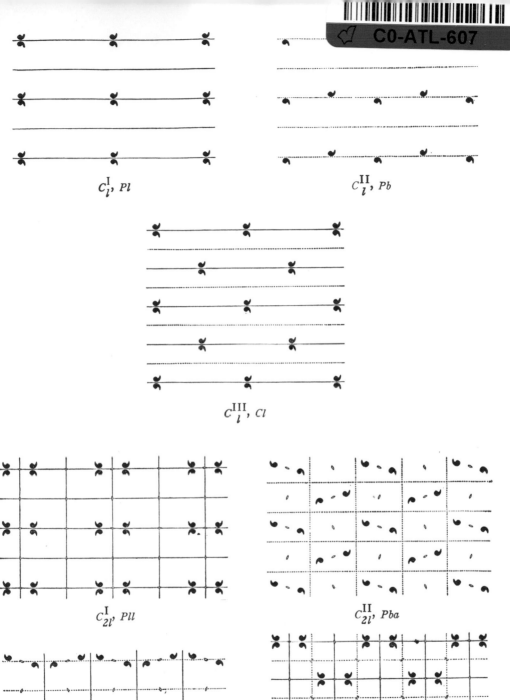

C_l^{I}, Pl

C_l^{II}, Pb

C_l^{III}, Cl

C_{2l}^{I}, Pll

C_{2l}^{II}, Pba

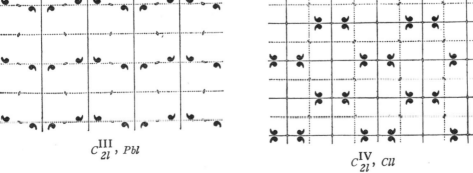

C_{2l}^{III}, Pbl

C_{2l}^{IV}, Cll

X-RAY CRYSTALLOGRAPHY

BOOKS BY M. J. BUERGER

Crystal-Structure Analysis
Elementary Crystallography. First Edition Revised
The Precession Method in X-Ray Crystallography
Vector Space
X-Ray Crystallography

X-Ray Crystallography

An Introduction to the Investigation of
Crystals by Their Diffraction of
Monochromatic X-Radiation

BY

M J. BUERGER

Professor of Mineralogy and Crystallography
Massachusetts Institute of Technology

JOHN WILEY & SONS, INC.

New York · London · Sydney

DEDICATED TO

WILLARD AND JENNIE WHITTEMORE

PREFACE

A crystal-structure investigation ordinarily proceeds through two distinct stages. The first consists of an investigation of the general geometry of the repetition or space pattern of the crystal. This leads to a knowledge of the crystal symmetry in the larger sense: the crystal class, the space lattice (its type and dimensions), and the space group. This stage of the structural investigation is a matter of pure geometry. It requires only a crystal and proceeds independently of any data or interpretation of data regarding the chemical constitution or nature of the crystal under investigation. Its results are truly the ultimate goal of the classical crystallographer. These results are easily achieved by employing the phenomenon of x-ray diffraction by the crystal. The science connecting x-ray diffraction and the desired crystallographic results might conveniently be called " x-ray crystallography."

X-ray crystallography is inadequately treated in current reference books. This is partly because the contents of such books usually include a wide selection of material from related fields such as x-ray physics, crystal-symmetry theory, the results of crystal-structure analysis, and crystal chemistry. This leaves the entire subject of the determination of crystal structures to be treated only in outline, and only part of this space can be given to x-ray crystallography. A student wishing to pursue structural studies by means of x-rays cannot learn through the aid of these books but must turn to original literature and possibly to an instructor for personal guidance. There is a definite need for a connected account of the theory and practice of structural investigations for such students.

The present book is intended to fill the x-ray crystallography part of this need. It is restricted in scope. The original intention was to discuss the moving-film methods only, first because they are covered so very superficially in all current books, and, second, because recent developments have shown their great importance and convenience. In order to discuss this phase of x-ray crystallography, it is desirable to develop a number of concepts ordinarily used in other diffraction methods, particularly the rotating-crystal method. It was therefore thought desirable to include chapters on the rotating- and oscillating-crystal methods, which are also ordinarily given all too little space in current books.   The subject matter of this book, therefore, is confined to methods utilizing a single crystal and monochromatic x-rays (in the sense of screened or unscreened characteristic radiation), and the theory necessary to the intelligent use of these methods.

Many users of this book will be those with comparatively little background in the field.   For their benefit, several of the elementary topics have been approached in a gradual manner in order to encourage a real understanding of the situation.   To this end, subjects such as x-ray diffraction and the reciprocal lattice, which require space geometry, have first been treated in a plane.   In this way, all the basic ideas involved can be developed with a minimum of extraneous mathematical fog.   Once the basic ideas are grasped, the student's attention may be readily broadened to include the additional dimension of space. In one other direction I have deliberately erred on the side of making things easy for the student who is not already a complete master of the subject.   Many textbooks and original papers require the reader to play detective with pencil and paper in an endeavor to find out how the author arrived at equation (2) after leaving equation (1).   In this book, an attempt has been made to avoid this by including the important intermediate steps.   This procedure will doubtless bore the expert, but I hope the plan will permit the novice to pay more attention to the subject matter, and less to the mystery of arriving at it.

Many crystallographic proofs are customarily given by vector algebraic methods.   For the benefit of the reader without training in vector algebra, alternative proofs have been given by geometrical, analytic geometrical, and algebraic methods.

A departure has been made in the usual vector algebra treatment of the reciprocal lattice.   The usual treatment has always appeared to me to be a classic example of seeing whether one could achieve an objective by the use of vector methods just because they were vector methods.   A little crystallography thrown in here and there eases the understanding of the treatment.   Furthermore, the usual treatment of the reciprocal lattice skips over the proof, except by implication, that the points reciprocal to the direct lattice constitute a lattice array, and focuses attention, rather, upon the cumbersome transformation equations connecting triclinic reciprocal and direct cells.   This highlights a comparatively unimportant quantitative aspect, of use chiefly in computation.   In the present treatment, the lattice aspect and the reciprocity aspect of the polar point system are emphasized first, the quantitative relations between the two point arrays being introduced only subsequently to impressing the more fundamental qualitative matters on the student.

The book contains a good deal of material appearing for the first time, or hitherto inadequately treated. Among such topics are the following:

A method of producing symmetry-true photographs (page 206).

Moving-film photographs taken with the x-ray beam inclined to the layers of the reciprocal lattice.

The Sauter and the Schiebold methods.

The de Jong and Bouman method.†

The choice of setting of a triclinic crystal.

Discussion of Bradley and Jay's absorption error.

Cohen's method extended to the refinement of the lattice constants of single crystals.

I am deeply grateful to the editor and publishers of the *Zeitschrift für Kristallographie*, not only for permission to reproduce, in slightly modified form, some of my articles which originally appeared there, but also for having had the kindness to permit reproduction of many of the illustrations. These include Figs. 136, 138, 142, 144, 147, 148, 155, 156, 221, 222, 223, 224, 225, 226, 227, 228, 229, 230, 231, 232, 233, 234, 236, 240, 243, 244, 245, 247, 252, and 253.

The editor of the *American Mineralogist* has also kindly permitted me to reproduce Figs. 1, 113, 214, 248, 249, 250, and 251. The plane-group patterns inside the covers are from blocks released to me by the *Technology Review*.

I am also indebted to the *Zeitschrift für Kristallographie* and to Dr. E. Schiebold for permission to reproduce Fig. 174; to the *Zeitschrift für Kristallographie* and to Dr. Linus Pauling for permission to reproduce Fig. 31; to Dr. Erwin Sauter for permission to reproduce Figs. 166 and 172; to the Royal Society of London and to Prof. J. D. Bernal for permission to reproduce Figs. 79, 80, and 86; and to Dr. I. Fankuchen for Fig. 105.

With the exception of the cuts acknowledged above, the illustrations have been prepared especially for this book. I am indebted to several assistants, particularly Mr. Howard Brooke Hindle, Jr., Mr. Herman William Ahrenholz, Jr., Mr. W. H. Dennen, and Mr. Joseph W. Mills, for transforming some of my pencil sketches of illustrations into draftings. Mr. Frank G. Chesley executed Fig. 60. I am indebted to my wife and to Miss Edna Howley for taking charge of transforming the most complicated parts of the manuscript into typescript.

† The chapter on the de Jong and Bouman method was written just after the first paper by these authors appeared. Their subsequent papers have covered ground almost identical with the material in Chapter 17.

My thanks are especially due to Dr. I. Fankuchen and to Mr. Joseph S. Lukesh who kindly offered to read the proof. Because of their efforts many misprints have been eliminated and not a few inaccuracies corrected.

I also wish to take this opportunity to thank the administration of the Massachusetts Institute of Technology as well as Dr. Warren J. Mead, Head of the Department of Geology, for their liberal policy of encouraging and providing facilities for undertaking work of the kind discussed in this book.

I hope that this account of x-ray crystallography will encourage the use of moving-film methods and level analysis methods not only among those who wish to employ them as a step in the complete determination of crystal structure, but also among those engaged in systematic crystallography.

M. J. BUERGER

CAMBRIDGE, MASSACHUSETTS

CONTENTS

CONTENTS

CHAPTER 10

CHAPTER 11

CHAPTER 12

CHAPTER 13

CHAPTER 16

CHAPTER 17

CHAPTER 18

CHAPTER 21

CHAPTER 22

CONTENTS

TABLES

CHAPTER 1

INTRODUCTION

From a certain point of view, a crystal structure (Fig. 1) is like a wallpaper pattern, except that it is three-dimensional and therefore more complex. In a wallpaper pattern (or in fact in any extended plane pattern such as is found, for example, in textile weaves, linoleums, tiling, etc.), the pattern as a whole has two aspects:

(1) a motif, such as a picture of a flower or other figure, a tile, or group of tiles, and

(2) a mechanism or scheme of repetition of this motif.

In a crystal structure, the same two general aspects can be recognized, only they occur in three dimensions, and they apply specifically to the packing of atoms. The motif is a cluster of atoms, which may or may not be a chemical molecule. (Molecules occur chiefly in crystals of organic material and, for the most part, cannot be recognized in the patterns of crystals of inorganic substances.) This motif cluster is repeated by a mechanism or scheme of repetition in three dimensions, the entire resulting pattern being the crystal structure.

FIG. 1. The space pattern of atoms in a crystal of marcasite, orthorhombic FeS_2. Dark balls represent iron atoms, light balls sulfur atoms.

In anticipation, it can be said that such a three-dimensional pattern acts as a diffraction grating to light having wavelengths of the same order of magnitude as the translation repeat period of the pattern. This period is of the order of 1×10^{-8} cm., and light having wavelengths of this order is x-radiation. In other words, x-rays may be used to explore the nature of crystal patterns through the phenomenon of diffraction by the crystal pattern. The discovery of this fact was made by von Laue and his collaborators. The technique of investigating this effect was quickly improved by Bragg, and subsequently by many others.

1

Again in anticipation, it can be said that the x-ray investigation of a crystal structure has two aspects:

(a) the investigation of the shape of the motif, and

(b) the investigation of the geometry of the repeating scheme.

Part (a), which corresponds with (1), above, depends upon the relative intensities and phases of the x-ray diffraction spectra. The intensities can be measured, but the phases cannot; hence it is not always possible to determine a crystal structure completely, although many have been solved by hunches and other indirect aids. Part (b), which corresponds with (2), above, depends, as we shall see, solely upon the geometry of the distribution of diffraction spectra, and this can be completely determined. It is thus always possible to determine the geometry of the repeat scheme of a crystal.

In actual practice a crystal-structure analysis starts with the determination of the geometry of the repeat scheme (i. e., with the determination of the *space group* and *unit cell* of the crystal). It is with this phase of crystal-structure analysis that the present volume is chiefly concerned. When the groundwork has been laid by this first analytical step, then *an attempt may be made* to determine the shape of the motif; this problem may or may not prove capable of solution. With this second phase of crystal-structure analysis, this particular volume is not immediately concerned.

The elements of repetition. It is not one of the purposes of this volume to discuss the theory of pattern repetition, i. e., *the theory of space groups*. This background should be acquired from one of the standard treatises, and at least a slight familiarity with the results of this theory is assumed. For the purposes of the development necessary here, certain general results of this theory are recapitulated in the following form:

Given any general space motif, there are 230, and only 230, ways of repeating it in space. These are the so-called *230 space groups* (there are only *17 plane groups*, i. e., 17 ways of repeating a plane motif in a plane). Each of these repeating schemes can be factored into a limited number of simpler, elementary repeat schemes which have, as their basis, repetition according to the following simple repeating devices, or elements of repetition:

(1) repetition by translation,

(2) repetition by rotation,

(3) repetition by screw motion,

(4) repetition by reflection,

(5) repetition by glide-reflection,

(6) repetition by roto-reflection.

By appropriately combining groups of each of the above, in all possible ways, the 230 space groups result.

When one has a crystal to be pattern-analyzed, he has an example of one of these 230 possible space patterns. He is then faced with the problem of determining systematically whether the crystal pattern contains groups of any of the above repetition elements and, if so, of determining how each is oriented with respect to the crystal axes. The sum total of such information leads to a knowledge of the space group of the crystal.

While a crystal may lack any of the elements of repetition (2), (3), (4), (5), or (6), it cannot lack (1). This is because the crystal is necessarily based upon an extended pattern, and this implies translation repetitions in three dimensions. Thus every crystal pattern has a basic three-dimensional translation group. It may or may not have additional repeating mechanisms.

LITERATURE

Non-mathematical introduction

M. J. Buerger and J. S. Lukesh. *Wallpaper and atoms.* *Technology Review* **39** (June, 1937), 338–342.

Excellent introduction and illustrations.

Stefan Kreutz. *Elemente der Theorie der Krystallstruktur.* (Wilhelm Engelmann, Leipzig, 1915.)

Standard Texts

Arthur Schoenflies. *Krystallsysteme und Krystallstructur.* (B. G. Teubner, Leipzig, 1891.)

Harold Hilton. *Mathematical crystallography and the theory of groups of movements.* (Clarendon Press, Oxford, 1903.)

Paul Niggli. *Geometrische Kristallographie des Diskontinuums.* (Gebrüder Borntraeger, Leipzig, 1919.)

Artur Schoenflies. *Theorie der Kristallstruktur, ein Lehrbuch.* (Gebrüder Borntraeger, Berlin, 1923.)

Georges Friedel. *Leçons de cristallographie.* (Berger-Levrault, Paris, 1926.)

H. Bouasse. *Cristallographie géométrique, groupes de déplacements.* (Librarie Delagrave, Paris, 1929.)

CHAPTER 2

SOME GEOMETRICAL ASPECTS OF LATTICES

ELEMENTARY LATTICE THEORY

Repetition by translation is a movement which can be represented by a vector. It should be remembered that, when a translation acts upon a motif, the translation does not act on or at any particular point in the motif, but upon the whole motif. In this sense, a translation, like other vectors, has no specific origin, but has only direction and magnitude. The origin of the conventional vector arrow representing the translation can be taken at any point suiting the convenience of the particular problem or discussion at hand. Suppose that it is taken at some arbitrary reference point in the pattern to be repeated. The repetitive action of the group of translations repeats this reference point (as well as all others of the motif) as a three-dimensional pattern of points in space (Fig. 2A). This point pattern is called a *point*

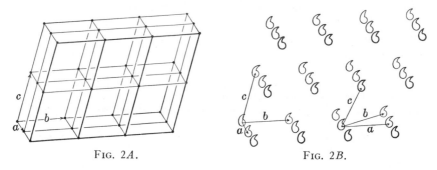

FIG. 2A.   FIG. 2B.

lattice (specifically a *point space lattice*). This nomenclature is in allusion to the fact that the points can be connected by a three-dimensional grid of lines, as shown in Fig. 2A. The line grid is a lattice, and is called, in distinction, a *line lattice* (specifically, a *line space lattice*). A line lattice determines a unique point lattice, but the points of a point lattice may be connected in various ways to form an infinite number of different line lattices.

It should again be emphasized that the space lattice of a crystal is a representation of its translation repetition. If the lattice scheme is shifted so that its origin point falls upon any desired point in the

4

motif, then there is an identical point at every one of the space lattice points, produced by the translation group of the crystal pattern.

The actual grid of the line lattice, Fig. 2A, can be described by specifying the directions and magnitudes of its three representative grid lines, a, b, and c. These are three non-coplanar translation vectors which, by repeated action, are sufficient to repeat the origin point at every point of the space lattice. They are conveniently known as the *three unit translations*. For a given pattern, Fig. 2B, they can be chosen in innumerable ways, and each way gives rise to a different line lattice gridwork of the type shown in Fig. 2A. The point lattice, or pattern of identical points, however, is always the same for a given pattern, no matter how the line lattice is chosen.

Current misuse of the term *lattice*. Many who have written about crystal structures, particularly chemists, physicists, and metallurgists unacquainted with pattern theory, have either used the term lattice incorrectly, or sought to extend its use to mean the actual material crystal structure of packed atoms. Not only is such usage incorrect, but also such an extension of meaning, if continued, is highly undesirable, for, if the word " lattice " is adapted to mean " structure," then what word shall be used to mean " lattice " in the original sense above discussed?

Unit cell. The small parallelepiped built upon the three translations selected as unit translations, Fig. 2A, is known as the unit cell. The unit cell, though imaginary, has an actual shape and a definite volume. The entire crystal structure is generated through the periodic repetition, by the three unit translations, of the matter contained within the volume of the unit cell. A unit cell does not necessarily have a definite, absolute origin or position, unless one is arbitrarily selected for a particular problem. It does have a definite orientation, since its edges are defined by the translation vectors. Thus, in the plane pattern of Fig. 3A, the solid lines block out a unit cell and the dotted lines block out an identical unit cell in another position. *A unit cell has shape and orientation, but it does not necessarily have position except for the convenience of a certain problem.*

Unit cells may be qualified as *primitive, doubly primitive, triply primitive, quadruply primitive*, etc., according as they contain 1, 2, 3, 4, etc., translation-equivalent chunks of pattern. This can be illustrated better in two dimensions than in three. In Fig. 3A, a pattern of commas is shown. This pattern is repeated both by twofold rotors (axes) and by translations. The chunk of pattern repeated by translation is the composite axially symmetrical cluster shown in Fig. 3B. In Fig. 3A, the cell outlined in solid lines contains one such cluster;

this can be proved to one's satisfaction by translating the cell outline until it occupies the dotted position.   Thus a *cell outline which has a cluster at each corner contains the equivalent of one total cluster*.   Such a cell is known as a *primitive* cell.   On the other hand, the cell shown

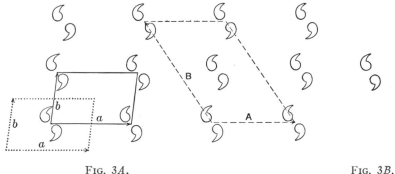

FIG. 3A.   FIG. 3B.

by dashed lines not only has a cluster at each corner but also has another whole cluster completely within the confines of the cell boundary.   It therefore contains a total of two translation-equivalent clusters and consequently is designated a *doubly primitive* cell.

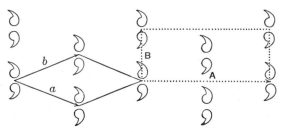

FIG. 4.

Usually, unit translations are chosen so that a primitive cell results. There are certain occasions, however, notably to gain advantages due to symmetry, where it is desirable to choose a doubly primitive cell. Such an occasion is illustrated in Fig. 4.   Here, because of symmetry, the primitive cell built upon translations a and b is diamond shaped, while the doubly primitive cell built upon translations A and B is rectangular.   The repetition of the translations of the rectangular cell builds up an orthogonal coordinate system, which is geometrically the more convenient of the two possibilities to use.

Lattice grid as a coordinate system.   The geometry of a crystal pattern is, of course, intimately connected with the pattern repetition. For this reason, the translation repetition, which is epitomized by the line lattice, is an ideal coordinate system for discussing the geometry of the pattern. This coordinate system may be an oblique one in certain cases, but it is nevertheless the natural one. For a given crystal pattern the translation identity pattern or point lattice is obvious. When three non-coplanar unit translations are chosen from this pattern, both the unit cell and the coordinate system gridwork are fixed. It has already been intimated that there are innumerable possible ways to choose these, and hence an innumerable set of unit cells and coordinate systems is available, all natural ones. The one selected for actual use is so chosen because it is the easiest to use for the problem at hand.

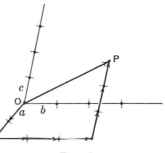

FIG. 5.

Indices of a lattice direction. The rational lattice direction, OP, Fig. 5, may be regarded as a vector. As such it may be resolved into its components along the coordinate (lattice) axes. Thus OP equals, vectorially, u units of translation $\mathbf{a}$, plus v units of translation $\mathbf{b}$, plus w units of translation $\mathbf{c}$, i. e.,

$$OP = u\mathbf{a} + v\mathbf{b} + w\mathbf{c}. \tag{1}$$

For any vector equation of this kind, only the numbers u, v and w change from direction to direction. These coefficients, enclosed in square brackets thus: $[uvw]$, are called the *indices* of the line OP; they completely specify it, for a given lattice coordinate system.

Indices of a point. The point P has coordinates u, v, w, referred to a coordinate system whose units are the unit translations a, b, c. To distinguish the coordinates of the point P from the indices of the line OP, the point coordinates u, v, w are written in double square brackets, thus: $[[uvw]]$.

The intercepts of a rational plane. A *rational plane* is one which contains lattice points (not all collinear). Unless otherwise indicated, the designation " plane," in pattern geometry, means a rational plane.

A plane may be specified by its equation. In Fig. 6, the rational plane is rational because it contains lattice points one of which happens to be located A units along the a-axis, a second B units along the b-axis, and a third C units along the c-axis. Since these intercepts are known, the equation of this plane can be put into its intercept form, namely,

$$\frac{x}{A} + \frac{y}{B} + \frac{z}{C} = 1. \tag{2}$$

The correctness of this equation can easily be verified by setting two variables at a time equal to zero; the third variable then equals the intercept along its axis.

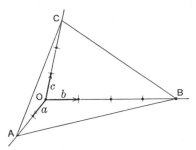

FIG. 6.

Repetition of a plane by translation. For every combination of three lattice points, one along each lattice axis, there is a rational plane which can be expressed in the form of an intercept equation (2). Each such plane, moreover, is repeated by the lattice translations so that any one equation implies a whole set of parallel planes. The question arises, how many planes are there from the origin to the rational intercept plane? This can best be solved first in two dimensions. In Fig. 7A, a simple " plane " is shown with intercepts of two translation units along a and three translation units along b. The b translation,

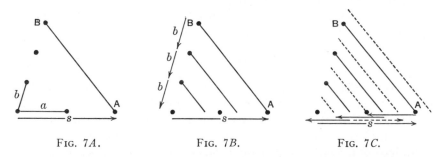

FIG. 7A.   FIG. 7B.   FIG. 7C.

Fig. 7B, repeats the one plane to become three, not including the origin plane. For each one of these three, the a translation repetition gives rise to two planes within the intercept interval s, Fig. 7C. There are thus 3·2, or, in general, $B·A$ planes between the intercept and the origin, provided B and A contain no common factor.

In a similar way, in three dimensions, if a plane has intercepts A, B, and C along the three coordinate axes, then, provided A, B, and C are prime to one another in pairs, the a translation repeats the intercept plane to become A planes, the b translation repeats these A planes B-fold to become AB planes, and the c translation repeats these AB planes C-fold to become a total of ABC planes.

Suppose, however, that A and B have the highest common factor, t. Then the first line from the origin having rational intercepts has intercepts A/t and B/t, not A and B. According to the discussion just given, there are $(A/t)(B/t)$ lines from the origin to this rational intercept line. There are t times as many lines to the line having inter-cepts A and B, that is, $(A/t) (B/t)t = AB/t$ lines. If, furthermore, the other two intercept pairs B, C, and A, C, have highest common factors r and s respectively, then the number of planes from the origin to the first rational intercept plane is ABC/rst. Equation (2) thus represents the ABC/rstth plane of its kind from the origin. The first plane from the origin evidently has intercepts $\dfrac{1}{ABC/rst}$ th as long as (2). The equation of this first plane is

$$\frac{x}{Arst/ABC} + \frac{y}{Brst/ABC} + \frac{z}{Crst/ABC} = 1,$$

or $\qquad\qquad (BC/rst)x + (AC/rst)y + (AB/rst)z = 1. \qquad\qquad (3)$

For convenience, now, let

$$\begin{cases} BC/rst = h, \\ AC/rst = k, \\ AB/rst = l. \end{cases} \qquad\qquad (4)$$

Equation (3) now becomes

$$hx + ky + lz = 1. \qquad\qquad (5)$$

This is the equation of the first plane from the origin. Every plane has an equation of this same form, only the coefficients h, k, and l differing from plane to plane. Therefore h, k, and l completely specify a plane; they are called the indices of the plane, and a set of three numbers is distinguished as indices of a plane by placing them in parentheses thus: (hkl).

Note that if the highest common factor of the pair A, B is t, and of A, C, is s, then A has the two factors s and t and an unfactored residue, say e. Thus

$$A = ste,$$

similarly, $\qquad\qquad B = rtf,$

and $\qquad\qquad C = rsg.$

If these factors are substituted for A, B, and C in (4), there results

$$\left.\begin{array}{l} h = BC/rst = rtf \cdot rsg/rst = rfg \\ k = AC/rst = ste \cdot rsg/rst = seg \\ l = AB/rst = ste \cdot rtf/rst = tef. \end{array}\right\} \qquad (4')$$

Since the right members of $(4')$ are products of integers, h, k, and l are integers.

The meaning of h, k, and l can be grasped best by recasting (5) in the form

$$\frac{x}{1/h} + \frac{y}{1/k} + \frac{z}{1/l} = 1. \qquad\qquad (6)$$

This can be recognized as the equation of a plane in intercept form, in which the intercepts are $\dfrac{1}{h}, \dfrac{1}{k}, \dfrac{1}{l}$. A series of planes of indices (hkl) thus intercepts the lattice axes in such a way as to divide the a-axis into h parts, the b-axis into k parts, and the c-axis into l parts.

The number of planes from the origin to the rational intercept plane. It was demonstrated in the last section that the number of planes from the origin to the first plane which makes rational intercepts on the three lattice axes is ABC/rst. It will be convenient to express this also in terms of indices. This can be done by transposing (4):

$$A/rst = \frac{k}{C}, \qquad B/rst = \frac{l}{A}, \qquad C/rst = \frac{h}{B}.$$

Hence,

$$\frac{ABC}{r^3 s^3 t^3} = \frac{k}{C} \cdot \frac{l}{A} \cdot \frac{h}{B},$$

thus

$$\frac{A^2 B^2 C^2}{r^2 s^2 t^2} = rst \cdot hkl,$$

from which

$$\frac{ABC}{rst} = \sqrt{rst \cdot hkl}. \tag{7}$$

There are $\sqrt{rst \cdot hkl}$† planes from the origin to the first plane which makes rational intercepts with the lattice axes.

TRANSFORMATION THEORY

Transformation of lattice axes. It has already been pointed out that, for a given point lattice, innumerable sets of the three unit trans-

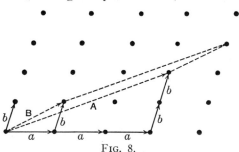

FIG. 8.

lations can be chosen resulting in as many unit cells. It often happens that one must start a problem with one such set of axes, and subsequently choose a new set. The question arises, how can the new axes be expressed in terms of the old? Figure 8 shows how these are related in the two-dimensional case. The old axes are a and b; the new ones are A and B. Evidently it is possible to express the new axes

---

† This can be shown to be equal to hkl/efg, where e, f, and g are the highest common factors of the pairs k and l, h and l, and h and k, respectively, according to (4′).

in terms of vector sums of the old. In this specific instance, the following vector additions hold:

$$\begin{cases} \mathbf{A} = 3\mathbf{a} + 2\mathbf{b}, \\ \mathbf{B} = \mathbf{a} + \mathbf{b}. \end{cases}$$

In any given two-dimensional case, this same form can be employed by changing the scalar coefficient used in this example for the ones appropriate to the problem. In general, the transformation is

$$\begin{cases} \mathbf{A} = u_1\mathbf{a} + v_1\mathbf{b}, \\ \mathbf{B} = u_2\mathbf{a} + v_2\mathbf{b}. \end{cases} \tag{8}$$

This result can easily be generalized to three dimensions. If $\mathbf{A}$, $\mathbf{B}$, and $\mathbf{C}$ are the new axes, they can be expressed in terms of the old axes, $\mathbf{a}$, $\mathbf{b}$, and $\mathbf{c}$, in the following form:

$$\begin{cases} \mathbf{A} = u_1\mathbf{a} + v_1\mathbf{b} + w_1\mathbf{c}, \\ \mathbf{B} = u_2\mathbf{a} + v_2\mathbf{b} + w_2\mathbf{c}, \\ \mathbf{C} = u_3\mathbf{a} + v_3\mathbf{b} + w_3\mathbf{c}. \end{cases} \tag{9}$$

The general form of (9) is always the same for any transformation of axes; for any particular transformation, the only individual character it has is the several coefficients u_1, v_1, w_1; u_2, v_2, w_2; and u_3, v_3, w_3. Relations (9) could therefore be expressed in abbreviated form by leaving out all but these important characteristic numbers, arranged in their proper order, thus:

$$\begin{matrix} u_1 & v_1 & w_1, \\ u_2 & v_2 & w_2, \\ u_3 & v_3 & w_3. \end{matrix}$$

This array, placed between double vertical bars, thus,

$$\left\| \begin{matrix} u_1 & v_1 & w_1 \\ u_2 & v_2 & w_2 \\ u_3 & v_3 & w_3 \end{matrix} \right\|, \tag{10}$$

to distinguish it from other, geometrically similar arrays, is known as the *matrix of the transformation*. It is an abbreviated expression of (9). This array has currently been written

$$u_1v_1w_1/u_2v_2w_2/u_3v_3w_3 \tag{11}$$

to save typographical effort. Form (10) is much more convenient for actual manipulation, as will be seen, because of its geometrical form.

Transformation of the indices of a plane. When one changes from one coordinate system to another, he naturally changes the intercepts which a given plane system makes with the axes. Accordingly, the indices of the plane also change. The nature of this change is illus-

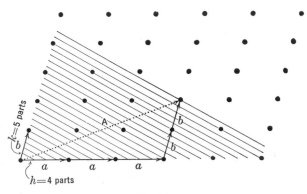

FIG. 9.

trated for two dimensions in Fig. 9. In this particular instance the " plane " system divides a into four parts and b into five parts. Thus $h = 4$ and $k = 5$. It is evident that the " planes " are independent of any choice of axes and divide any transverse line into equal segments. The particular line of the new axis is the vector sum

$$\mathbf{A} = 3\mathbf{a} + 2\mathbf{b}.$$

It is also evident that the division lines of the vector $3\mathbf{a}$ and the vector $2\mathbf{b}$ project along the " plane " directions on line $\mathbf{A}$. Line $\mathbf{A}$ is therefore divided into the same number of parts as the sum of the number of parts in the vector $3\mathbf{a}$ plus the number of parts in the vector $2\mathbf{b}$, i. e., into $3 \cdot 4 + 2 \cdot 5 = 12 + 10 = 22$ parts. The number of parts into which an axis is divided is the index corresponding with that axis. Thus:

$$H = 3 \cdot 4 + 2 \cdot 5 = 22.$$

Using the general values rather than numbers, the vector $\mathbf{A}$ is

$$\mathbf{A} = u_1 \mathbf{a} + v_1 \mathbf{b}.$$

The number of parts into which it is divided is the sum of the number of parts into which its projections on $\mathbf{a}$ and $\mathbf{b}$ are divided, namely,

$$H = u_1 h + v_1 k. \tag{12}$$

A second new axis, **B**, in terms of the old axes, is

$$B = u_2\mathbf{a} + v_2\mathbf{b},$$

and similarly the index of the " plane " on the new **B** axis is

$$K = u_2h + v_2k. \tag{13}$$

The entire two-dimensional index transformation is then given by the pair (12) and (13):

$$\begin{cases} H = u_1h + v_1k, \\ K = u_2h + v_2k, \end{cases} \tag{14}$$

whose matrix is

$$\begin{Vmatrix} u_1 & v_1 \\ u_2 & v_2 \end{Vmatrix}. \tag{15}$$

This entire discussion can be generalized to three dimensions. Here, a new axis in terms of the old, according to (9), is

$$\mathbf{A} = u_1\mathbf{a} + v_1\mathbf{b} + w_1\mathbf{c},$$

A plane, indexed (hkl) on the old axes, divides the new **A** axis into H parts, given by

$$H = u_1h + v_1k + w_1l. \tag{16}$$

By analogy, the entire index transformation is

$$\begin{cases} H = u_1h + v_1k + w_1l, \\ K = u_2h + v_2k + w_2l, \\ L = u_3h + v_3k + w_3l. \end{cases} \tag{17}$$

The matrix of this transformation is

$$\begin{Vmatrix} u_1 & v_1 & w_1 \\ u_2 & v_2 & w_2 \\ u_3 & v_3 & w_3 \end{Vmatrix}. \tag{18}$$

Note that this matrix of *plane index* transformation is identical with (10), the matrix for the axial transformation. Thus, if one decides upon a certain change of axes, he can immediately write down the new indices of any plane, (hkl), by filling in the u, v, w coefficients of (17) with the numbers taken from the axial transformation matrix (10).

Determination of the transformation matrix from particular index transformations. It frequently happens that an early investigator has assigned indices to a certain limited number of planes of a crystal species and a later investigator has assigned different indices to these same planes. The question arises, how will *any* index transform from the first system to the second? This problem can easily be solved by filling in the particular H, K, L, and h, k, l, quantities of (17) for the limited number of planes, and solving for the u's, v's, and w's. An example of this procedure is as follows:

Three planes, ①, ②, and ③, in old and new notation, are as follows:

	Old (hkl)	New (HKL)
①	(110)	(230)
②	(210)	(340)
③	(001)	(001)

This gives the index transformation of these three particular planes. How do the indices of any plane transform, or, more generally, what is the transformation matrix? To solve this question, note that the transformation must have the general form of (17), namely,

$$H = u_1 h + v_1 k + w_1 l.$$

The H, h, k, and l values listed in the table above give, for planes ①, ②, and ③,

$$
\begin{array}{ll}
① & 2 = u_1 \cdot 1 + v_1 \cdot 1 + w_1 \cdot 0, \\
② & 3 = u_1 \cdot 2 + v_1 \cdot 1 + w_1 \cdot 0, \\
③ & 0 = u_1 \cdot 0 + v_1 \cdot 0 + w_1 \cdot 1.
\end{array}
$$

These three equations can be easily solved for u_1, v_1, and w_1, as follows:

$$
\begin{array}{ll}
① & u_1 + v_1 = 2 \\
② & 2u_1 + v_1 = 3 \\
\hline
& u_1 \quad\quad = 1.
\end{array}
$$

Substituting this in ①:

$$1 + v_1 = 2;$$
$$v_1 = 1.$$

Substituting both of these in ③:

$$0 = 1 \cdot 0 + 1 \cdot 0 + w_1 \cdot 1;$$
$$w_1 = 0.$$

This determines u_1, v_1, and w_1. To find u_2, v_2, and w_2, substitute the values for planes ①, ②, and ③ from the above table in the second equation of (17):

$$K = u_2 h + v_2 k + w_2 l$$

① $\quad 3 = u_2 \cdot 1 + v_2 \cdot 1 + w_2 \cdot 0$
② $\quad 4 = u_2 \cdot 2 + v_2 \cdot 1 + w_2 \cdot 0$
③ $\quad 0 = u_2 \cdot 0 + v_2 \cdot 0 + w_2 \cdot 1$

① $\quad u_2 + v_2 = 3$
② $\quad 2u_2 + v_2 = 4$
$$\overline{\quad u_2 \qquad\qquad = 1.}$$

Substituting this in ①:

$$1 + v_2 = 3,$$
$$v_2 = 2.$$

Substituting both of these in ③:

$$0 = 1\cdot 0 + 2\cdot 0 + w_2 \cdot 1$$
$$w_2 = 0.$$

Finally, u_3, v_3, and w_3 may be found by substituting from the values for planes ①, ②, and ③ into the third equation of (17):

$$L = u_3 h + v_3 k + w_3 l$$

① $\quad 0 = u_3 \cdot 1 + v_3 \cdot 1 + w_3 \cdot 0$
② $\quad 0 = u_3 \cdot 2 + v_3 \cdot 1 + w_3 \cdot 0$
③ $\quad 1 = u_3 \cdot 0 + v_3 \cdot 0 + w_3 \cdot 1$

① $\quad u_3 + v_3 = 0$
② $\quad 2u_3 + v_3 = 0$
$$\overline{\quad u_3 \qquad\qquad = 0.}$$

Substituting this in ②:

$$2\cdot 0 + v_3 = 0,$$
$$v_3 = 0.$$

Substituting both of these in ③:

$$1 = 0\cdot 0 + 0\cdot 0 + w_3 \cdot 1,$$
$$w_3 = 1.$$

Assembling the values of u_1, v_1, w_1, u_2, v_2, w_2, u_3, v_3, w_3 in their appropriate places gives the matrix of the transformation:

$$\begin{Vmatrix} 1 & 1 & 0 \\ 1 & 2 & 0 \\ 0 & 0 & 1 \end{Vmatrix}.$$

Finally, these particular matrix elements, substituted for the coefficients in (17), give the general transformation which is consistent with the index transformation of planes ①, ②, and ③:

$$\begin{cases} H = h + k \\ K = h + 2k \\ L = \qquad l. \end{cases}$$

From this general transformation, the index change for any plane whatever can be determined.

Two transformations in sequence. If a transformation is made from a first set of axes to a second, and then from this second set of axes to a third, it is possible to pass directly from the first to the third by a manipulation of matrices known as matrix multiplication. This will be developed first for the two-dimensional case.

Suppose a new set of axes $\mathbf{A}$, $\mathbf{B}$ is chosen from the pattern. These new axes in terms of the old are

$$\begin{cases} \mathbf{A} = u_1\mathbf{a} + v_1\mathbf{b}, \\ \mathbf{B} = u_2\mathbf{a} + v_2\mathbf{b}. \end{cases} \tag{19}$$

The matrix of this transformation is

$$\left\| \begin{matrix} u_1 & v_1 \\ u_2 & v_2 \end{matrix} \right\|. \tag{20}$$

A " plane " whose indices were (hk) referred to the first set of axes now acquires new indices (HK). The second indices may be solved for in terms of the first by substituting the elements of the matrix (20) as coefficient of the transformation equation

$$\begin{cases} H = u_1h + v_1k, \\ K = u_2h + v_2k. \end{cases} \tag{21}$$

So far, this follows the discussion in the last section. Suppose, now, that one chooses a third set of axes $\mathbf{A}'$, $\mathbf{B}'$. Expressed in terms of the second set, these are

$$\begin{cases} \mathbf{A}' = U_1\mathbf{A} + V_1\mathbf{B}, \\ \mathbf{B}' = U_2\mathbf{A} + V_2\mathbf{B}. \end{cases} \tag{22}$$

The matrix of this transformation is

$$\left\| \begin{matrix} U_1 & V_1 \\ U_2 & V_2 \end{matrix} \right\|. \tag{23}$$

A " plane " whose indices were (HK), referred to the second set of axes, now becomes $(H'K')$ referred to the third set of axes. The third indices may be solved for in terms of the *second* by substituting the elements of matrix (23) as coefficients of the transformation equation

$$\begin{cases} H' = U_1H + V_1K, \\ K' = U_2H + V_2K. \end{cases} \tag{24}$$

We now have the third indices expressed in terms of the second (24), and the second indices expressed in terms of the first (21). It is evidently possible to express the third in terms of the first by properly combining relations (24) and (21). This can easily be done by sub-stituting, in (24), for H and K, the values given in (21). Relation (24) then develops thus:

$$\begin{cases} H' = U_1(u_1h + v_1k) + V_1(u_2h + v_2k), \\ K' = U_2(u_1h + v_1k) + V_2(u_2h + v_2k), \end{cases}$$

$$\begin{cases} H' = U_1u_1h + U_1v_1k + V_1u_2h + V_1v_2k, \\ K' = U_2u_1h + U_2v_1k + V_2u_2h + V_2v_2k. \end{cases}$$

Collecting coefficients of h and k,

$$\begin{cases} H' = (U_1u_1 + V_1u_2)h + (U_1v_1 + V_1v_2)k, \\ K' = (U_2u_1 + V_2u_2)h + (U_2v_1 + V_2v_2)k. \end{cases} \tag{25}$$

(25) represents the combined transformations (21) followed by (24). It has the standard transformation form, and its matrix can be written

$$\left\| \begin{array}{cc} (U_1u_1 + V_1u_2) & (U_1v_1 + V_1v_2) \\ (U_2u_1 + V_2u_2) & (U_2v_1 + V_2v_2) \end{array} \right\|. \tag{26}$$

The only quantities which have been manipulated in the production of (25) are the coefficients, which are collected in matrix form in (26). It should evidently be possible to pass from matrices (20) and (23) directly to the combined matrix (26) without the formal manipulation of equations (21) and (24). This is possible by a simple scheme which can be derived by a study of these three matrices. In order to follow conventional practice, the matrix of the second transformation (called **b**) is written first, thus:

$$\begin{matrix} \textbf{b} & \text{“ multi-} & \textbf{a} & & \textbf{c} \\ \text{(matrix} & \text{plied by”} & \text{(matrix} & = & \text{(matrix (26))} \\ \text{(23))} & & \text{(20))} & & \end{matrix}$$

$$\left\| \begin{array}{cc} U_1 & V_1 \\ U_2 & V_2 \end{array} \right\| \quad . \quad \left\| \begin{array}{cc} u_1 & v_1 \\ u_2 & v_2 \end{array} \right\| = \left\| \begin{array}{cc} (U_1u_1 + V_1u_2) & (U_1v_1 + V_1v_2) \\ (U_2u_1 + V_2u_2) & (U_2v_1 + V_2v_2) \end{array} \right\| \tag{27}$$

Careful study of (27) will show that, to fill in any desired element of matrix (26), one has only to note the row and column of this element;

for example, for the upper left element:

$$(26): \left\| \; \boxminus \; \right\|.$$

To form this element from matrices (23) and (20), take the elements of the corresponding row of (23) (the matrix in first position), thus:

$$(23): \left\| \; ① \;\; ② \; \right\|,$$

and the elements in the corresponding column of (20) (the matrix in second position), thus:

$$(20): \left\| \begin{matrix} ①' \\ ②' \end{matrix} \; \right\|,$$

then multiply ① × ①′ and add it to the product ② × ②′. The result is the desired matrix element of (26). This *row-by-column* multiplication is continued for the other desired matrix elements of (26). The entire process may be symbolized in the following form:

(b)   (a)   (c)

$$\left\| \; \rightarrow \; \right\| \cdot \left\| \downarrow \; \right\| = \left\| \; \square \; \right\| \qquad (27A)$$

$$\left\| \; \rightarrow \; \right\| \cdot \left\| \; \downarrow \right\| = \left\| \;\; \square \right\| \qquad (27B)$$

$$\left\| \; \rightarrow \; \right\| \cdot \left\| \downarrow \; \right\| = \left\| \; \square \;\; \right\| \qquad (27C)$$

$$\left\| \; \rightarrow \; \right\| \cdot \left\| \; \downarrow \right\| = \left\| \;\; \square \; \right\| \qquad (27D)$$

This process is known as the multiplication of matrices **b·a**. It should be emphasized that in order to apply it to two transformations taken in sequence, as above, the matrix of the second transformation is placed first, and the first second. Matrix multiplication is not commutative, i. e.,

$$\mathbf{b \cdot a \neq a \cdot b}$$

Thus, in (27), **b·a** gave the result

$$\mathbf{b \cdot a} = \left\| \begin{array}{cc} (U_1 u_1 + V_1 u_2) & (U_1 v_1 + V_1 v_2) \\ (U_2 u_1 + V_2 u_2) & (U_2 v_1 + V_2 v_2) \end{array} \right\|, \tag{28}$$

but the same multiplication rule applied to **a·b** gives,

$$\mathbf{a \cdot b} = \left\| \begin{array}{cc} (u_1 U_1 + v_1 U_2) & (u_1 V_1 + v_1 V_2) \\ (u_2 U_1 + v_2 U_2) & (u_2 V_1 + v_2 V_2) \end{array} \right\|. \tag{29}$$

a·b gives a combined transformation resulting from making the **b** transformation first, then following it by the **a** transformation. These two results are quite different; consequently the order of multiplication must be carefully observed.

Two transformations in sequence in three dimensions. The discussion of two three-dimensional transformations in sequence follows the two-dimensional case closely but is more complicated owing to the additional dimension.

Starting with an original set of axes, a new set is chosen, which is represented in terms of the original by the transformation

$$\begin{cases} \mathbf{A} = u_1 \mathbf{a} + v_1 \mathbf{b} + w_1 \mathbf{c}, \\ \mathbf{B} = u_2 \mathbf{a} + v_2 \mathbf{b} + w_2 \mathbf{c}, \\ \mathbf{C} = u_3 \mathbf{a} + v_3 \mathbf{b} + w_3 \mathbf{c}. \end{cases} \tag{30}$$

A plane, whose indices are $(h\,k\,l)$ referred to the first axes, receives a new set of indices $(H\,K\,L)$ referred to the second set, and these indices are given by

$$\begin{cases} H = u_1 h + v_1 k + w_1 l, \\ K = u_2 h + v_2 k + w_2 l, \\ L = u_3 h + v_3 k + w_3 l. \end{cases} \tag{31}$$

A third set of axes is now chosen from the second, such that

$$\begin{cases} \mathbf{A'} = U_1 \mathbf{A} + V_1 \mathbf{B} + W_1 \mathbf{C}, \\ \mathbf{B'} = U_2 \mathbf{A} + V_2 \mathbf{B} + W_2 \mathbf{C}, \\ \mathbf{C'} = U_3 \mathbf{A} + V_3 \mathbf{B} + W_3 \mathbf{C}. \end{cases} \tag{32}$$

Accordingly, a plane of index $(H\,K\,L)$ referred to the second axes becomes $(H'\,K'\,L')$ referred to the third set of axes, the new indices being given by the relation

$$\begin{cases} H' = U_1 H + V_1 K + W_1 L, \\ K' = U_2 H + V_2 K + W_2 L, \\ L' = U_3 H + V_3 K + W_3 L. \end{cases} \tag{33}$$

The relation of the third set of indices to the second is given by (33), and the relation of the second to the first is given by (31). It is now desired to know in what way the third set of indices $(H' \, K' \, L')$ is related to the first set $(h \, k \, l)$. This can evidently be found by substituting the values of H, K, and L given by (31) in (33). This gives

$$\begin{cases} H' = U_1u_1h + U_1v_1k + U_1w_1l + V_1u_2h + V_1v_2k + V_1w_2l + W_1u_3h + W_1v_3k + W_1w_3l, \\ K' = U_2u_1h + U_2v_1k + U_2w_1l + V_2u_2h + V_2v_2k + V_2w_2l + W_2u_3h + W_2v_3k + W_2w_3l, \\ L' = U_3u_1h + U_3v_1k + U_3w_1l + V_3u_2h + V_3v_2k + V_3w_2l + W_3u_3h + W_3v_3k + W_3w_3l. \end{cases}$$

Collecting the coefficients of the h's, k's, and l's gives

$$\begin{cases} H' = (U_1u_1 + V_1u_2 + W_1u_3)h + (U_1v_1 + V_1v_2 + W_1v_3)k + (U_1w_1 + V_1w_2 + W_1w_3)l, \\ K' = (U_2u_1 + V_2u_2 + W_2u_3)h + (U_2v_1 + V_2v_2 + W_2v_3)k + (U_2w_1 + V_2w_2 + W_2w_3)l, \quad (34) \\ L' = (U_3u_1 + V_3u_2 + W_3u_3)h + (U_3v_1 + V_3v_2 + W_3v_3)k + (U_3w_1 + V_3w_2 + W_3w_3)l. \end{cases}$$

This is in the form of the standard transformation equation. The matrices of (31), (33), and (34) may now be assembled, reversing the order of (31) and (33):

$$\begin{matrix} \mathbf{b} & & \mathbf{a} \\ (33) & & (31) \end{matrix}$$

$$\left\| \begin{matrix} U_1 & V_1 & W_1 \\ U_2 & V_2 & W_2 \\ U_3 & V_3 & W_3 \end{matrix} \right\| \cdot \left\| \begin{matrix} u_1 & v_1 & w_1 \\ u_2 & v_2 & w_2 \\ u_3 & v_3 & w_3 \end{matrix} \right\| =$$

$$\begin{matrix} \mathbf{c} \\ (34) \end{matrix}$$

$$\left\| \begin{matrix} (U_1u_1 + V_1u_2 + W_1u_3) & (U_1v_1 + V_1v_2 + W_1v_3) & (U_1w_1 + V_1w_2 + W_1w_3) \\ (U_2u_1 + V_2u_2 + W_2u_3) & (U_2v_1 + V_2v_2 + W_2v_3) & (U_2w_1 + V_2w_2 + W_2w_3) \\ (U_3u_1 + V_3u_2 + W_3u_3) & (U_3v_1 + V_3v_2 + W_3v_3) & (U_3w_1 + V_3w_2 + W_3w_3) \end{matrix} \right\| . (35)$$

A study of (35) shows that the rules for matrix multiplication $\mathbf{b \cdot a = c}$ hold also for three-dimensional matrices. The scheme is complicated only by the fact that, after selecting the row of the first and the column of the second, there are three elements in each. To derive each element of the composite matrix three products are formed, in order, and their sum placed in the correct position of the product matrix.

In order to fix the ideas of matrix multiplication the following concrete example is given: Axes were assigned to the triclinic crystal, axinite, in 1892 by Dana. In 1897, Goldschmidt chose a new set of axes, which could be described in terms of Dana's by means of the transformation

DANA TO GOLDSCHMIDT

$$\begin{cases} \mathbf{A} = 1\mathbf{a} + 1\mathbf{b} + 0\mathbf{c}, \\ \mathbf{B} = \bar{2}\mathbf{a} + 0\mathbf{b} + 0\mathbf{c}, \quad \text{matrix:} \\ \mathbf{C} = 0\mathbf{a} + 0\mathbf{b} + 2\mathbf{c} \end{cases} \left\| \begin{matrix} 1 & 1 & 0 \\ \bar{2} & 0 & 0 \\ 0 & 0 & 2 \end{matrix} \right\| .$$

Subsequently, in 1926, Friedel chose new axes for axinite which were related to Goldschmidt's axes by the following transformation:

GOLDSCHMIDT TO FRIEDEL

$$\begin{cases} A' = \bar{1}A + \bar{1}B + \bar{1}C \\ B' = 1A + 0B + 0C, \\ C' = 0A + 0B + 1C \end{cases} \quad \text{matrix:} \quad \begin{Vmatrix} \bar{1} & \bar{1} & \bar{1} \\ 1 & 0 & 0 \\ 0 & 0 & 1 \end{Vmatrix}.$$

We wish to know how Dana's axes and indices transform directly to Friedel's. To determine this, we multiply the second matrix by the first, thus:

GOLDSCHMIDT TO FRIEDEL	DANA TO GOLDSCHMIDT		DANA TO FRIEDEL

$$\begin{Vmatrix} \bar{1} & \bar{1} & \bar{1} \\ 1 & 0 & 0 \\ 0 & 0 & 1 \end{Vmatrix} \cdot \begin{Vmatrix} 1 & 1 & 0 \\ \bar{2} & 0 & 0 \\ 0 & 0 & 2 \end{Vmatrix} = \begin{Vmatrix} (\bar{1}\cdot 1 + \bar{1}\cdot\bar{2} + \bar{1}\cdot 0) & (\bar{1}\cdot 1 + \bar{1}\cdot 0 + \bar{1}\cdot 0) & (\bar{1}\cdot 0 + \bar{1}\cdot 0 + \bar{1}\cdot 2) \\ (1\cdot 1 + 0\cdot\bar{2} + 0\cdot 0) & (1\cdot 1 + 0\cdot 0 + 0\cdot 0) & (1\cdot 0 + 0\cdot 0 + 0\cdot 2) \\ (0\cdot 1 + 0\cdot\bar{2} + 1\cdot 0) & (0\cdot 1 + 0\cdot 0 + 1\cdot 0) & (0\cdot 0 + 0\cdot 0 + 1\cdot 2) \end{Vmatrix}$$

$$= \begin{Vmatrix} 1 & \bar{1} & \bar{2} \\ 1 & 1 & 0 \\ 0 & 0 & 2 \end{Vmatrix}$$

With the aid of the new matrix elements, it is possible to fill in the coefficients of both the axial and plane index transformation equations for a direct transformation from Dana to Friedel, as follow:

$$\begin{matrix} \text{Axial} \\ \text{transformation} \end{matrix} \quad \begin{cases} A' = 1a + \bar{1}b + \bar{2}c \\ B' = 1a + 1b + 0c \\ C' = 0a + 0b + 2c, \end{cases}$$

$$\begin{matrix} \text{Plane} \\ \text{index} \\ \text{transformation} \end{matrix} \quad \begin{cases} H' = 1h + \bar{1}k + \bar{2}l \\ K' = 1h + 1k + 0l \\ L' = 0h + 0k + 2l. \end{cases}$$

The identical transformation. When a set of axes, or a set of plane indices, is transformed to itself, this is called *the identical transformation*, whose symbol is usually given as **I** or **1**, sometimes as **E**. The identical transformation is evidently no transformation at all, and has the following unique set of coefficients:

$$\begin{cases} A = 1a + 0b + 0c \\ B = 0a + 1b + 0c, \\ C = 0a + 0b + 1c \end{cases} \quad \text{matrix:} \quad \begin{Vmatrix} 1 & 0 & 0 \\ 0 & 1 & 0 \\ 0 & 0 & 1 \end{Vmatrix}. \quad (36)$$

Any transformation, followed by the identical transformation, is evidently the original transformation itself, and this is confirmed in the multiplication of matrices:

$$\begin{Vmatrix} 1 & 0 & 0 \\ 0 & 1 & 0 \\ 0 & 0 & 1 \end{Vmatrix} \cdot \begin{Vmatrix} u_1 & v_1 & w_1 \\ u_2 & v_2 & w_2 \\ u_3 & v_3 & w_3 \end{Vmatrix} = \begin{Vmatrix} u_1 & v_1 & w_1 \\ u_2 & v_2 & w_2 \\ u_3 & v_3 & w_3 \end{Vmatrix}, \quad (37)$$

or

$$\mathbf{I} \quad \cdot \quad \mathbf{a} \quad = \quad \mathbf{a}.$$

The inverse transformation. If a transformation is made, and then a second transformation is made bringing the system back to its original condition, the net result is no transformation, i. e., the identical transformation. The return transformation is known as the *inverse transformation*. If the direct transformation is designated by **a**, it is customary to designate the inverse transformation by $\mathbf{a}^{-1}$. Thus,

$$\mathbf{a} \cdot \mathbf{a}^{-1} = \mathbf{I}. \quad (38)$$

The matrix of the inverse transformation is useful in recovering an original system of axes or plane indices from the final one if the direct transformation is known. In order to determine how the matrices of the direct and inverse transformations are related, suppose that the original transformation is

$$\begin{cases} H = u_1 h + v_1 k + w_1 l, \\ K = u_2 h + v_2 k + w_2 l, \\ L = u_3 h + v_3 k + w_3 l. \end{cases} \quad (39)$$

This is a system of simultaneous equations, and the inverse transformation is simply the solutions of these for the original indices h, k, and l. These solutions may be accomplished by direct elimination, but this procedure is very tedious. The use of determinants in this elimination gives rise to a comparatively compact solution. The determinant of (39) is

$$\Delta = \begin{vmatrix} u_1 & v_1 & w_1 \\ u_2 & v_2 & w_2 \\ u_3 & v_3 & w_3 \end{vmatrix}. \quad (40)$$

Provided that Δ is not zero, the solution of (39) may be immediately written in the following determinant form:

$$
\left\{
\begin{array}{l}
h = \dfrac{\begin{vmatrix} H & v_1 & w_1 \\ K & v_2 & w_2 \\ L & v_3 & w_3 \end{vmatrix}}{\Delta} \\[3em]
k = \dfrac{\begin{vmatrix} u_1 & H & w_1 \\ u_2 & K & w_2 \\ u_3 & L & w_3 \end{vmatrix}}{\Delta} \\[3em]
l = \dfrac{\begin{vmatrix} u_1 & v_1 & H \\ u_2 & v_2 & K \\ u_3 & v_3 & L \end{vmatrix}}{\Delta}
\end{array}
\right.
\tag{41}
$$

Each of the numerator determinants may be expanded by minors of H, K, and L to give the following form of (41):

$$
\left\{
\begin{array}{l}
h = \dfrac{\begin{vmatrix} v_2 & w_2 \\ v_3 & w_3 \end{vmatrix}}{\Delta} H - \dfrac{\begin{vmatrix} v_1 & w_1 \\ v_3 & w_3 \end{vmatrix}}{\Delta} K + \dfrac{\begin{vmatrix} v_1 & w_1 \\ v_2 & w_2 \end{vmatrix}}{\Delta} L \\[3em]
k = -\dfrac{\begin{vmatrix} u_2 & w_2 \\ u_3 & w_3 \end{vmatrix}}{\Delta} H + \dfrac{\begin{vmatrix} u_1 & w_1 \\ u_3 & w_3 \end{vmatrix}}{\Delta} K - \dfrac{\begin{vmatrix} u_1 & w_1 \\ u_2 & w_2 \end{vmatrix}}{\Delta} L \\[3em]
l = \dfrac{\begin{vmatrix} u_2 & v_2 \\ u_3 & v_3 \end{vmatrix}}{\Delta} H - \dfrac{\begin{vmatrix} u_1 & v_1 \\ u_3 & v_3 \end{vmatrix}}{\Delta} K + \dfrac{\begin{vmatrix} u_1 & v_1 \\ u_2 & v_2 \end{vmatrix}}{\Delta} L
\end{array}
\right.
\tag{42}
$$

(42) is a transformation in standard form, whose matrix is

$$
\left\|
\begin{array}{ccc}
\dfrac{\begin{vmatrix} v_2 & w_2 \\ v_3 & w_3 \end{vmatrix}}{\Delta} & -\dfrac{\begin{vmatrix} v_1 & w_1 \\ v_3 & w_3 \end{vmatrix}}{\Delta} & \dfrac{\begin{vmatrix} v_1 & w_1 \\ v_2 & w_2 \end{vmatrix}}{\Delta} \\[3em]
-\dfrac{\begin{vmatrix} u_2 & w_2 \\ u_3 & w_3 \end{vmatrix}}{\Delta} & \dfrac{\begin{vmatrix} u_1 & w_1 \\ u_3 & w_3 \end{vmatrix}}{\Delta} & -\dfrac{\begin{vmatrix} u_1 & w_1 \\ u_2 & w_2 \end{vmatrix}}{\Delta} \\[3em]
\dfrac{\begin{vmatrix} u_2 & v_2 \\ u_3 & v_3 \end{vmatrix}}{\Delta} & -\dfrac{\begin{vmatrix} u_1 & v_1 \\ u_3 & v_3 \end{vmatrix}}{\Delta} & \dfrac{\begin{vmatrix} u_1 & v_1 \\ u_2 & v_2 \end{vmatrix}}{\Delta}
\end{array}
\right\| .
\tag{43}
$$

The matrix of the direct transformation, (39), is

$$\left\| \begin{array}{ccc} u_1 & v_1 & w_1 \\ u_2 & v_2 & w_2 \\ u_3 & v_3 & w_3 \end{array} \right\| . \tag{44}$$

These two matrices are related in a very simple manner. To understand this, attention is directed to the coefficients of (42). The numerators of these coefficients are minors of the determinant (40). More specifically, they are cofactors, i. e., minors with their correct signs. In developing the determinants of (41) by minors, the minors of H, K, and L were written down as coefficients of a row of terms. These cofactors consequently appear in the final matrix (43) as a row of matrix elements. However, they are cofactors of a *column* of H, K, L terms in (41) and these terms replace a *column* of terms in the original determinant (40) or its matrix (44). Thus each numerator term of a row of the inverse matrix (43) is the cofactor of the terms in the corresponding column of the direct matrix (44).

Any element of the inverse matrix, (43), is a fraction having for its denominator the determinant of the direct matrix, and having for its numerator the cofactor of the term in the direct matrix which is symmetrical with itself across the principal diagonal (upper left–lower right).

To illustrate the use of the inverse transformation, assume the direct transformation already given for axinite, namely, "Dana to Goldschmidt":

$$\begin{cases} \mathbf{A} = 1\mathbf{a} + 1\mathbf{b} + 0\mathbf{c} \\ \mathbf{B} = \bar{2}\mathbf{a} + 0\mathbf{b} + 0\mathbf{c}, \\ \mathbf{C} = 0\mathbf{a} + 0\mathbf{b} + 2\mathbf{c} \end{cases} \quad \text{matrix:} \quad \left\| \begin{array}{ccc} 1 & 1 & 0 \\ \bar{2} & 0 & 0 \\ 0 & 0 & 2 \end{array} \right\| .$$

We wish to know the inverse matrix and transformation. The determinant of this matrix is:

$$\left| \begin{array}{ccc} 1 & 1 & 0 \\ \bar{2} & 0 & 0 \\ 0 & 0 & 2 \end{array} \right| = 1 \left| \begin{array}{cc} 0 & 0 \\ 0 & 2 \end{array} \right| - \bar{2} \left| \begin{array}{cc} 1 & 0 \\ 0 & 2 \end{array} \right| + 0 \left| \begin{array}{cc} 1 & 0 \\ 0 & 0 \end{array} \right|$$

$$\begin{aligned} &= 1(0-0) + 2(2-0) + 0(0-0) \\ &= \quad\quad 0 \quad + \quad 4 \quad + \quad 0 \\ &= \quad\quad 4. \end{aligned}$$

Perhaps the easiest way of applying the rules for forming the inverse matrix is to take the cofactors of the direct matrix elements by *columns* and write them in the inverse matrix by corresponding *rows*, then divide each term by the determinant, which is 4 in this case.   Following this system, the matrix of the inverse transformation develops as follows:

$$
\begin{Vmatrix}
\dfrac{(0\cdot2 - 0\cdot0)}{4} & \dfrac{-(1\cdot2 - 0\cdot0)}{4} & \dfrac{(1\cdot0 - 0\cdot0)}{4} \\[2mm]
\dfrac{-(\bar{2}\cdot2 - 0\cdot0)}{4} & \dfrac{(1\cdot2 - 0\cdot0)}{4} & \dfrac{-(1\cdot0 - \bar{2}\cdot0)}{4} \\[2mm]
\dfrac{(\bar{2}\cdot0 - 0\cdot0)}{4} & \dfrac{-(1\cdot0 - 1\cdot0)}{4} & \dfrac{(1\cdot0 - \bar{2}\cdot1)}{4}
\end{Vmatrix}
$$

$$
= \begin{Vmatrix}
\dfrac{0}{4} & \dfrac{-2}{4} & \dfrac{0}{4} \\[2mm]
\dfrac{4}{4} & \dfrac{2}{4} & \dfrac{0}{4} \\[2mm]
\dfrac{0}{4} & \dfrac{0}{4} & \dfrac{2}{4}
\end{Vmatrix}
$$

$$
= \begin{Vmatrix}
0 & -\tfrac{1}{2} & 0 \\[1mm]
1 & \tfrac{1}{2} & 0 \\[1mm]
0 & 0 & \tfrac{1}{2}
\end{Vmatrix}.
$$

The inverse transformation, " Goldschmidt to Dana," therefore, is

$$
\begin{cases}
\mathbf{A} = 0\mathbf{a} - \tfrac{1}{2}\mathbf{b} + 0\mathbf{c} \\
\mathbf{B} = 1\mathbf{a} + \tfrac{1}{2}\mathbf{b} + 0\mathbf{c} \\
\mathbf{C} = 0\mathbf{a} + 0\mathbf{b} + \tfrac{1}{2}\mathbf{c}.
\end{cases}
$$

Changes of cell volume with transformation.   When one selects a new set of axial vectors in terms of an original set, the new cell may or

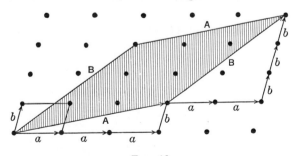

Fig. 10.

may not differ in volume.   The discussion of this can best start with the two-dimensional case:

Suppose that the geometry of a two-dimensional lattice, Fig. 10, was first referred to a set of axes, $\mathbf{a}$, $\mathbf{b}$, and subsequently a new cell based upon other vectors, $\mathbf{A}$, $\mathbf{B}$, was chosen. The new vectors in terms of the old are given by the transformation

$$\begin{cases} \mathbf{A} = u_1\mathbf{a} + v_1\mathbf{b}, \\ \mathbf{B} = u_2\mathbf{a} + v_2\mathbf{b}. \end{cases} \tag{45}$$

The area of the new cell is the vector (cross) product $\mathbf{A} \times \mathbf{B}$. By substituting from (45), this area can be developed thus:

$$\text{Area}_{A,B} = \mathbf{A} \times \mathbf{B} \tag{46}$$

$$= (u_1\mathbf{a} + v_1\mathbf{b}) \times (u_2\mathbf{a} + v_2\mathbf{b}) \tag{47}$$

$$= u_1u_2(\mathbf{a} \times \mathbf{a}) + u_1v_2(\mathbf{a} \times \mathbf{b}) + u_2v_1(\mathbf{b} \times \mathbf{a}) + v_1v_2(\mathbf{b} \times \mathbf{b}).$$

Remembering that, in vector algebra, $\mathbf{a} \times \mathbf{a} = 0$ and $\mathbf{b} \times \mathbf{b} = 0$, and that $\mathbf{b} \times \mathbf{a} = -\mathbf{a} \times \mathbf{b}$, this simplifies to

$$\text{Area}_{A,B} = 0 + u_1v_2(\mathbf{a} \times \mathbf{b}) - u_2v_1(\mathbf{a} \times \mathbf{b}) + 0$$
$$= (u_1v_2 - u_2v_1)\mathbf{a} \times \mathbf{b}. \tag{48}$$

The scalar part of this is evidently an expanded determinant, and the vector product, $\mathbf{a} \times \mathbf{b}$, is the area of a cell built upon the unit translations $\mathbf{a}$ and $\mathbf{b}$. Hence

$$\text{Area}_{A,B} = \begin{vmatrix} u_1 & v_1 \\ u_2 & v_2 \end{vmatrix} \text{area}_{a,b}. \tag{49}$$

Hence, *the transformation from one set of two-dimensional axes to another is accompanied by a multiplication of the cell area, and the multiplication factor is the determinant of the transformation. The multiplication factor is known as the modulus of the transformation.*

This discussion can easily be generalized to three dimensions. In this case the transformation of axes is

$$\begin{cases} \mathbf{A} = u_1\mathbf{a} + v_1\mathbf{b} + w_1\mathbf{c}, \\ \mathbf{B} = u_2\mathbf{a} + v_2\mathbf{b} + w_2\mathbf{c}, \\ \mathbf{C} = u_3\mathbf{a} + v_3\mathbf{b} + w_3\mathbf{c}. \end{cases} \tag{50}$$

The volume of the parallelepiped cell built upon the translations $\mathbf{A}$, $\mathbf{B}$, and $\mathbf{C}$ is

$$\text{Vol.}_{A,B,C} = \mathbf{A} \cdot \mathbf{B} \times \mathbf{C}$$
$$= (u_1\mathbf{a} + v_1\mathbf{b} + w_1\mathbf{c}) \cdot (u_2\mathbf{a} + v_2\mathbf{b} + w_2\mathbf{c}) \times (u_3\mathbf{a} + v_3\mathbf{b} + w_3\mathbf{c})$$

$$= \{\ u_1u_2\mathbf{a}\cdot\mathbf{a} + u_1v_2\mathbf{a}\cdot\mathbf{b} + u_1w_2\mathbf{a}\cdot\mathbf{c}$$
$$+ v_1u_2\mathbf{b}\cdot\mathbf{a} + v_1v_2\mathbf{b}\cdot\mathbf{b} + v_1w_2\mathbf{b}\cdot\mathbf{c}$$
$$+ w_1u_2\mathbf{c}\cdot\mathbf{a} + w_1v_2\mathbf{c}\cdot\mathbf{b} + w_1w_2\mathbf{c}\cdot\mathbf{c}\} \times (u_3\mathbf{a} + v_3\mathbf{b} + w_3\mathbf{c})$$

$$= \{u_1u_2\mathbf{a}\cdot\mathbf{a} + v_1v_2\mathbf{b}\cdot\mathbf{b} + w_1w_2\mathbf{c}\cdot\mathbf{c}$$
$$+(u_1v_2 + v_1u_2)\mathbf{a}\cdot\mathbf{b}$$
$$+(v_1w_2 + w_1v_2)\mathbf{b}\cdot\mathbf{c}$$
$$+(u_1w_2 + w_1u_2)\mathbf{a}\cdot\mathbf{c}\} \times (u_3\mathbf{a} + v_3\mathbf{b} + w_3\mathbf{c}). \tag{51}$$

Remembering that if two members of the triple scalar product, [rst], are the same the product vanishes, (51) reduces to

$$\text{Vol.}_{A,B,C} = 0 + 0 + 0 + 0 + 0 + 0 + 0 + 0 + 0$$
$$+(u_1v_2 + v_1u_2)(\quad 0 \quad + \quad 0 \quad + \quad w_3[\mathbf{abc}])$$
$$+(v_1w_2 + w_1v_2)(u_3[\mathbf{bca}] \quad + \quad 0 \quad + \quad 0 \quad)$$
$$+(u_1w_2 + w_1u_2)(\quad 0 \quad + v_3[\mathbf{acb}] + \quad 0 \quad)$$

$$= u_3(v_1w_2 + w_1v_2)[\mathbf{abc}] - v_3(u_1w_2 + w_1u_2)[\mathbf{abc}] + w_3(u_1v_2 + v_1u_2)[\mathbf{abc}]. \tag{52}$$

The scalar terms in parentheses are evidently expanded determinants, so (52) may be written more compactly thus:

$$\text{Vol.}_{A,B,C} = \left\{ u_3 \begin{vmatrix} v_1 & w_1 \\ v_2 & w_2 \end{vmatrix} - v_3 \begin{vmatrix} u_1 & w_1 \\ u_2 & w_2 \end{vmatrix} + w_3 \begin{vmatrix} u_1 & v_1 \\ u_2 & v_2 \end{vmatrix} \right\} [\mathbf{abc}] \tag{53}$$

It is further evident that the scalar part of (53) represents an expansion of a determinant by minors. It may be compacted to the form

$$\text{Vol.}_{A,B,C} = \begin{vmatrix} u_1 & v_1 & w_1 \\ u_2 & v_2 & w_2 \\ u_3 & v_3 & w_3 \end{vmatrix} [\mathbf{abc}] \tag{54}$$

and, finally, [abc] is the volume of the parallelepiped cell built upon the translation **a**, **b**, and **c**, hence

$$\text{Vol.}_{A,B,C} = \begin{vmatrix} u_1 & v_1 & w_1 \\ u_2 & v_2 & w_2 \\ u_3 & v_3 & w_3 \end{vmatrix} \text{Vol.}_{a,b,c}. \tag{55}$$

Consequently, *a transformation from one set of axes to another is accompanied by a multiplication of the cell volume, and the multiplication factor, or modulus, is the determinant of the transformation.* When the modulus is 1, the transformation is said to be unimodular.

Suppose that the transformation is from a set of primitive axes to some new axes:

$$\begin{cases} \mathbf{A} = u_1\mathbf{a} + v_1\mathbf{b} + w_1\mathbf{c}, \\ \mathbf{B} = u_2\mathbf{a} + v_2\mathbf{b} + w_2\mathbf{c}, \\ \mathbf{C} = u_3\mathbf{a} + v_3\mathbf{b} + w_3\mathbf{c}. \end{cases}$$

Then the u's, v's, and w's are all integral, and hence their determinant and modulus, which is a sum of the products of the u's, v's, and w's, must be integral.   The new cell is primitive, doubly primitive, triply primitive, quadruply primitive, etc., according as the modulus of the transformation is 1, 2, 3, 4, etc., respectively.   It should be observed, however, that the modulus (volume change factor) need not necessarily be integral; it can also be a simple fraction.   This occurs, for example, when transforming from a multiply primitive cell to one of lower multiplicity.   For example, in the transformation from doubly primitive, triply primitive, quadruply primitive, back to a primitive cell, the modulus (volume change factor) is $\frac{1}{2}$, $\frac{1}{3}$, and $\frac{1}{4}$, respectively. In the transformation of axinite axes from " Dana to Goldschmidt," the modulus is 4, but from " Goldschmidt to Dana " it was $\frac{1}{4}$.   The modulus of any inverse transformation, in general, is the reciprocal of the modulus of the direct transformation.

It should also be observed that, if the transformation is made from a primitive cell to another cell, the number of lattice points associated with the new cell is the modulus of the transformation.   This is because there is one lattice point associated with the primitive cell, and the new cell has a volume factor increase equal to the modulus. It accordingly contains the modulus times as many lattice points. Thus, in Fig. 10, **a** and **b** outline a primitive cell.   The specific transformation indicated is

$$\begin{cases} \mathbf{A} = 3\mathbf{a} + 1\mathbf{b}, \\ \mathbf{B} = 2\mathbf{a} + 3\mathbf{b}. \end{cases}$$

The modulus here is

$$\Delta = \begin{vmatrix} 3 & 1 \\ 2 & 3 \end{vmatrix} = 3 \cdot 3 - 2 \cdot 1 = 9 - 2 = 7.$$

In accordance with this, the illustration shows one lattice point at the cell corners, plus six completely enclosed by the cell.   The new cell is sevenfold primitive.

LITERATURE

Matrices
   MAXIME BÔCHER.   *Introduction to higher algebra.*   (Macmillan, New York, 1936.)   See especially pages 60–78.
   A. C. AITKIN.   *Determinants and matrices.*   (Oliver and Boyd, London, 1939.)

Application of matrices to coordinate transformations
   J. D. H. DONNAY.   Transformation of co-ordinates.   *Am. Mineralogist* **22** (1937), 621–624.

CHAPTER 3

THE DIFFRACTION OF X-RAYS BY CRYSTALS

THE DIFFRACTION OF X-RAYS BY SIMPLE LATTICE ARRAYS OF ATOMS

Scattering of x-rays by individual atoms. Like visible light, x-rays may be regarded as an electromagnetic wave disturbance radiating from the point of origin. The wavelengths of x-rays useful in crystal-structure determinations vary between about 0.5 and 3 Ångström units. This unit of length, abbreviated Å, is equal to 1×10^{-8} cm.

If an electron happens to be in the path of an x-ray wave, Fig. 11, it is set into forced vibrations by the periodically changing electric field of the x-ray waves passing by it. Such oscillation involves acceleration and deceleration of the electron. Now an accelerating electrically charged particle is itself a source of electromagnetic disturbance; hence, since the electron is oscillating in phase with the x-ray wave, it is the source of an electromagnetic wave of the same frequency and wavelength as the original x-ray wave. By this interaction the electron is said to *scatter* the original x-ray wave.

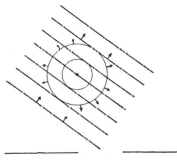

Fig. 11.

An atom consists of a swarm of electrons about a positively charged nucleus. The nucleus may be neglected in connection with scattering because of its relatively great mass. Each of the electrons in the atom scatters x-rays. The atom as a whole, therefore, scatters x-rays to an extent dependent on the number of electrons it has, i.e., dependent on its atomic number. The intensity of scattered radiation, however, varies with direction, falling off as the angle between the directions of the scattered radiation and original x-ray beam increases. This occurs for reasons brought out later. For present purposes, an atom in an x-ray beam may be considered as a point source of scattered x-radiation.

Cooperative scattering by a row of atoms. Figure 12 shows a series of wavefronts of x-rays impinging on a row of regularly spaced atoms.

29

Each atom scatters the x-radiation as discussed in the last section, producing about itself a new set of spherical wave envelopes. Any line-up of envelopes constitutes a combined wave moving in the direction of the common tangent, as shown in the figure. When it is near the atoms which produce it, such a wavefront is somewhat crenulated,

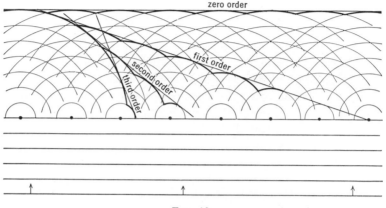

Fig. 12.

in detail, but at long distances from this source, each spherical wavefront has such a large radius of curvature that the crenulations vanish, and the wavefront becomes a straight line coinciding with the common tangent to the spherical wave envelopes.

This cooperative combining of scattered wavelets is known as *diffraction*. It is evident, from the above discussion, that a diffracted wavefront develops along the tangent to any line-up of spherical wavefronts. Perhaps the most obvious tangent is parallel with the original wavefront. This wave is known as the *zero-order diffraction wave*, or the *direct beam*. The figure also shows that there are other ways of developing tangents to the spherical wave surfaces. The next simplest one is a tangent starting at the innermost spherical envelope of one atom, continuing through the second nearest envelope of the next atom, through the third nearest envelope of the next atom, and so on. Between neighboring atoms the envelope difference is 1. The wave built up along this front is known as the *first-order diffraction wave*. In a similar way, a tangent which connects envelope differences of 2 between neighboring atoms builds up the *second-order diffraction wave*, and in general, an *nth order diffraction wave* is built up by the cooperation of wavelets of neighboring atoms differing by n envelopes or wavelengths phase difference.

Figure 12 showed the special case of a wave traveling exactly at right angles to a row of atoms. The more general case of a wave striking a row of atoms at some oblique angle is illustrated in Fig. 13. The two figures are analogous; the second lacks some of the special features of the first which are dependent on symmetry.

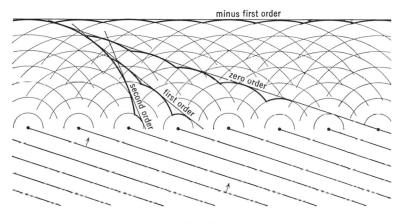

FIG. 13.

Conditions for diffraction by a row of atoms. It has just been seen that a diffracted wavefront develops along any common tangent to the spherical wave envelopes scattered by a row of atoms. The diffracted wave is propagated as a beam normal to this tangent. This can be restated in another way: A diffracted beam develops in any direction such that the envelope difference in the direction of propagation is 1, 2, 3, · · ·, Fig. 14A. In other words, a diffracted beam develops in any direction such that the wavelet contributions from two neighboring atoms are in phase in that direction. This formulation of diffraction is illustrated in Fig. 14B, which shows a wavefront, MN, advancing upon a row of atoms, and a wavefront ST produced by cooperative scattering by the row of atoms. For simplicity, attention is directed to the two particular atoms O and R. In order that the radiation scattered by O and R be in phase in the direction OS, path $MOQS$ must be an integral number of wavelengths longer than $NPRT$. Since $MO = NP$, and $QS = RT$, this requires that OQ be m whole wavelengths longer than PR; i.e., the following formulation must hold:

$$OQ - PR = m\lambda. \tag{1}$$

If a is the spacing of atoms along the row, then

$$\cos \bar{\nu} = \frac{OQ}{a} \qquad (2)$$

and

$$\cos \bar{\mu} = \frac{PR}{a}. \qquad (3)$$

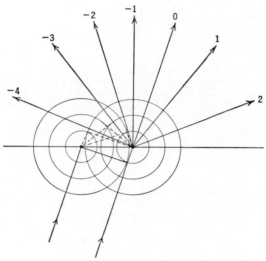

Fig. 14A.

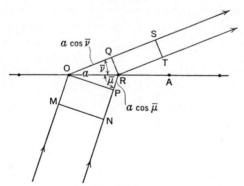

Fig. 14B.

Substituting the values of OQ and PR indicated by (2) and (3) in (1) gives

$$a \cos \bar{\nu} - a \cos \bar{\mu} = m\lambda,$$

or

$$a(\cos \bar{\nu} - \cos \bar{\mu}) = m\lambda. \qquad (4)$$

This relation can also be interpreted in the following way: Given a row of atoms separated by translation a, and a beam of x-radiation of wavelength λ, inclined to the row by angle $\bar{\mu}$, then a diffracted beam will be developed by cooperative scattering by the row of atoms, and its direction will be

$$\cos \bar{\nu} = \cos \bar{\mu} + \frac{m\lambda}{a},\qquad(5)$$

where m is some integer.

Equation (5) provides the angle, $\bar{\nu}$, of the mth-order beam of diffracted radiation, given the angle $\bar{\mu}$, which the incident radiation makes with the row of atoms. The directions satisfying (5) for a given set of conditions lie on a cone of angular half opening, $\bar{\nu}$. The loci of the several orders of diffraction from a line of atoms thus form a series of cones coaxial with the line of atoms and having a common apical point at the position where the incident x-ray beam meets the line of atoms, as shown in Fig. 15.

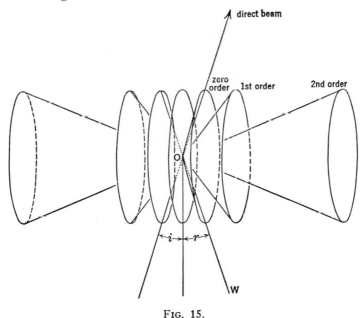

FIG. 15.

Several interesting points should be noticed in connection with diffractions from a line of atoms. When $m = 0$, it follows from (5) that $\bar{\nu} = \bar{\mu}$; i.e., the zero-order cone of diffraction has an internal cone half-angle equal to the angle between incident beam and the line of atoms. The incident beam is therefore a generator line on the surface of the zero-order cone. The generator line, OW, on the diametrically opposite side of the zero-order cone acts as if it were reflected from the line of atoms, for the angle of incidence equals the angle of reflection. In a subsequent section it will be seen that diffraction from a three-dimensional lattice array of points is a generalization of this reflection effect.

It should be pointed out that fractional orders of diffraction do not exist. The derivation of (5) depended on the scattered radiation being in phase in the diffraction direction, and therefore the phase interval between adjacent rays was exactly $0, 1, 2, 3, \cdots$ wavelengths; i. e., only integral orders exist. In connection with this, a question which may arise in the student's mind is why a diffracted beam is not built up by cooperation of alternate atoms. If this were the case, then the first order from alternate atoms is composed of radiation from alternate atoms $360°$ out of phase. The radiation of the intermediate atoms is exactly $180°$ out of phase with this and so destroys it. A similar proof can be given for the non-existence of cooperation by every m atoms, or what amounts to the same thing, for the non-existence of any fractional order.

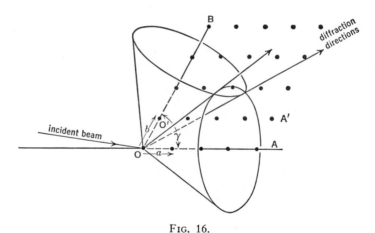

F<small>IG</small>. 16.

Diffraction by a plane lattice array of atoms. A plane lattice array is shown in Fig. 16. It may be defined in terms of two translations, a and b (which correspond with the regular spacing of atoms along the line in the one-dimensional line of atoms), and also the interaxial angle γ, the angle between the two primitive translations. The plane lattice may be conveniently thought of as being generated by the line lattice, OA, repeated by the translation, b.

For the axial lattice directions OA, the locus of directions for which the scattered rays are in phase for the mth order is the surface of a cone coaxial with OA. In a similar way for the axial direction OB, the locus of directions for which the scattered rays are in phase for the nth order is a cone of rays coaxial with OB. The two lines where these two cones intersect are directions along which all the atoms in rows OA and OB are simultaneously in phase. If another line, $O'A'$, is paired with $O'B$ and similarly considered, then two identical cones and two identical cone intersections define the diffraction maxima. Since the latter two maxima are identical in *direction* with the first two derived, the diffraction maxima of rays scattered by all the atoms in the first two horizontal rows extend in the same direction. Now one atom of $O'A'$, namely O', is also contained in the line OB, all of whose atoms are scattering radiation in phase with the lines OA and $O'A'$ along the two cone intersection directions. Hence all the atoms in lines OA and $O'A'$ are scattering in phase along these two lines. By considering additional rows parallel with OA in the same manner, it can be shown that the entire plane lattice array is scattering radiation in phase in the two cone intersection directions.

Two interesting points about the geometry of diffraction by a plane lattice should be noticed. In the first place, the two cone intersection directions are symmetrically disposed on both sides of the plane of the plane lattice regardless of the direction of the incident beam. This follows from the fact that the diffraction maximum of each axial line is a cone coaxial with the line. In the second place, only two diffraction cones have been illustrated and discussed for the sake of clearness, one corresponding with an mth order for the lattice row OA and the other corresponding with the nth order for the row OB. All orders, within the limits discussed in the last section, are possible, however, and the cone of any order about axial line OA may (possibly) intersect a cone of any order about axial line OB within certain limits. There fore when a beam of x-rays is incident upon a plane lattice there are produced, in general, a large number of cone intersections, symmetrically disposed on both sides of the lattice plane and related as mirror images of one another, each intersection representing a combination of some order of diffraction of one axial line with some order of diffraction of the second chosen axial line.

A more quantitative picture of diffraction by a plane lattice is the following: The condition that the line OA scatters in phase is

$$a(\cos \bar{\nu}_1 - \cos \bar{\mu}_1) = m\lambda. \tag{6}$$

The condition that the second line, OB, scatters in phase is similarly

$$b(\cos \bar{\nu}_2 - \cos \bar{\mu}_2) = n\lambda. \tag{7}$$

When conditions (6) and (7) are simultaneously satisfied, then the entire plane lattice scatters in phase. Equations (6) and (7) may be rearranged in the form,

$$\begin{cases} \cos \bar{\nu}_1 = \cos \bar{\mu}_1 + \dfrac{m}{a}\lambda, \\[2ex] \cos \bar{\nu}_2 = \cos \bar{\mu}_2 + \dfrac{n}{b}\lambda. \end{cases} \quad (8)$$

The two left members of these simultaneous conditions are the direction cosines of the diffracted ray referred to the oblique coordinate axes, OA and OB, respectively; these are separated by the interaxial angle,

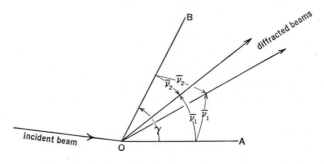

Fig. 17.

γ, as shown in Fig. 17. Now, given a pair of direction cosines referred to two different axes in space, Fig. 17, a pair of directions is uniquely defined except when the sum of the angles corresponding to the direction cosines is less than the angle between the coordinate axes. It follows, therefore, that with a given value of the wavelength λ, given values of the integers m and n, there will exist for any two-dimensional lattice in any orientation (within certain limits) a pair of direction cosines for the diffracted radiation which define a pair of real directions. This is the direction of the diffracted radiation for the given values of m and n under consideration. Similar solutions are also possible for all other combinations of integral values of m and n within certain limits. These integers cannot increase without limit because on increasing m, for example, a point is eventually reached when an increase by another whole number so increases the last term that the whole right member of (8) becomes greater than unity and a solution for $\cos \bar{\nu}$ consequently becomes impossible.

The integers m and n are known as the *indices* of the diffracted beam. For a given set of conditions a, b, λ, $\bar{\mu}_1$, and $\bar{\mu}_2$, a certain *field of indices*, m, n is possible. If λ is too large, or a and b too small, the field may be non-existent because the entire right expression of (8) exceeds unity, for which there is no solution for the left member.

Diffraction by a plane lattice array of atoms in plane space. The easiest appreciation of the more complicated theory of the diffraction of x-rays by actual three-dimensional crystals is through a study of the simpler analogous situation of the diffraction of x-rays in two-dimensional space by a plane lattice.

In Fig. 16 it is obvious that the two diffraction cones, in general, intersect outside the plane of the plane lattice. It is therefore evident that, for a given direction of incident beam, there is, in general, no diffraction maximum *in the plane of the plane lattice*. If, however, the orientation of the lattice and beam with respect to each other is altered, the cone angles change and the lines of cone intersection consequently migrate along the cone surfaces. A special case of this intersection develops when the cones are tangent to one another, when the pair of diffraction directions degenerates to a single line in the plane of the lattice. Under such circumstances, but under no other, the two-dimensional crystal diffracts x-rays in its own two-dimensional space.

As pointed out in the last section, there are a number of cones of different orders for each axial lattice direction, and each cone on the first lattice line is potentially able to form a pair of intersections with each of the cones of the other lattice line. Each such pair of intersections migrates along the cone surfaces if the crystal is rotated with respect to the x-ray beam, and, as each pair degenerates to a straight line in the plane of the lattice, diffraction within the plane of the plane lattice takes place instantaneously. Diffraction in a plane thus takes place only at definite orientations of the lattice with respect to the incident x-ray beam.

Another way of looking at the phenomenon of diffraction in a plane is as follows: In the last section it was shown that a simultaneous solution of the two relations (8) always exists between certain limits, because in three-dimensional space it is possible to specify two direction cosines independently. If the diffraction is limited to a plane, however, then only one direction cosine may be independently specified, for the other is related to it by the more confined geometry of a plane. Thus, in Fig. 18A, it is evident that, if angle $\bar{\nu}_2$ is specified, and the interaxial angle is γ, then $\bar{\nu}_1$ is fixed by the relation

$$|\gamma| = |\bar{\nu}_1| + |\bar{\nu}_2|. \tag{9}$$

Given a beam of x-rays incident upon a plane lattice array of atoms, therefore, the scattering in phase for one axial row occurs in a direction indicated by the upper member of (8) while scattering in phase for the other axial row occurs in the direction indicated by the lower member of (8). The variables of angle and spacing are independent in these two rows, and consequently the in-phase scattering angles, $\bar{\nu}$, of the

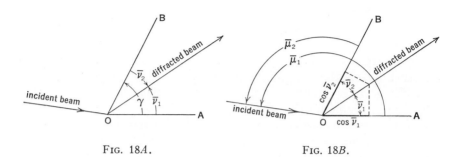

FIG. 18A.   FIG. 18B.

two rows are independent of one another. Unless these two directions are fortuitously related by (9) there is no *simultaneous* scattering in phase in the plane and no diffraction occurs. If the lattice is rotated, however, the first part of the right-hand members of (8) continually changes, and whenever this variation causes the left members to be related by (9), diffraction occurs in the plane. The geometrical interpretation of (8) is then as diagrammatically shown in Fig. 18B.

The " reflection " of x-rays.   Diffraction maxima may be expressed not only in terms of combinations of orders of diffraction from axial lines, but also in terms of reflection from lattice lines (in three dimensions, from lattice planes):

In Fig. 19, an x-ray beam is incident on a plane lattice array of atoms, only a few important points of which are shown, and diffraction is actually taking place in the plane of the lattice. The lattice line OA is diffracting radiation in the mth order, and the line OB is diffracting in the nth order. This means that, between the waves scattered by adjacent atoms M and O, m wavelengths' difference exists, and between N and O, n wavelengths' difference exists. It follows, therefore, that, n lattice points out along the line OA, a wave is scattered by an atom, S, differing in phase by $n(m)$ wavelengths from that scattered by the origin lattice point. Similarly, m lattice points out along the line OB, a wave is scattered by an atom, T, differing in phase by $m(n)$ wavelengths from that scattered by the origin lattice point. Since S and T scatter waves differing in phase from the origin by the same number of wavelengths, namely mn, they therefore scatter waves differing by zero wavelengths from each other.

The two atoms S and T lie on, and determine, a rational lattice line having intercepts n and m respectively, and the line accordingly has lattice indices of $\left(\dfrac{1}{n}, \dfrac{1}{m}\right) \backsimeq (mn)$. Since the paths from a wavefront before diffraction to a wavefront after diffraction differ by zero wavelengths for the rays scattered by S and T and, indeed, for rays scat-

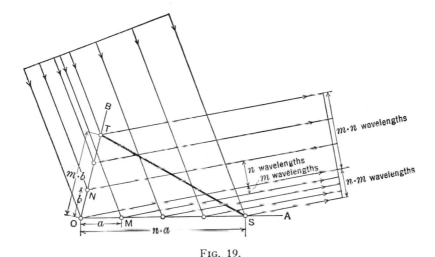

Fɪɢ. 19.

tered by all atoms on the line ST, these rays diffracted by line ST suffer no change in total length and the line therefore behaves as if it were "reflecting" x-rays. Referring back to Fig. 15, it is also evident that if diffraction takes place simultaneously, as above described, in the mth order from line OA and the nth order from line OC, this is equivalent to diffraction taking place in the 0th order from lines of index (mn).

Diffraction by a three-dimensional lattice array of atoms. The diffraction of x-rays by an actual three-dimensional lattice array of atoms may be understood by a generalization of the development already given for diffraction by a two-dimensional lattice. Any three rows may be chosen in the three-dimensional lattice as axial rows or coordinate reference lines. For the sake of clearness only these three rows, OA, OB, and OC, of the lattice are illustrated in Fig. 20. When an x-ray beam is incident upon the lattice array, each of these rows diffracts x-rays along a nest of cones coaxial with the respective row. Consider only the cone of order m for row OA, the cone of order n for row OB, and the cone of order p for row OC. In general, these cones may or may not intersect one another. In the event that each mutually intersects both of the others, six intersections are produced, OP, OQ, OR, OS, OT, and OU. Along the directions of these six lines, the diffracted radiation from two rows of atoms at a time, i. e., a plane of atoms, is being scattered in phase. All the atoms of the lattice array are *not* simultaneously scattering in phase in any

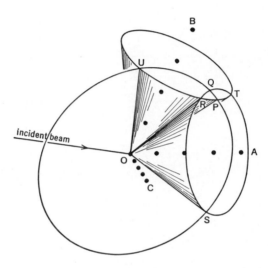

Fig. 20.

direction unless the three lines OP, OQ, and OR chance to coincide, that is, unless the three cones have a single line in common. For any given direction of the incident x-ray beam, therefore, the whole lattice array does not diffract x-rays. Since, however, angular opening of a cone is a function of the angle between the incident x-ray beam and the lattice row, it is possible to change the three cone angles by an appropriate shift in x-ray-beam angle in such a way that OP, OQ, and OR come to coincide. When this is accomplished, every row and thus every atom in the lattice array is scattering in phase in the common direction of cone intersection, and a diffracted beam of x-rays is developed traveling along this line.

Diffraction in a three-dimensional lattice may also be treated analytically by the method used for the two-dimensional lattice: If a, b, and c are the spacings of atoms along the three axial rows, OA, OB, and OC, then the conditions that each of these rows scatters as rows are similar to (6) and (7), namely:

For row OA: $a(\cos \bar{\nu}_1 - \cos \bar{\mu}_1) = m\lambda,$
For row OB: $b(\cos \bar{\nu}_2 - \cos \bar{\mu}_2) = n\lambda,$ (10)
For row OC: $c(\cos \bar{\nu}_3 - \cos \bar{\mu}_3) = p\lambda.$

These are the *Laue equations* of x-ray diffraction. For an x-ray diffraction beam to develop, these three conditions must be satisfied
along the same direction simultaneously. Relations (10) may be
conveniently rearranged as follows:

$$
\left\{
\begin{aligned}
\cos \bar{\nu}_1 &= \cos \bar{\mu}_1 + \frac{m\lambda}{a}, \\
\cos \bar{\nu}_2 &= \cos \bar{\mu}_2 + \frac{n\lambda}{b}, \\
\cos \bar{\nu}_3 &= \cos \bar{\mu}_3 + \frac{p\lambda}{c}.
\end{aligned}
\right.
\qquad (11)
$$

The direction of the diffracted beam is defined by the direction
cosines, $\cos \bar{\nu}_1$, $\cos \bar{\nu}_2$, and $\cos \bar{\nu}_3$, which are referred to the oblique

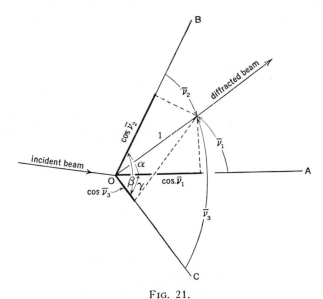

FIG. 21.

lattice coordinate directions OA, OB, and OC, Fig. 21. For cooperative scattering to occur in some direction, i. e., for x-ray diffraction
actually to occur, it is necessary that each of the directions defined

by (11) be the same. Now, in three dimensions, if two direction cosines are given, the third is necessarily fixed. It therefore follows that any two of the equations of (11) may have a simultaneous solution and define a common direction, but this direction, in general, is incompatible with the direction defined by the third equation, because each direction is conditioned by the independent variables of x-ray beam direction, $\bar{\mu}$, order m, n, or p, and lattice row distance, a, b, or c. In general, therefore, diffraction does not occur. Since, however, the angle, $\bar{\mu}$, between the beam and the lattice rows, is variable, it is possible so to arrange this variable that the three cos $\bar{v}$'s define an identical direction. The entire lattice array then scatters x-radiation in phase in this direction, and diffraction occurs.

It can be shown in this three-dimensional case, just as for the two-dimensional lattice array, that, when diffraction occurs in the mth order for the row OA, in the nth order for the row OB, and the pth order for the row OC, this is equivalent to a reflection of the incident x-ray beam by the atoms of the planes (mnp). The proof of this is simply a generalization of that given on pages 38–39: An atom one translation interval out on row OA scatters x-rays m wavelengths out of phase with the origin atom; an atom np translation intervals out on this same row scatters x-rays $(np)m$ wavelengths out of phase with the origin atom. A similar relation holds for each co-ordinate row of atoms, thus:

Atom row	Order of diffraction for row (i. e., number of wavelengths' difference in phase between wave scattered by origin atom and first atom of row)	Number of translations from origin to atom which is scattering a wave whose phase differs by mnp wavelengths with that scattered by the origin atom
OA	m	np
OB	n	mp
OC	p	mn

All the atoms in the right column scatter waves with the same number of wavelengths' difference from the wave scattered by the origin atom, O, namely mnp wavelength. The atoms in the right column also define a plane, whose intercepts are np, mp, and mn, along the a-, b-, and c-axes, respectively. According to the general discussion on pages 7–8, the intercept form of the equation of this plane is

$$\frac{x}{np} + \frac{y}{mp} + \frac{z}{mn} = 1. \tag{12}$$

Multiplying by mnp gives

$$\frac{mnp\,x}{np} + \frac{mnp\,y}{mp} + \frac{mnp\,z}{mn} = mnp,$$

which reduces to

$$mx + ny + pz = mnp. \tag{13}$$

This represents a plane of index (mnp), but it is not the nearest plane to the origin; it is removed from the origin plane by $m \cdot n \cdot p$ planes. Every atom in the plane scatters in exactly the same absolute phase. This is equivalent to saying that the path from an incoming wavefront, to an atom of the plane, and on to a wavefront of the outgoing, diffracted wave, is identical for all atoms in the plane. This condition is identical with the condition for reflection from this plane.

From the above discussion, the following conclusion can be drawn: *If a lattice array of atoms diffracts x-rays so that the a-axis row is diffracting in the mth order, the b-axis row in the nth order, and the c-axis row in the pth order, then this is geometrically equivalent to a reflection of the x-ray beam by the plane (mnp) referred to these axes.*

Bragg's law. In the foregoing development of the theory of diffraction of x-rays by a lattice array of atoms, the viewpoint has been that of treating the phenomenon as a formal problem in diffraction, i.e., that the diffraction from a lattice array of atoms is the cooperative diffraction from three non-coplanar rows of atoms. This way of looking at the problem is both formal and cumbersome. The discussion of the section just preceding, however, has proved that the whole phenomenon is equivalent to cooperative reflection from planes of atoms. This is a geometrically much more compact way of regarding it, and because of the simplicity of this approach, the present-day tendency is to speak of the diffraction of x-rays by a crystal as the cooperative reflection of the x-rays by the planes of the crystal. It is the purpose of the present section to look anew at the phenomenon of x-ray diffraction from this viewpoint.

Figure 22 shows a lattice array of atoms, and attention is drawn to the alignment of the atoms into planes, for example, those indicated in Fig. 23. Several of these planes, designated sheets 1, 2, and 3, will be referred to in subsequent development. Attention is first directed to any of these planes, say sheet 1, by itself. Every point of this plane reflects x-rays regardless of the angle, θ, which the x-ray beam makes with the plane. That this is so is evident from the following considerations: In Fig. 24, the x-ray beam, in phase along the wavefront, ABC, is scattered by the atoms Q, R, and S. The several rays

AQ, BR, and CS can combine, after scattering, to form a wavefront, provided that they are in phase with each other. This means that the total paths AQA', BRB', and CSC', each contain the same number

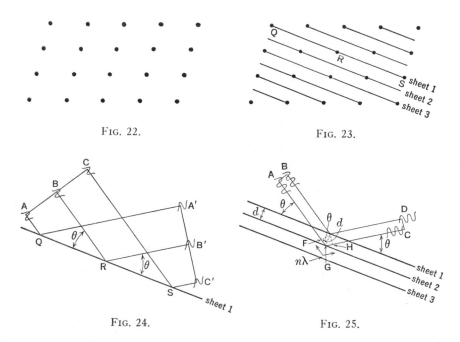

FIG. 22.   FIG. 23.

FIG. 24.   FIG. 25.

of waves, that is, that the absolute distances $AQA' = BRB' = CSC'$. This can occur in only two ways. One of these corresponds with a continuation of the direct beam and can therefore be neglected for this purpose. The other corresponds with a reflection from the plane QRS as shown. Thus, if x-rays strike a single plane of atoms at any glancing angle, θ, they are diffracted in such a way as to obey the standard laws of optical reflection. (It should be observed in passing that the equality of optical paths $AQA' = BRB' = CSC'$ holds not only for the particular points Q, R, and S in the plane, but for all points; i.e., any point in the plane would act as a reflecting point contributing to the building-up of wavefront $A'B'C'$ if there were only an atom there to scatter the x-radiation. In other words, the ability of a plane of atoms to reflect x-rays does not depend upon the regular arrangement of the atoms in a plane-lattice array in it. However, if the atoms are not in plane-lattice array in the plane, this would be the only *plane* of atoms in the assemblage, and there would consequently be no other planes to reflect x-rays.)

It is not sufficient, however, to know that the atoms of any single plane reflects x-rays at all angles, θ. Not only must each of the planes of atoms, sheet 1, sheet 2, sheet 3, etc., Fig. 23, reflect x-rays but also they must *cooperatively* reflect x-rays, i. e., they must not give rise to destructive interference. The condition that the reflections from each of the planes of atoms do not thus annul one another is that their individual reflected waves be in phase. The way this can occur is illustrated in Fig. 25. Each plane farther down the stack offers a longer path to the x-rays it reflects. If this additional path is exactly an integral number of wavelengths, then all the reflected waves are in phase again along the front DC, and the waves from each plane reinforce one another. The additional path for sheet 2 is evidently FGH, and this must equal some integral number of wavelengths. If n is an integer, this can be expressed:

$$n\lambda = FGH. \tag{14}$$

Since the path AGC is a reflected ray, the incident and reflected parts of the path are symmetrical and equal; the additional path can therefore be split in half, giving

$$\frac{n\lambda}{2} = FG. \tag{15}$$

In Fig. 25 it is evident that the length FG is related to the interplanar spacing d, and to the glancing angle θ, by the relation

$$\sin\theta = \frac{FG}{d}. \tag{16}$$

Substituting the value of FG from (15) into (16) gives the very important relation

$$\sin\theta = \frac{n\lambda}{2d}. \tag{17}$$

This is known as *Bragg's law* and is usually stated in the rearranged form

$$n\lambda = 2d\sin\theta. \tag{18}$$

The form given in (17) is most convenient for the discussion of many problems in crystal diffraction. It provides the angle, θ, at which x-rays are cooperatively reflected by all sheets of a given plane system of spacing d. Note that, for a given experimental set-up with monochromatic radiation, the wavelength λ is fixed, and therefore (17) has only a particular set of solutions, namely,

$$\theta = \sin^{-1} 1\left(\frac{\lambda}{2d}\right),$$

$$\sin^{-1} 2\left(\frac{\lambda}{2d}\right),$$

$$\sin^{-1} 3\left(\frac{\lambda}{2d}\right) \cdots \text{etc.}, \tag{19}$$

corresponding with 1, 2, 3, etc., wavelengths' path difference for adjacent sheets, Fig. 26. These reflections are known as the first-, second-, third-, etc., order reflection, according as n is 1, 2, 3, etc.

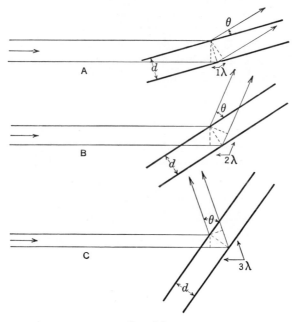

FIG. 26.

This means that the crystal cannot give rise to reflections at any angle, but only at those certain discrete angles indicated by (19). Furthermore, the total number of these solutions is distinctly limited because, as the order n increases, the fraction $\dfrac{n\lambda}{2d}$ increases; but since the upper limit of the sine is 1, it cannot increase beyond this value.

Thus, for a particular set of planes of spacing d, n can only have values from zero *up to, but not including*, the number which makes the term $n \dfrac{\lambda}{2d}$ just exceed unity. This limitation can also be grasped geometrically from Fig. 26. Reflection can occur only so long as the path difference $n\lambda$ remains less than $2d$.

All this may be briefly epitomized by saying that, for a given experimental set-up in which the wavelength is fixed, the planes of spacing d in a lattice array will give rise to a series of distinct reflections only when the lattice is turned so that it makes certain definite angles, Fig. 26. The lattice planes do not act like an ordinary mirror in giving rise to a continuous reflection at all angles, but give rise to reflections only at angles characteristic of the spacing d.

Now every lattice array has an infinite number of planes of different d spacings. The d's, of course, are dependent on the lattice dimensions, and, therefore, for each specific lattice there is an infinite series of d's characteristic of the lattice. Only a few of the larger values of these d's, however, make the value of the sine in (17) less than unity, and consequently there is always a limited total number of reflections for a given lattice array and a given wavelength, but the angles which these reflections make with the original x-ray beam are specific and characteristic of the lattice dimensions. A systematic treatment of all these reflections will be given in later chapters.

THE DIFFRACTION OF X-RAYS BY ACTUAL CRYSTAL STRUCTURES

The relation between diffraction by a crystal and by a simple lattice array. Up to this point the diffraction of x-rays by a highly simplified type of crystal structure has been considered, namely, a structure consisting of a lattice array of atoms. In order to see what modifications must be made in this theory of x-ray scattering for a real crystal structure, it is first necessary to see in what way a real crystal structure is related to a lattice array of atoms.

A simple illustration of a real crystal pattern is given in Fig. 27. This particular example of a pattern has a pair of atoms as its motif, and the entire pattern may be thought of as this atom group repeated by the lattice translations. The two atoms of the motif group may or may not be related by additional symmetry operations; for example, they might be related by a center of symmetry half way between them, or by a twofold rotation axis or twofold screw axis half way between them. Whether this is the case or not, the motif group is repeated by the lattice translations, and therefore a separate lattice grid could be drawn through each kind of atom, Fig. 28. In this way, this particular crystal pattern can be decomposed into two simple lattice arrays of atoms, the two lattices being identical in dimensions and identically oriented, but slightly displaced from each other. In the same way *any* crystal pattern, no matter how complex, can be decomposed into simple lattice arrays. If the pattern motif contains N different atoms not related by lattice translations, then the pattern can be broken down into N lattice arrays of atoms.

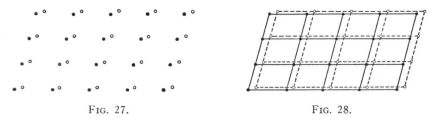

FIG. 27.   FIG. 28.

Each such lattice array is dimensionally identical with the point lattice of the crystal, Fig. 22. Each array reflects x-rays at angles according to Bragg's law,

$$\theta = \sin^{-1}\left(\frac{n\lambda}{2d}\right),$$

and any such reflection occurs simultaneously for each lattice array of atoms. This is illustrated for the pattern of Fig. 27 in Fig. 29. It

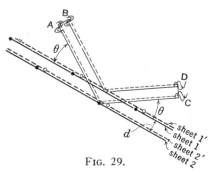

FIG. 29.

should be observed that the reflections from the sheets of black atoms are in phase with each other, because it was on this basis that Bragg's law, above, was derived. The reflections from the sheets of white atoms are also in phase with one another, but the reflections from the black lattice and the reflection from the white lattice are not mutually in phase. They are out of phase by an amount equal to the separation of sheet 1 and sheet 1′, based upon the spacing d as a full 360° phase difference. In other words, since the two lattice arrays have identical dimensions and orientations, they reflect at the same angle, θ. These two reflections differ merely in phase. The resulting reflected wave is thus the composite of two waves of the same wavelength, not quite in phase. This composite is a wave of the same wavelength but different amplitude.

It may therefore be said that, in general, a real crystal structure reflects x-rays at exactly the same angles as does a simple lattice array of atoms of identical lattice dimensions. The amplitudes of the reflections are controlled by the phase relations of the reflections of the separate lattice arrays of the pattern. Since these phase relations depend upon the displacements of the several lattice arrays from one another, this means that *the reflection intensities are controlled by the shape of the motif of the pattern, while the reflection angles depend only on the lattice dimensions.*

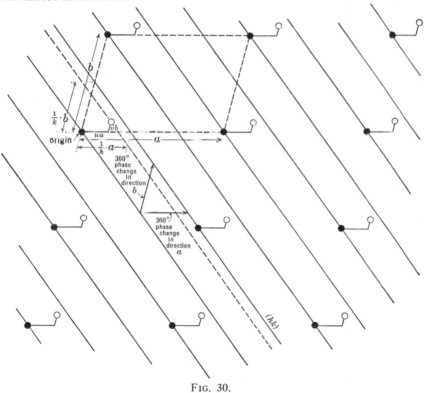

FIG. 30.

The scattering phase of a pattern point. A qualitative picture has just been given of the way several lattice arrays of atoms in a crystal structure contribute to the diffraction by the structure. This picture can easily be made quantitative:

In Fig. 30 there is shown a pattern consisting of two lattice arrays of atoms. To facilitate the discussion, a cell with edges a and b is chosen with its origin at a black atom.   In this two-dimensional illustration, the " planes " (hk) scatter x-rays with a phase difference of 360° between neighboring planes.   In particular, if the phase scattered by the " plane " through the origin is 0°, then the phase scattered by the first " plane " away from the origin is 360°.   The problem is, what is the phase scattered by the (dashed) " plane " of white atoms, or, what amounts to the same thing, what is the phase scattered by the white atom itself?

The white atom lattice array is displaced from the black atom lattice array by a vector having components along the a and b axes of $u\mathbf{a}$ and $v\mathbf{b}$, respectively, i. e., the coordinates of the white atom with respect to the origin and cell chosen are $[[uv]]$.   In this case, u and v are *fractional* parts of the unit translations $\mathbf{a}$ and $\mathbf{b}$, respectively.

The " planes " (hk) divide a into h parts, and therefore the intercept of the first plane on the a-axis is $(1/h)\cdot a$.   This distance corresponds with a 360° phase change.   The distance ua corresponds with an undetermined phase change, p_a.   These facts can be formulated thus:

$$360° \infty \frac{1}{h}\cdot a, \tag{20}$$

$$p_a \infty ua. \tag{21}$$

Therefore

$$\frac{p_a}{360°} = \frac{ua}{(1/h)\cdot a}, \tag{22}$$

from which

$$p_a = 360°\cdot hu. \tag{23}$$

Similarly, for the b axial direction,

$$360° \infty \frac{1}{k}\cdot b, \tag{24}$$

$$p_b \infty vb. \tag{25}$$

Therefore

$$\frac{p_b}{360°} = \frac{vb}{(1/k)\cdot b}, \tag{26}$$

from which

$$p_b = 360°\cdot kv. \tag{27}$$

The total phase change, P, due to the displacement of the white atom from the origin, is equal to the sum of the phase changes p_a and p_b, due to the displacement of the atom in the a and b directions:

$$P = p_a + p_b. \tag{28}$$

Substituting the values of p_a and p_b given by (23) and (27), this becomes

$$P = 360° \cdot hu + 360° \cdot kv. \qquad (29)$$

This may readily be generalized to the three-dimensional case by adding to the right member of (28) the additional phase change for the third dimension:

$$P = p_a + p_b + p_c; \qquad (30)$$

then,

$$\begin{aligned} P &= 360° \cdot hu + 360° \cdot kv + 360° \cdot lw \\ &= 360° \, (hu + kv + lw) \\ &= 2\pi \, (hu + kv + lw). \end{aligned} \qquad (31)$$

It should be noted that u, v, and w are fractions referring to fractional parts of the a, b, and c translational distances. To express (31) in absolute form, let x, y, and z represent absolute distance components of the white atom in the a, b, and c directions. Then,

$$\begin{cases} x = ua, \\ y = vb, \\ z = wc. \end{cases} \qquad (32)$$

Substituting the values of u, v, and w provided by (32) into (31) gives an alternative form of (31), namely,

$$P = 2\pi \left(h\frac{x}{a} + k\frac{y}{b} + l\frac{z}{c} \right). \qquad (33)$$

The form previously given in (31) is simpler and to be preferred, but form (33) is in frequent use.

It should be remarked that the phase, P, is an angle, specifically the phase angle between two waves, one scattered by an atom having cell coordinates $[[uvw]]$, and the other scattered by an atom at the origin. There need not actually be an atom at the origin. Relation (31) thus also provides the phase of a wave scattered by any atom whose coordinates, referred to any desired unit cell and origin, are $[[uvw]]$.

The compounding of waves diffracted by the several lattice arrays of atoms in a crystal structure. The scattering efficiency of an atom varies with the glancing angle θ, as illustrated in Fig. 31. The figure shows that the scattering efficiency falls off with $\dfrac{\sin \theta}{\lambda}$. The decrease with angle is due to the fact that in scattering, Fig. 32, the several electrons in the various parts of the atom do not scatter in phase, but, in general, somewhat out of phase. This partial interference weakens

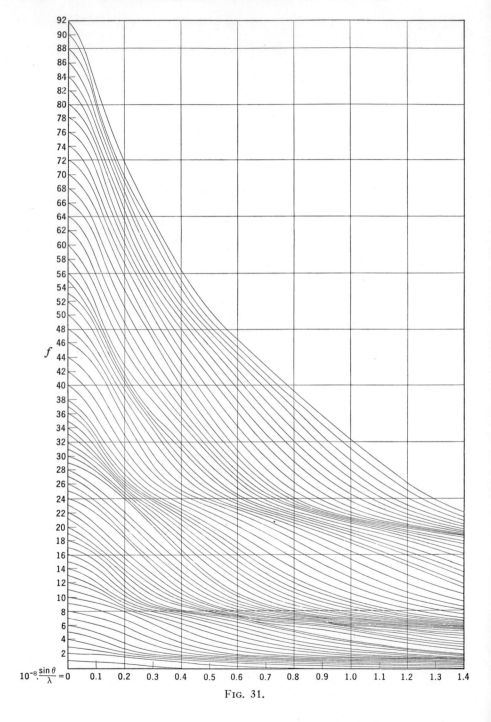

FIG. 31.

the net amplitude of scattering from any single atom. The net amplitude is designated f, as given in Fig. 31. It represents the scattering power of an atom expressed in terms of the scattering power of a single

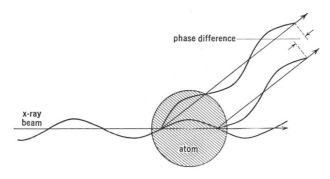

phase difference

x-ray
beam

atom

FIG. 32.

electron as unity. Each curve of Fig. 31 represents the variation of scattering power of a particular atomic species. For example, Mo (atomic number 42) scatters directly forward with all 42 electrons scattering in phase with one another. The scattering power of Mo as a function of $\dfrac{\sin \theta}{\lambda}$ is accordingly indicated in Fig. 31 by the particular curve starting at the left of the diagram at $f = 42$.

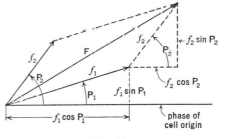

FIG. 33.

The amplitudes and phases of waves scattered by several atoms in the same direction are compounded vectorially. This is diagrammatically illustrated in Fig. 33 for two atoms, 1 and 2, having scattering phases and amplitudes of P_1, P_2, f_1 and f_2, respectively. The resulting composite amplitude, F, is vectorially

$$\mathbf{F} = \mathbf{f}_1 + \mathbf{f}_2. \tag{34}$$

By using complex quantities, this resulting amplitude may be compactly expressed as

$$F = f_1 e^{P_1 i} + f_2 e^{P_2 i}. \tag{35}$$

Figure 33 indicates that (34) can also be expressed with the aid of sines and cosines of the phase angles, thus:

$$|F| = \sqrt{(f_1 \cos P_1 + f_2 \cos P_2)^2 + (f_1 \sin P_1 + f_2 \sin P_2)^2}. \tag{36}$$

Substituting the values of P given by (31), this becomes

$$|F| = [\{f_1 \cos 2\pi(hu_1 + kv_1 + lw_1) + f_2 \cos 2\pi(hu_2 + kv_2 + lw_2)\}^2 \\ + \{f_1 \sin 2\pi(hu_1 + kv_1 + lw_1) + f_2 \sin 2\pi(hu_2 + kv_2 + lw_2)\}^2]^{\frac{1}{2}}. \tag{37}$$

Since the intensity is proportional to the square of the amplitude, the expression for the intensity is

$$I \propto |F|^2 = \{f_1 \cos 2\pi(hu_1 + kv_1 + lw_1) \quad + \quad \{f_1 \sin 2\pi(hu_1 + kv_1 + lw_1) \\ +f_2 \cos 2\pi(hu_2 + kv_2 + lw_2) \qquad +f_2 \sin 2\pi(hu_2 + kv_2 + lw_2) \\ +\dots\dots\dots\dots \}^2 \qquad +\dots\dots\dots\dots \}^2. \tag{38}$$

If additional atoms are present, additional lines are added to equation (38) with subscripts 3, 4, 5, $\cdots$ etc.

A simplifying feature occurs if the space group of the crystal pattern contains groups of symmetry centers, and if the origin is taken at one of these centers. In this case, an atom having coordinates [[uvw]] is always accompanied by a centrosymmetrically equivalent atom at [[$\bar{u}\bar{v}\bar{w}$]]. If these atoms are designated 1 and 2, then the terms for which they are responsible occur in lines 1 and 2, respectively, of (38). All f's then are identical. $\sin 2\pi(h\bar{u}_2 + k\bar{v}_2 + l\bar{w}_2)$ can be expressed $-\sin 2\pi(hu_2 + kv_2 + lw_2)$, and therefore the sine terms cancel, leaving only

$$I \propto |F|^2 = 2^2 f^2 \cos^2 2\pi(hu + kv + lw) \tag{39}$$
(for a pair of centrosymmetrical atoms).

The vanishing of the sine terms is graphically illustrated in Fig. 34. If the structure contains more than one centrosymmetrical pair of atoms, the more general expression corresponding to (39) is

$$I \propto |F|^2 = \quad 2^2 f_1^2 \cos^2 2\pi(hu_1 + kv_1 + lw_1) \\ +2^2 f_3^2 \cos^2 2\pi(hu_3 + kv_3 + lw_3) \\ +2^2 f_5^2 \cos^2 2\pi(hu_5 + kv_5 + lw_5) \\ +\dots\dots\dots\dots\dots \tag{40}$$

In all space groups except one, there are elements of repetition other than translations. The unit cells of such space groups consequently contain sets of identical atoms whose coordinates are related by space-group operations. If, for a given space group, the coordinates of an atom as repeated by the space-group operations are substituted for the several sets of u, v, w, of (38), an expression is obtained which can usually be much simplified by combining terms with the

aid of standard rules of trigonometric combination.   This results in a
compacted form of (38), known as the *structure factor* of the space
group.   In actually carrying out x-ray diffraction intensity computations, which are important in locating the positions of the atoms in
crystals, the use of this compact expression greatly simplifies the pro-

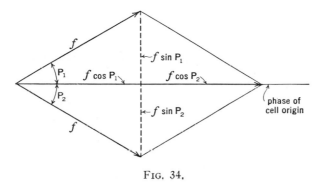

FIG. 34.

cedure.   For a fuller discussion of this phase of structural determination, the reader is referred to *Numerical structure factor tables*, noted
in the literature at the end of this chapter.

Computations of the intensities of x-ray reflection may usually be
simplified in such a way as to involve products of terms of the form

$$\cos 2\pi(hu),$$
$$\sin 2\pi(hu).$$

Tables of these functions are available for convenience in computing
intensities.

Before such computations can be compared with observed intensities, they must be corrected by various factors.   The two most
important of these are the Lorentz and the polarization factors.   The
appropriate combined form of these for use with equatorial reflections
recorded by rotating a single crystal in a beam of monochromatic
radiation is

$$\frac{1 + \cos^2 2\theta}{\sin 2\theta}.$$

Tables of this function (used in computing intensities) and tables of
its square root (used in computing corresponding amplitudes, rather
than intensities), and also tables of the reciprocals of these two functions, are also available (see *Numerical structure factor tables*).

Diffraction symmetry.   For any given crystal structure, the surface
patterns of two different planes $(h_1k_1l_1)$ and $(h_2k_2l_2)$, which are not

related by symmetry, are quite different, and the stacking of sheets, such as shown in Fig. 29, is also different for such planes. It follows that two planes $(h_1k_1l_1)$ and $(h_2k_2l_2)$, not related by symmetry, give rise to reflections which are, in general, of different intensities. On the other hand, if two planes are equivalent by symmetry, they give rise to identical reflections. For example, in the orthorhombic normal class, $D_{2h}(=mmm)$, the two planes (111) and $(1\bar{1}1)$ are symmetrically equivalent; they therefore give rise to identical reflections. However, in the triclinic normal class, $C_i (=\bar{1})$, the two planes (111) and $(1\bar{1}1)$ are not symmetrically equivalent and consequently they give rise to reflections which are, in general, of different intensity.

This suggests that the intensities of the x-ray reflections may be used to furnish a key to the symmetry of the crystal. This information would be very valuable if the symmetry of the crystal were

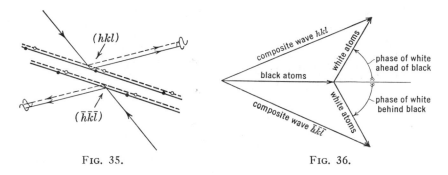

Fig. 35.   Fig. 36.

unknown. There is, however, an important relation which modifies this conclusion, namely, that the intensities of reflection from (hkl) and $(\bar{h}\bar{k}\bar{l})$ are identical. Why this is so is easily demonstrated with the aid of Fig. 35, which illustrates, for simplicity, a pattern composed of two different atoms. The reflection from the top of the stack of (hkl) planes is a composite wave made up of two components. The reflection from the bottom of the stack of (hkl) planes is equivalent to the reflection from the top of the stack of $(\bar{h}\bar{k}\bar{l})$ planes. It is a composite wave made up of the same two components. The only difference between reflections hkl and $\bar{h}\bar{k}\bar{l}$ is that, for hkl, the phase of the white atom component is *ahead* of that of the black atom component, while for $\bar{h}\bar{k}\bar{l}$ the phase of the white atom component is *behind* that of the black atom component. If the phase of the black atom component is adopted as the zero for phase measurement, then the composite waves hkl and $\bar{h}\bar{k}\bar{l}$ are compounded as shown in Fig. 36. Note that the two composite waves are equal in amplitude and have phases

TABLE 1

DIFFRACTION SYMMETRY

Crystal symmetry				Symmetry of x-ray diffraction effects	
Triclinic		C_1	1	C_i	$\bar{1}$
		C_i	$\bar{1}$		
Monoclinic		C_2	2	C_{2h}	$2/m$
		C_v	m		
		C_{2h}	$2/m$		
Orthorhombic		D_2	222	D_{2h}	mmm
		C_{2v}	$2mm$		
		D_{2h}	mmm		
Tetragonal		S_4	$\bar{4}$	C_{4h}	$4/m$
		C_4	4		
		C_{4h}	$4/m$		
		D_{2d}	$\bar{4}2m$	D_{4h}	$4/mmm$
		C_{4v}	$4mm$		
		D_4	422		
		D_{4h}	$4/mmm$		
Hexagonal	rhombohedral	C_3	3	C_{3i}	$\bar{3}$
		C_{3i}	$\bar{3}$		
		C_{3v}	$3m$	D_{3d}	$\bar{3}m$
		D_3	32		
		D_{3d}	$\bar{3}m$		
	hexagonal	C_6	6	C_{6h}	$6/m$
		C_{3h}	$\bar{6}$		
		C_{6h}	$6/m$		
		D_{3h}	$\bar{6}m2$	D_{6h}	$6/mmm$
		C_{6v}	$6mm$		
		D_6	622		
		D_{6h}	$6/mmm$		
Isometric		T	23	T_h	$m3$
		T_h	$m3$		
		T_d	$\bar{4}3m$	O_h	$m3m$
		O	432		
		O_h	$m3m$		

of opposite sign.   The intensities of hkl and $\bar{h}\bar{k}\bar{l}$ are equal to the squares of their amplitudes, and are accordingly equal; the phases cannot be observed.

Since the quality of reflections from hkl and $\bar{h}\bar{k}\bar{l}$ cannot be distinguished, it follows that all x-ray diffraction effects are centrosymmetrical.   Whether a crystal does or does not have a center of symmetry, its diffraction effects *do* have a center of symmetry, and hence it is impossible to judge by diffraction effects whether the crystal has a center of symmetry or not.   This insertion of a symmetry center into the apparent symmetry of a crystal as judged by x-ray diffraction symmetry was first pointed out by Friedel, and is known as *Friedel's law*.†

There are 11 centrosymmetrical symmetry classes.   Each of these is a possible symmetry combination observable by x-ray diffraction effects, and x-ray diffraction effects can show no other symmetry. Each class of diffraction symmetry, however, can be produced by crystals of several different crystallographic symmetries.   Table 1 shows the 11 classes of diffraction symmetry and the crystals which produce them.   The table is prepared by inserting a symmetry center into each of the 32 point groups.   The resulting symmetry combination is the diffraction symmetry displayed by the crystal.

LITERATURE

Diffraction of x-rays

W. FRIEDRICH, P. KNIPPING, and M. LAUE.   Interferenz-Erscheinungen bei Röntgenstrahlen. *Sitzungsberichte der mathematisch-physikalischen Klasse der Königlich Bayerischen Akademie der Wissenschaften zu München*, 1912, 303–322.

M. LAUE.   Eine quantitative Prüfung der Theorie für die Interferenz-Erscheinungen bei Röntgenstrahlen. *Sitzungsberichte der mathematisch-physikalischen Klasse der Königlich Bayerischen Akademie der Wissenschaften zu München*, 1912, 363–373.

M. LAUE.   Röntgenstrahlinterferenzen. *Physik. Z.*, **14** (1913), 1075–1079.

W. L. BRAGG.   The diffraction of short electromagnetic waves by a crystal. *Proc. Cambridge Phil. Soc.*, **17** (1913) 43–57.

W. H. BRAGG.   The reflection of x-rays by crystals. *Proc. Roy. Soc. London* (*A*), **88** (1913), 428–438.

W. L. BRAGG.   The structure of some crystals as indicated by their diffraction of x-rays. *Proc. Roy. Soc. London* (*A*), **89** (1913), 248–277.

W. H. BRAGG and W. L. BRAGG.   The structure of diamond. *Proc. Roy. Soc. London* (*A*), **89** (1913), 277–291.

W. LAWRENCE BRAGG.   The analysis of crystals by the x-ray spectrometer. *Proc. Roy. Soc. London* (*A*), **89** (1913), 468–489.

† For conditions which cause exceptions to Friedel's law, see literature at end of chapter, especially the paper by Coster, Knol, and Prins.

W. H. BRAGG and W. L. BRAGG. *X-rays and crystal structure.* (G. Bell and Sons, London, 1915.) Especially 8–21.

WILLIAM BRAGG. *An introduction to crystal analysis.* (G. Bell and Sons, London, 1928). Especially 9–26.

W. H. BRAGG and W. L. BRAGG. *The crystalline state.* Vol. 1. (G. Bell and Sons, London, 1933.) Especially 12–21.

CHARLES F. MEYER. *The diffraction of light, x-rays, and material particles.* (University of Chicago Press, Chicago, 1934.) Especially 279–341.

Symmetry of diffraction effects

G. FRIEDEL. Sur les symétries cristallines que peut révéler la diffraction des rayons Röntgen. *Compt. rend.*, **157** (1913), 1533–1536.

F. M. JAEGER. On a new phenomenon accompanying the diffraction of Röntgen-rays in birefringent crystals. *Proc. of the Section of Sciences, Koninklijke Akademie van Wetenschappen te Amsterdam,* **17** (1915), 1204–1236.

H. HAGA and F. M. JAEGER. On the symmetry of the Röntgen-patterns of trigonal and hexagonal crystals and on normal and abnormal diffraction-images of birefringent crystals in general. *Proc. of the Section of Sciences, Koninklijke Akademie van Wetenschappen te Amsterdam,* **18** (1916), 542–558.

H HAGA and F. M. JAEGER. On the symmetry of the Röntgen-patterns of rhombic crystals. *Proc. of the Section of Sciences, Koninklijke Akademie van Wetenschappen te Amsterdam,* **18** (1916), 559–571.

M. v. LAUE. Über die Symmetrie der Kristall-Rontgenogramme. *Ann. Physik,* **50** (1916), 433–446.

P. P. EWALD. Über die Symmetrie der Röntgeninterferenzen. *Physica,* **5** (1925), 363–369.

P. P. EWALD and C. HERMANN. Gilt der Friedelsche Satz über die Symmetrie der Röntgeninterferenzen? *Z. Krist. (A),* **63** (1927), 251–260.

D. COSTER, K. S. KNOL, and J. A. PRINS. Unterschiede in der Intensität der Röntgenstrahlenreflexion an der beiden 111-Flächen der Zinkblende. *Z. Physik,* **63** (1930), 345–369. *[Experimental proof of the deviation from Friedel's law caused by using x-radiation in the region of an absorption edge.]*

J. TER BERG and F. M. JAEGER. On the possibility of distinguishing righthanded and lefthanded structures in crystals by means of their Laue-patterns. *Proc. of the Section of Sciences, Koninklijke Akademie van Wetenschappen te Amsterdam,* **40** (1937), 406–410.

Tables for computing x-ray diffraction intensities

M. J. BUERGER. Numerical structure factor tables. *Geol. Soc. America Special Paper* 33.

CHAPTER 4

SPACE-GROUP EXTINCTIONS

In Chapter 3, the tacit assumption was made that crystals could be considered as simple patterns of atoms repeated only by the space-lattice translations, and the diffraction of x-rays by crystals was discussed from this simple viewpoint.   It is true that in triclinic crystals the pattern is this simple, but in any of the more symmetrical crystals the group of operations which repeats the atomic motif is more complex and contains, in general, rotation and reflection operations or both, in addition to pure translation.   The rotation and reflection operations may be pure rotation and pure reflections, in which case no additional considerations are necessary, or they may be composite rotation-translation (screw motion) or composite reflection-translation (glide reflection).   If this is the case, the additional translations contained in these composite operations involve a new translation periodicity in the pattern.

It is the purpose of this chapter to develop the theory of what modifications these additional translations entail in the diffraction of x-rays. It will be shown that each symmetry operation in the space group involving translations in addition to the three primitive lattice translations has the important effect of *extinguishing* a certain class of x-ray diffraction spectra.   These x-ray *extinctions* are exceptionally important in structural crystallography, because each is characteristic of the repeating operation which produced it.   A list of x-ray extinctions of a given crystal is thus (within certain limits) characteristic of the repeating operations of the pattern of that crystal, i. e., of the space group of the crystal in hand.

The repeating devices which are concerned with space-group extinctions may be divided into three categories: lattice centerings, glide planes, and screw axes.   Each of these is responsible for causing extinctions in a different class of reflections.

EXTINCTIONS DUE TO NON-PRIMITIVE LATTICES

Introduction.   It is convenient to refer a crystal to whatever coordinate system appears to give the simplest geometry consistent with its symmetry.   For example, the obvious kind of coordinate system to which to refer an orthorhombic crystal is one based upon a set of three mutually orthogonal axes. Unfortunately, the axes of a coordinate system, so chosen for the convenience of orthogonality, do not necessarily correspond with the axes of a primitive cell, but may correspond, instead, with the axes of a doubly primitive or quadruply primitive cell. An insight into this situation and the results it entails may be most readily seen by reference to a two-dimensional example:

FIG. 37A.   FIG. 37B.

In Fig. 37A, the natural primitive coordinate system furnished by the diamond plane lattice is the diamond net itself, and the obvious primitive cell is the one whose axes are the diamond sides, a and b, which are inclined to one another at the oblique angle, γ. The geometry of the pattern can be simplified, however, by reference to an orthogonal coordinate system, Fig. 37B. This can be blocked out into rectangular cells, which, however, are not primitive, but doubly primitive. There are penalties for using doubly primitive cells, which will appear presently.

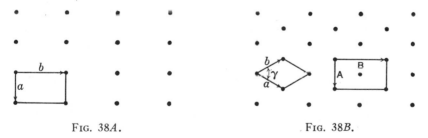

FIG. 38A.   FIG. 38B.

Now, in the beginning stages of the analysis of a crystal structure, the only information one has about the pattern is, possibly, its symmetry, as indicated by the symmetry of the external form development of the crystal. For sake of definiteness and simplicity, suppose that one is dealing with a particular two-dimensional crystal and its crystal form indicates that the pattern probably has the symmetry $C_{2i} = mm$. This symmetry is consistent with either a primitive rectangular lattice, Fig. 38A, or a primitive diamond lattice, Fig. 38B. Which of these alternatives, however, is the correct one, is not known in advance of a study of the x-ray results, and, indeed, it is one of the functions of the x-ray work to provide criteria for a decision between them.   If the primitive rectangular lattice should ultimately prove to be the correct alternative, then there would be no need to transform the coordinate system to get a rectangular cell, for the primitive cell is rectangular already.   If, however, the diamond lattice should prove to be the correct alternative, then, in order to gain the geometrical advantages of orthogonality, it would be necessary to transform to a doubly primitive rectangular coordinate system, Fig. 38*B*.   In either case, one wishes to end with a rectangular coordinate system for the sake of convenience. In the first case the rectangular cell will prove to be a primitive rect- angular cell, and in the second it will prove to be a doubly primitive (centered) rectangular cell.   Knowing of the two possibilities in advance, one therefore preselects the geometrically most convenient cell consistent with the known symmetry of the crystal, and after studying the x-ray results (about to be discussed) he decides whether this cell is either centered or primitive.   In the present illustration, if the cell subsequently turns out to be centered, it simply means that a non-primitive cell was chosen, and that the shape of the true primitive cell is a diamond.   Attention is now directed to the nature of the penalty which is imposed if the cell chosen was non-primitive.

In Chapter 3 (pages 38 to 43) it was shown that, when a crystal diffracts x-rays, each diffracted beam is the equivalent of a reflection from a rational plane in the crystal.   It was also shown that, pro- vided that the planes of the crystal are referred to a primitive cell — and this was tacitly assumed in the discussion — then the Laue equa- tion indices of each reflection correspond with the indices of the reflecting plane.   This implies that to each real crystal plane, (*hkl*), there corresponds an x-ray reflection, *hkl*.   If a new reference system of coordinates is adopted, then the existence of the rational planes and their corresponding reflections is unaffected, but these receive new indices in accordance with the scheme of transformation from one axial system to another.

If, after a transformation of axes in this fashion, the new reference cell should be non-primitive, then not all possible index combinations referred to the new axial system have a real meaning.   This may be illustrated by the two-dimensional lattice shown in Fig. 39*B*.   In this illustration, the natural primitive cell consistent with the symmetry of the pattern is the one based upon the primitive translations *a* and *b*. If one wishes to refer the geometry of the pattern to orthogonal axes, which is also permitted by symmetry, then he can do this by choosing a cell based upon *A* and *B*.   It is evident that **A** and **B** are related to **a** and **b** by the transformation

$$\begin{cases} \mathbf{A} = \mathbf{a} - \mathbf{b}, \\ \mathbf{B} = \mathbf{a} + \mathbf{b}. \end{cases} \tag{1}$$

The matrix of this transformation is

$$\left\| \begin{matrix} 1 & \bar{1} \\ 1 & 1 \end{matrix} \right\|. \tag{2}$$

According to the theory developed in Chapter 2, the indices of lines also transform with the aid of equations having this same matrix. If the indices of a line referred to the old coordinates are (hk), then the indices (HK), referred to the new orthogonal coordinates, are given by the transformation:

$$\begin{array}{cc} \text{New,} & \text{Old,} \\ \text{centered-cell} & \text{primitive-cell} \\ \text{indices} & \text{indices} \end{array}$$

$$\begin{cases} H & = & h - k. \\ K & = & h + k. \end{cases} \tag{3}$$

If the new indices, H and K, are added, there results

$$H + K = 2h. \tag{4}$$

This new composite relation entails interesting consequences: Each of the original indices, h and k, must be integers, so that the term on the right of (4) must be an integer divisible by 2, i.e., it must be an even integer. The sum $H + K$, therefore, must be even. The meaning of this is that only such planes exist as have indices whose sum $H + K$ is even. If one arbitrarily chooses any pair of numbers, H and K, for which the sum $H + K$ is *not* even, then $2h$ could not have been even, i.e., $2h$ is odd, and the only way this could be accounted for is by h being fractional, which is contrary to the conventional usage of the index symbol.

The geometrical significance of all this is illustrated in Fig. 39A and B. Lines of index $(HK) = (12)$ are first shown in an uncentered rectangular lattice in Fig. 39A. The same lines are shown as full lines in Fig. 39B, which is the corresponding centered lattice. The full line, MM, correctly described by the index (12) in the uncentered case, no longer has this index in the second case, because the index number $K = 2$ means that the *first* line of the system designated (12) has an intercept on the B-axis of $\frac{1}{2}$. *But MM is not the first line; it is the second,* because the primitive but centering translation, b, carries the line MM to NN, which becomes the *first* line. Also, in terms of indices (hk) referred to primitive coordinates, the line MM has indices

$(\frac{3}{2}\,\frac{1}{2})$. The system of index designation disallows these indices because of their fractional nature, and the meaning of this is that MM is not the *first* line, but the second, of the line system. The *first* line is NN, and its primitive indices are (31).

In any centered pattern, the index (HK), where $H + K$ is odd, is thus a kind of fictitious index, and since such a " plane " does not really exist, no x-ray reflection could possibly arise from it. The reflections corresponding with such non-existent planes, or to non-permissible index combinations, are said to be *extinguished*. The *extinction rule*

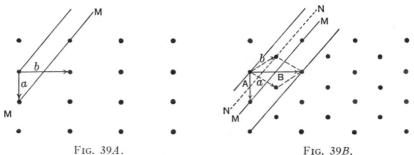

FIG. 39A.   FIG. 39B.

for a centered plane lattice is that the entire class of reflections, (HK), where $H + K$ is odd, is extinguished. Note, in Fig. 39B, that extinction really means the shifting of the position of first-line-from-the-origin in the uncentered lattice case, Fig. 39A, to a position which is some submultiple of it in the centered lattice case. This is caused directly by the additional translation involved in the centered lattice and is characteristic of it. The extinction rule is thus characteristic of submultiple translations and provides a specific criterion for the nature of such translations. X-ray diffraction, of course, is the experimental means of determining extinctions in patterns of atomic dimensions, for, as already shown in Chapter 3, to each primitive line (hk) or plane (hkl) there corresponds a real x-ray reflection hk or hkl, respectively.

The method of regarding centered lattices as non-primitive axial systems derived from some primitive system can, of course, be extended to three-dimensional lattices. This procedure is carried out for the several non-primitive lattices in the following pages. In seeking extinctions, the scheme is followed which has just been developed in the two-dimensional pattern, namely, seeking classes of (HKL) planes in which, with the aid of the transformation equations, some index combination is expressible in terms of multiples of the primitive indices h, k, and l.   Since these must be integers in all instances, the class of indices (HKL) must conform to that multiple, or else be fictitious, in which case the corresponding reflections are extinguished.

To anticipate the results to be derived, it may be said that the translations inherent in non-primitive lattices provide extinctions in the most general class of reflections, HKL; that the translation components of glide planes provide extinctions in more limited classes of reflections like $HK0$, and that the translation components of screw axes provide extinctions in the even more limited classes of reflections like $H00$.

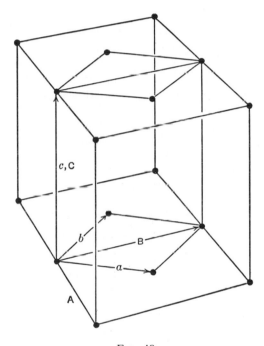

FIG. 40.

The end-centered lattice.   If, from a primitive (prismatic) lattice, a new lattice which is end-centered on its C- or (001)-pinacoid, Fig. 40, is chosen, the axial transformation, according to Fig. 40, is

$$\begin{cases} A = a - b \\ B = a + b \\ C = \qquad c, \end{cases} \qquad \text{matrix:} \quad \begin{Vmatrix} 1 & \bar{1} & 0 \\ 1 & 1 & 0 \\ 0 & 0 & 1 \end{Vmatrix} \quad \text{(modulus 2).} \quad (5)$$

The new cell is doubly primitive.   The new indices (HKL) of a plane may be obtained from the primitive indices by means of the transformation

New, C-centered lattice indices		Old, primitive lattice indices

$$\begin{cases} H &=& h - k, \\ K &=& h + k, \\ L &=& \qquad l. \end{cases} \qquad (6)$$

The following sum is expressible in terms of multiples of the primitive indices:

$$H + K = 2h, \text{ an even number.} \qquad (7)$$

If, therefore, a lattice is C-centered, the only planes which exist and the only reflections which can appear are those for which the sum $H + K$ is even. Reflections for which the sum $H + K$ is odd are extinguished. For centering on the A and B pinacoids, the sums $K + L$ and $H + L$, respectively, substitute for the $H + K$ in C-centering.

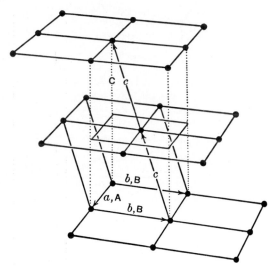

Fig. 41.

The body-centered lattice. If, from a primitive lattice, a new lattice which is body-centered, Fig. 41, is chosen, the axial transformation, according to Fig. 41, is

$$\begin{cases} \mathbf{A} = \mathbf{a} \\ \mathbf{B} = \qquad \mathbf{b} \\ \mathbf{C} = \mathbf{a} + \mathbf{b} + 2\mathbf{c} \end{cases}, \quad \text{matrix:} \quad \begin{Vmatrix} 1 & 0 & 0 \\ 0 & 1 & 0 \\ 1 & 1 & 2 \end{Vmatrix} \quad \text{(modulus 2).} \qquad (8)$$

The new cell is doubly primitive. The new indices (HKL) may be obtained from the primitive indices by means of the transformation

$$\begin{cases} H &=& h, \\ K &=& k, \\ L &=& h + k + 2l. \end{cases} \qquad (9)$$

New, body-centered lattice indices — Old, primitive lattice indices

The following sum is expressible in terms of multiples of the primitive indices:

$$H + K + L = 2(h + k + l), \qquad (10)$$

or

$$H + K + L = 2n, \qquad (11)$$

where n is an integer. If, therefore, a lattice is body-centered, the only planes which exist and the only reflections which can appear are those for which the sum $H + K + L$ is even. Reflections for which the sum $H + K + L$ is odd are extinguished.

The face-centered lattice. If, from a primitive lattice, a new lattice which is face-centered is chosen, the axial transformation, according to Fig. 42, is

$$\begin{cases} \mathbf{A} = \mathbf{a} - \mathbf{b} \\ \mathbf{B} = \mathbf{a} + \mathbf{b} \\ \mathbf{C} = \mathbf{a} + \mathbf{b} + 2\mathbf{c} \end{cases}, \quad \text{matrix:} \quad \begin{Vmatrix} 1 & \bar{1} & 0 \\ 1 & 1 & 0 \\ 1 & 1 & 2 \end{Vmatrix} \quad \text{(modulus 4)}. \qquad (12)$$

The new indices, (HKL), may be obtained from the primitive indices by means of the transformation

$$\begin{cases} H &=& h - k, \\ K &=& h + k, \\ L &=& h + k + 2l. \end{cases} \qquad (13)$$

New, face-centered lattice indices — Old, primitive lattice indices

The following *three* sums are expressible in terms of multiples of the primitive indices:

$$\begin{aligned} H + K &= 2h, && \text{an even number,} \\ H \quad + L &= 2h && + 2l, \text{ an even number,} \\ K + L &= 2h + 2k + 2l, \text{ an even number.} \end{aligned} \qquad (14)$$

If, therefore, a lattice is face-centered, the only planes which exist and the only reflections which can appear are those for which the three sums, $H + K, H + L,$ and $K + L,$ are all even. A rule derivable from this is that $H, K,$ and L must be all even or all odd, a condition known as *unmixed indices*. Reflections for which any of the sums $H + K,$ $H + L,$ or $K + L$ are odd, or for which $H, K,$ and L are *mixed*, are extinguished.

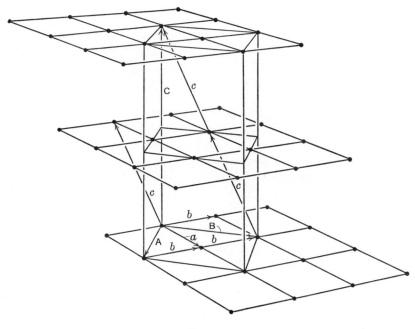

Fig. 42.

Rhombohedral centering. A rhombohedral pattern may be referred to a hexagonal coordinate system by choosing hexagonal axes from the rhombohedral lattice as shown in Fig. 43. The axial transformation is

$$\begin{cases} \mathbf{A}_1 = \mathbf{a}_1 - \mathbf{a}_2 \\ \mathbf{A}_2 = \quad\ \mathbf{a}_2 - \mathbf{a}_3, \quad \text{matrix:} \\ \mathbf{C} = \mathbf{a}_1 + \mathbf{a}_2 + \mathbf{a}_3 \end{cases} \quad \begin{Vmatrix} 1 & \bar{1} & 0 \\ 0 & 1 & \bar{1} \\ 1 & 1 & 1 \end{Vmatrix} \quad \text{(modulus 3).} \quad (15)$$

The new cell is triply primitive. The new hexagonal indices, $(HK\cdot L),$ of a plane may be obtained from the primitive rhombohedral indices by means of the following transformation:

New, (centered) hexagonal lattice indices		Old, rhombohedral lattice indices	
H	$=$	$h - k,$	
K	$=$	$k - l,$	(16)
L	$=$	$h + k + l.$	

The following algebraic sum is expressible in terms of multiples of the primitive indices:

$$-H + K + L = 3k, \text{ a number divisible by 3.} \qquad (17)$$

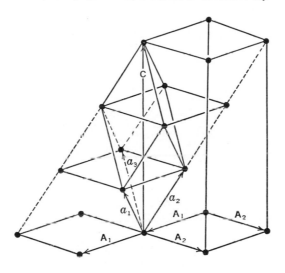

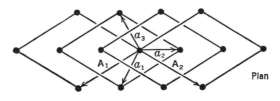

Plan

FIG. 43.

If, therefore, a hexagonal lattice is rhombohedral-centered, i.e., if a rhombohedral cell is indexed on a multiply primitive hexagonal lattice, the only planes which exist and the only reflections which can appear are those for which the algebraic sum $-H + K + L$ is divisible by 3.

Reflections for which the sum $-H + K + L$ is not divisible by 3 are extinguished.

Hexagonal centering. A hexagonal pattern may be referred to a rhombohedral coordinate system by choosing rhombohedral axes from

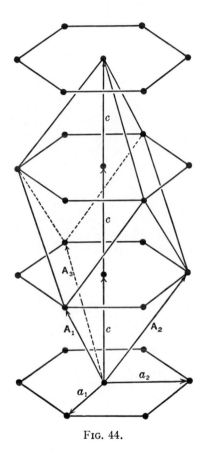

FIG. 44.

the hexagonal lattice as shown in Fig. 44.   The axial transformation is

$$
\begin{cases}
\mathbf{A}_1 = \mathbf{a}_1 \qquad\;\; + \mathbf{c} \\
\mathbf{A}_2 = \qquad \mathbf{a}_2 + \mathbf{c}, \\
\mathbf{A}_3 = -\mathbf{a}_1 - \mathbf{a}_2 + \mathbf{c}
\end{cases}
\quad \text{matrix:} \quad
\begin{Vmatrix}
1 & 0 & 1 \\
0 & 1 & 1 \\
\bar{1} & \bar{1} & 1
\end{Vmatrix}
\quad \text{(modulus 3).} \qquad (18)
$$

The new cell is triply primitive.   The new rhombohedral indices of a plane may be obtained from the primitive hexagonal indices by means of the following transformation:

$$\begin{cases} H = & h & + l, \\ K = & k + l, \\ L = -h - k + l. \end{cases} \qquad (19)$$

The following algebraic sum is expressible in terms of multiples of the primitive indices:

$$H + K + L = 3l, \text{ a number divisible by 3.} \qquad (20)$$

If, therefore, a hexagonal lattice is indexed on a multiply primitive rhombohedral lattice, the only planes which exist and the only reflections which can appear are those for which the algebraic sum $H + K + L$ is divisible by 3. Reflections for which the sum $H + K + L$ is not divisible by 3 are extinguished.

EXTINCTIONS DUE TO GLIDE PLANES

Introduction. The effect of a glide plane on the repetition of a pattern is illustrated in Fig. 45A. It will be observed that the projection of the repeated motif on a plane parallel with the glide plane is a pattern with half the true identity period, t. This comes about because:

(a) A glide plane symmetry operation involves two components: (1) a translation component parallel with the symmetry plane, having a magnitude of half the true space identity period, and (2) a reflection component, x, normal to the plane.

(b) The translation component projects on the parallel glide plane as its full magnitude, $\dfrac{t}{2}$, while

(c) The reflection component, x, being normal to the plane of projection, *vanishes for this particular plane of projection.*

Because of (c), above, the degenerate, halved identity period of the projected motif applies only to projections on the glide plane, or planes parallel with it. For any other projection plane, Fig. 45A, the reflection component in (c) does not vanish, and the projected identity period is a direct projection of the true period.

It should also be observed that, for a reflection mechanism not involving a glide component, the projection on the reflection plane of the repetition, Fig. 45B, is a pattern whose identity period is the projection of the true identity period. This projection is the same as if there were no symmetry plane of any sort present, Fig. 45C. Of the two kinds of symmetry planes, a glide plane is uniquely characterized

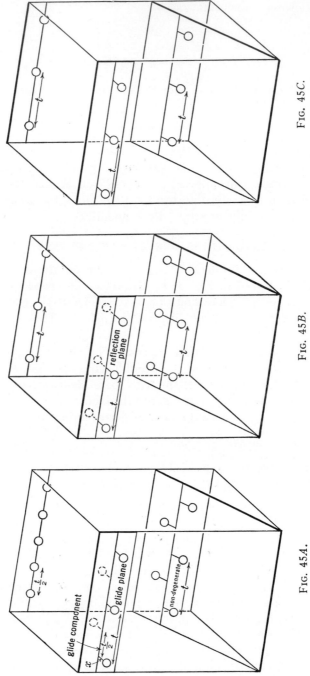

Fig. 45C.

Fig. 45B.

Fig. 45A.

by a projection pattern of half the true space identity period on the symmetry plane. If there is any way by which this degenerate projected identity period can be detected, a glide plane in a crystal pattern can also be detected.

The relation between the projected pattern with its rational lines, and the space pattern and its corresponding rational planes, is shown

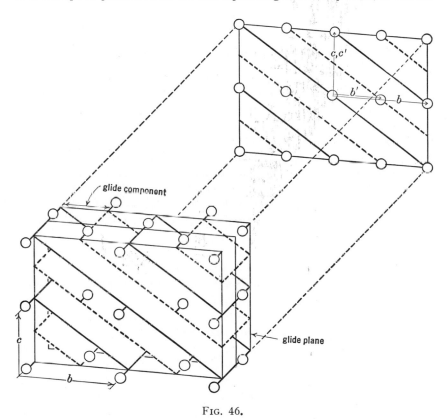

FIG. 46.

in Fig. 46. It is apparent in this illustration that the glide plane operates on the true lattice planes perpendicular to itself, shown in full lines, and produces, from them, another set interleaved between them. These two sets of planes differ only by being right- and left-handed equivalents of one another, i. e., by being associated with right- and left-handed equivalents of the pattern motif. The true identity period is evidently from full-lined plane to full-lined plane. Note, however, that, because of the geometry of reflection, the reflection components, x, *lie in the plane.* If, therefore, no distinction is made between right- and left-handed planes, and if no importance attaches to displacement, x, within the plane, then the plane system normal to the glide plane is referable to the same degenerate coordinate system as the projected pattern. According to the development of x-ray diffraction given on pages 43–47, it is evident that diffraction makes no distinction between right-handed and left-handed planes and *is not concerned with displacements within the plane. Consequently, so far as x-ray diffraction effects and the indexing of x-ray reflections are concerned, the correct coordinate system of this set of planes is that provided by the projection on the glide plane.* X-rays, therefore, provide a way of detecting the degenerate projected identity period and hence a means of detecting the presence of a glide plane in the space group of the crystal.

It should be observed that, if the planes shown in Fig. 46 are given any slope with respect to the glide plane other than 90°, then the right and left planes are no longer parallel, are no longer interleaved, and the reflection components, x, lie outside the planes, and therefore the projections of the planes no longer have a degenerate identity period. This follows for the same reason that the projection of the pattern on any plane other than the glide plane, Fig. 45A, does not have a degenerate identity period.

In investigating the effect of the glide plane upon the extinguishing of certain spectra, the same procedure may be followed as used in the investigation of the extinction effects of additional lattice translations, with this exception: *additional lattice translations change the identity periods of every plane in the pattern*, but *glide planes affect only the diffraction identity periods of the planes normal to the glide plane.* Thus, if the glide plane is (100), only planes normal to this, or (0kl), will have their x-ray identity periods disturbed. The absence of reflections due to this effect must therefore be confined to the reflections of these planes, namely, the reflections 0kl. The general procedure in investigating these planes will be to see what new indexing is necessary on the basis of the degenerate identity period caused by the translation component of the glide.

Kinds of glide planes. Symmetry planes can occur only parallel with pinacoidal or prism planes. Consequently these are also the only positions which could be occupied by glide planes. Furthermore, a glide operation repeated twice in succession is equivalent to a translation to identity, Fig. 45A. Consequently the gliding direction of a glide plane is always along a simple lattice line. If this lattice line is a cell edge, the glide is known as an axial glide. If it is parallel with a cell diagonal, then two cases arise according as a twice repeated glide becomes equal to a full cell diagonal or a half cell diagonal. The last case, known as a diamond glide because of its occurrence in the diamond structure, occurs only in multiply primitive cells, specifically:

(*a*) parallel with *B* in a simple monoclinic lattice in which a *B*-centered cell is chosen;

(*b*) parallel with pinacoids in a face-centered lattice;

(*c*) parallel with (110) in (tetragonal and isometric) body-centered lattices.

The three kinds of glide planes have the characteristics shown in Table 2.

TABLE 2

KINDS OF GLIDE PLANES AND THEIR DESIGNATIONS

Kind of glide	Designation, in orientation space-group symbol		Glide components
Axial glide	*a*		$\dfrac{a}{2}$
	b		$\dfrac{b}{2}$
	c		$\dfrac{c}{2}$
Diagonal glide	*n*		$\dfrac{a}{2} + \dfrac{b}{2}$,
		or	$\dfrac{a}{2} + \dfrac{c}{2}$,
		or	$\dfrac{b}{2} + \dfrac{c}{2}$.
Diamond glide	*d*		$\dfrac{a}{4} + \dfrac{b}{4}$,
		or	$\dfrac{a}{4} + \dfrac{c}{4}$,
		or	$\dfrac{b}{4} + \dfrac{c}{4}$.

Extinctions due to an axial glide. In Fig. 46, the glide plane is (100) and the glide component is $\dfrac{b}{2}$. This may be taken as the representative case of axial glide and others may be derived from this one by appropriate interchanges of axes. The plane pattern projection on the symmetry plane has a b' identity period equal to half the b identity period of the space pattern. The relation between the cell axes of the space pattern and those of its projection on the glide plane may be expressed by means of the transformation

$$
\begin{matrix} \text{Space} \\ \text{pattern} \end{matrix} \quad \begin{matrix} \text{Pattern projected on} \\ \text{(100) glide plane} \end{matrix}
$$

$$
\begin{cases} \mathbf{b} = 2\mathbf{b}' \\ \mathbf{c} = \mathbf{c}' \end{cases}, \qquad \text{matrix:} \quad \left\| \begin{matrix} 2 & 0 \\ 0 & 1 \end{matrix} \right\|.
$$

The relation between the indices of the same plane on the two patterns is, therefore,

$$
\begin{cases} k = 2k', \\ l = l'. \end{cases}
$$

The index, k, is thus expressible in multiples of the projection index, namely,

$$k = 2k', \text{ an even number.}$$

This result can be interpreted as follows: If the plane pattern were referred to axes b and c, then " planes " (kl) would be present only for k even, and absent for k odd. The geometry of the reflection of x-rays from the planes of the space structure which are normal to the symmetry plane, namely, planes $(0kl)$, is exactly the same as the geometry of reflection from the corresponding " planes " (kl) of the projection on the symmetry plane. Since reflections kl, when k is odd, would be extinguished for the projected pattern, reflections $0kl$, when k is odd, are extinguished for the space pattern. The other examples of extinction due to axial glide, listed in Table 3, can be derived by similar reasoning.

Extinctions due to a diagonal glide. Figure 47 shows the cell of a space pattern having a (100) glide plane with diagonal glide components $\frac{b}{2} + \frac{c}{2}$. This pattern projects on the symmetry plane as a centered plane pattern having the same b and c dimensions as the space pattern. The relation between the cell axes of the space pattern and those of its projection on the glide plane may be expressed by means of the transformation

$$
\begin{matrix} \text{Space} \\ \text{pattern} \end{matrix} \qquad \begin{matrix} \text{Pattern projected on} \\ \text{(100) glide plane} \end{matrix}
$$

$$
\begin{cases} \mathbf{b} = \mathbf{b}' + \mathbf{c}' \\ \mathbf{c} = -\mathbf{b}' + \mathbf{c}' \end{cases}, \qquad \text{matrix:} \quad \left\| \begin{matrix} 1 & 1 \\ 1 & 1 \end{matrix} \right\|.
$$

The relation between the indices of the same plane on the two patterns is, therefore,

$$\begin{cases} k = & k' + l', \\ l = & -k' + l'. \end{cases}$$

The index sum, $k + l$, is expressible in multiples of the projection pattern index,

$$k + l = 2l', \text{ an even number.}$$

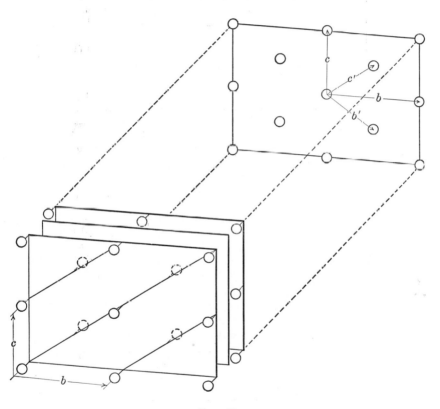

FIG. 47.

By the interpretation given for extinctions due to axial glide, this means that reflections kl are present only when $k + l$ is even, and absent when $k + l$ is odd. Hence reflections $0kl$ are possible from the space pattern only when $k + l$ is even, and are extinguished when $k + l$ is odd. The other examples of extinction due to diagonal glide, listed in Table 4, can be derived by similar reasoning.

TABLE 3

EXTINCTIONS DUE TO GLIDE PLANES WITH AXIAL GLIDE

Glide plane	Glide components	Extinctions			
		P setting	C setting	I setting	F setting
(100)	$\dfrac{b}{2}$	$0kl$ when k is odd	$0kl$ when k is odd		
	$\dfrac{c}{2}$	$0kl$ when l is odd	$0kl$ when l is odd	$0kl$ when l is odd	
(1$\bar{1}$0)	$\dfrac{c}{2}$	hhl when l is odd	hhl when l is odd		hhl when l is odd

TABLE 4

EXTINCTIONS DUE TO GLIDE PLANES WITH DIAGONAL GLIDE

Glide plane	Glide components	Extinctions	
		P setting	C setting
(100)	$\dfrac{b}{2} + \dfrac{c}{2}$	$0kl$ when $k + l$ is odd	————
(1$\bar{1}$0)	$\dfrac{a}{2} + \dfrac{b}{2}$	No extinctions	hhl when h is odd
	$\dfrac{a}{4} + \dfrac{b}{4} + \dfrac{c}{4}$	————	hhl when $h + l$ is odd

TABLE 5

EXTINCTIONS DUE TO GLIDE PLANES WITH DIAMOND GLIDE

Glide plane	Glide components	Extinctions		
		B setting	F setting	I setting
(010)	$\dfrac{a}{4} + \dfrac{c}{4}$	$h0l$ when $h + l$ is divisible by 4		
(100)	$\dfrac{b}{4} + \dfrac{c}{4}$		$0kl$ when $k + l$ is divisible by 4	————
(1$\bar{1}$0)	$\dfrac{a}{4} + \dfrac{b}{4} + \dfrac{c}{4}$		————	hhl when $2h + l$ is divisible by 4

Extinctions due to a diamond glide.   Figure 48 shows a cell with a (010) glide plane having diamond glide components $\frac{a}{4} + \frac{c}{4}$. The projection of this pattern on the symmetry plane (010) is also shown. Taking the primitive cell based upon a' and c' for the plane pattern,

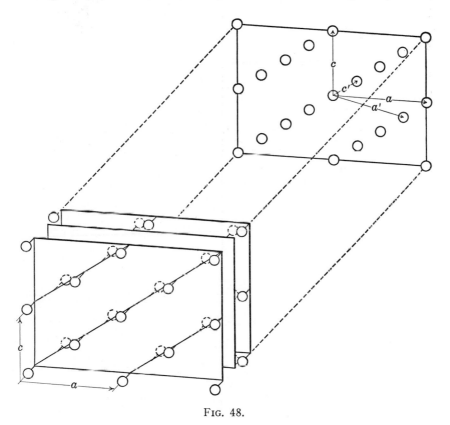

FIG. 48.

it is evident from the illustration that the axes of the space pattern and those of the plane pattern are related by the transformation

$$\begin{cases} \mathbf{a} = \mathbf{a'} + \mathbf{c'} \\ \mathbf{c} = -\mathbf{a'} + 3\mathbf{c'} \end{cases} \quad \text{matrix:} \quad \begin{Vmatrix} 1 & 1 \\ \bar{1} & 3 \end{Vmatrix}.$$

The relations between the indices of the same plane on the two patterns is, evidently,

$$\begin{cases} h = h' + l', \\ l = -h' + 3l'. \end{cases}$$

The index sum, $h + l$, is expressible as multiples of the projection pattern indices,

$$h + l = 4l'.$$

Following the interpretation given for extinctions due to axial glide, this means that reflections hl are present only when $h + l$ is divisible by 4, and absent when $h + l$ is not divisible by 4. Hence, for the space pattern, reflections $h0l$ are possible only for values of h and l such that $h + l$ is divisible by 4, and are extinguished for values of h and l such that $h + l$ is not divisible by 4. The other examples of extinction due to diamond glide, listed in Table 5, can be derived by similar reasoning.

EXTINCTIONS DUE TO SCREW AXES

The repetition of a point or atom by a screw axis is illustrated in Fig. 49A. The coordinates of the successive repeated points are such that they involve

    (a) identical radial components, r, measured from the screw axis as origin,

    (b) uniformly increasing angular components, ω, measured from the first point as origin, and

    (c) uniformly increasing translation components, t/n.

Following the general plan developed in the investigation of extinctions due to glide planes, it will be observed that the entire pattern projects on the screw axis as a linear pattern with an identity period, t', equal to $\dfrac{t}{n}$, where t is the identity period of the space pattern in the direction of the screw axis, and n is the period of the axis. (This, of course, is because r and ω have no component on the axis.) It should also be observed that duplicate planes of the space pattern, Fig. 49C, which are normal to the screw axis are interleaved with a spacing of $\dfrac{t}{n}$, while planes of all other slopes have a less regular arrangement whose identity period along the screw axis is non-degenerate and equal to true identity period, t. This difference is due to the way r and ω project on planes of these slopes.

Suppose that the screw axis is parallel with [001]. Then the space axes and the degenerate axis along the line of the screw, Fig. 49B, are related by the transformation

Space axis		Line axis
c	$=$	$n\mathbf{c}'$,

and the indices of planes normal to the screw axis transform:

Space index   Line index
$$l = nl'.$$

This means that, if the " planes " along the line axis, Fig. 49B, were indexed with respect to " cell edge," c, all " planes," except those for

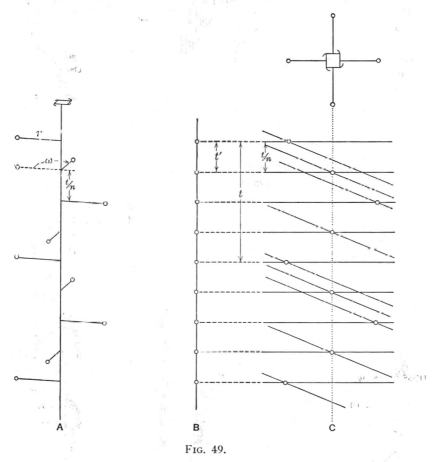

FIG. 49.

which l was divisible by n, would be fictitious, absent, and represented by an extinguished reflection. Since the geometry of the reflection from planes normal to the screw axis of the space pattern, Fig. 49C, is the same as that of the reflection from the points of the line pattern, Fig. 49B, only such x-ray reflections of the class 00l appear for which l is divisible by n. Reflections 00l, where l is not divisible by n, are extinguished.

THE DETERMINATION OF SPACE-GROUP SYMMETRY ELEMENTS FROM EXTINCTIONS

This chapter has been devoted to showing that those space-group symmetry elements involving any translations in addition to the primitive lattice translatiors entail extinctions of certain classes of reflections. The symmetry elements have also been examined individually, and it has been shown that each brings about extinction in a class characteristic of itself. These are summarized in Table 6.

A survey of Table 6 shows that an extinction rule *uniquely characterizes the symmetry element which produces it*. Therefore a knowledge of the x-ray extinctions of a crystal gives information for detection and location of symmetry elements in the crystal pattern, i. e., for the identification of the space group of the crystal pattern. Two fundamental precautions, however, should always be kept in mind in the application of extinction information to space-group symmetry content of a crystal:

1. An extinguished class implies the presence of a symmetry-element repetition operation with a translation component other than the cell-edge translations. *The absence of an extinguished class implies the absence of the corresponding symmetry element with translation component; it does not imply the presence of some alternative symmetry element.* For example, the extinction of $h0l$ when h is odd implies the presence of a glide plane parallel to (010) with glide component $\frac{a}{2}$. The non-extinction of this class does *not* imply that the crystal contains a reflection plane, (010), which is an alternative kind of symmetry plane for this position in the space group.

Suppose that, from other information available (diffraction symmetry, for example, see pages 55–58), the crystal is known to be monoclinic. Further, suppose, as above, that it shows only the extinction of $h0l$ when h is odd. Then the space group of the crystal definitely contains a glide plane, but additional information cannot be obtained with regard to the presence of other possible monoclinic symmetry elements. It is not known whether it contains a twofold rotation axis or not, because a twofold rotation axis involves no additional translations and therefore produces no characteristic extinctions. The space group may therefore be either† Pa or $P2/a$, although it cannot be

(*Text is continued on page 89.*)

† The symbols used here, and listed in the last two columns of Table 7, show, at a glance, the important repetition characteristics of their respective space groups. The symbols for symmetry planes and symmetry axes are set forth on page 507. A full explanation of space-group nomenclature is given in the *International Tables* (see literature at end of chapter).

TABLE 6

Symmetry Interpretations of Extinctions

Class of reflection	Condition for non-extinction (n = an integer)	Interpretation of extinction	Symbol of symmetry element
hkl	$h + k + l = 2n$	Body-centered lattice	I
	$h + k = 2n$	C-centered lattice	C
	$h + l = 2n$	B-centered lattice	B
	$k + l = 2n$	A-centered lattice	A
	$\left\{\begin{matrix} h + k = 2n \\ h + l = 2n \\ k + l = 2n \end{matrix}\right\}$	Face-centered lattice	F
	$\backsim h, k, l,$ all even or all odd		
	$-h + k + l = 3n$	Rhombohedral lattice indexed on hexagonal reference system	R
	$h + k + l = 3n$	Hexagonal lattice indexed on rhombohedral reference system	H
$0kl$	$k = 2n$	(100) glide plane, component $\dfrac{b}{2}$	b (P, B, C)
	$l = 2n$	" " " " $\dfrac{c}{2}$	c (P, C, I)
	$k + l = 2n$	" " " " $\dfrac{b}{2} + \dfrac{c}{2}$	n (P)
	$k + l = 4n$	" " " " $\dfrac{b}{4} + \dfrac{c}{4}$	d (F)
$h0l$	$h = 2n$	(010) glide plane, component $\dfrac{a}{2}$	a (P, A, I)
	$l = 2n$	" " " " $\dfrac{c}{2}$	c (P, A, C)
	$h + l = 2n$	" " " " $\dfrac{a}{2} + \dfrac{c}{2}$	n (P)
	$h + l = 4n$	" " " " $\dfrac{a}{4} + \dfrac{c}{4}$	d $(F), (B)$
$hk0$	$h = 2n$	(001) glide plane, component $\dfrac{a}{2}$	a (P, B, I)
	$k = 2n$	" " " " $\dfrac{b}{2}$	b (P, A, B)
	$h + k = 2n$	" " " " $\dfrac{a}{2} + \dfrac{b}{2}$	n (P)
	$h + k = 4n$	" " " " $\dfrac{a}{4} + \dfrac{b}{4}$	d (F)
hhl	$l = 2n$	(110) glide plane, component $\dfrac{c}{2}$	c (P, C, F)
	$h = 2n$	" " " " $\dfrac{a}{2} + \dfrac{b}{2}$	b (C)
	$h + l = 2n$	" " " " $\dfrac{a}{4} + \dfrac{b}{4} + \dfrac{c}{4}$	n (C)
	$2h + l = 4n$	" " " " $\dfrac{a}{2} + \dfrac{b}{4} + \dfrac{c}{4}$	d (I)
$h00$	$h = 2n$	[100] screw axis, component $\dfrac{a}{2}$	$2_1, 4_2$
	$h = 4n$	" " " " $\dfrac{a}{4}$	$4_1, 4_3$
$0k0$	$k = 2n$	[010] screw axis, component $\dfrac{b}{2}$	$2_1, 4_2$
	$k = 4n$	" " " " $\dfrac{b}{4}$	$4_1, 4_3$
$00l$	$l = 2n$	[001] screw axis, component $\dfrac{c}{2}$	$2_1, 4_2, 6_3$
	$l = 3n$	" " " " $\dfrac{c}{3}$	$3_1, 3_2, 6_2, 6_4$
	$l = 4n$	" " " " $\dfrac{c}{4}$	$4_1, 4_2$
	$l = 6n$	" " " " $\dfrac{c}{6}$	$6_1, 6_2$
$hh0$	$h = 2n$	[110] screw axis, component $\dfrac{a}{2} + \dfrac{b}{2}$	2_1

TABLE 7

Space-Group Symbols

Schoenflies symbol		Complete Mauguin symbol	International symbol	International symbol for other orientations
C_1^1		$P1$	$P1$	$A1, B1, C1, F1, I1. \ldots$
$C_i^1,$	S_2^1	$P\bar{1}$	$P\bar{1}$	$A\bar{1}, B\bar{1}, C\bar{1}, F\bar{1}, I\bar{1}. \ldots$
$C_s^1,$	C_{1h}^1	Pm	Pm	Bm
$C_s^2,$	C_{1h}^2	Pc	Pc	Pa, Pn, Ba, Bd
$C_s^3,$	C_{1h}^3	Cm	Cm	Am, Im, Fm
$C_s^4,$	C_{1h}^4	Cc	Cc	Aa, Ia, Fd
C_2^1		$P2$	$P2$	$B2$
C_2^2		$P2_1$	$P2_1$	$B2_1$
C_2^3		$C2$	$C2$	$A2, I2, F2$
C_{2h}^1		$P2/m$	$P2/m$	$B2/m$
C_{2h}^2		$P2_1/m$	$P2_1/m$	$B2_1/m$
C_{2h}^3		$C2/m$	$C2/m$	$A2/m, I2/m, F2/m$
C_{2h}^4		$P2/c$	$P2/c$	$P2/a, P2/n, B2/a, B2/c, B2/d$
C_{2h}^5		$P2_1/c$	$P2_1/c$	$P2_1/a, P2_1/n, B2_1/a, B2_1/c, B2_1/d$
C_{2h}^6		$C2/c$	$C2/c$	$A2/a, I2/a, F2/d$
C_{2v}^1		$Pmm2$	Pmm	$P2mm; \ Pm2m$
C_{2v}^2		$Pmc2_1$	Pmc	$Pcm(2); P2ma; P2am; Pb2m; Pm2b$
C_{2v}^3		$Pcc2$	Pcc	$P2aa; \ Pb2b$
C_{2v}^4		$Pma2$	Pma	$Pbm(2); P2mb; P2cm; Pc2m; Pm2a$
C_{2v}^5		$Pca2_1$	Pca	$Pbc(2); \ P2ab; \ P2ca; \ Pc2b; \ Pb2a$
C_{2v}^6		$Pcn2$	Pnc	$Pcn(2); P2na; P2an; Pb2n; Pn2b$
C_{2v}^7		$Pmn2_1$	Pmn	$Pnm(2); P2mn; P2nm; Pn2m; Pm2n$
C_{2v}^8		$Pba2$	Pba	$P2cb; \ Pc2a$
C_{2v}^9		$Pna2_1$	Pna	$Pbn(2); \ P2nb; \ P2cn; \ Pc2n; \ Pn2a$
C_{2v}^{10}		$Pnn2$	Pnn	$P2nn; \ Pn2n$
C_{2v}^{11}		$Cmm2$	Cmm	$A2mm; \ Bm2m$
C_{2v}^{12}		$Cmc2_1$	Cmc	$Ccm(2); A2ma; A2am; Bb2m; Bm2b$
C_{2v}^{13}		$Ccc2$	Ccc	$A2aa; \ Bb2b$
C_{2v}^{14}		$Amm2$	Amm	$Bmm(2); B2mm; C2mm; Cm2m; Am2m$
C_{2v}^{15}		$Abm2$	Abm	$Bma(2); \ B2am; \ C2ma; \ Cm2a; \ Ab2m$
C_{2v}^{16}		$Ama2$	Ama	$Bbm(2); B2mb; C2cm; Cc2m; Am2a$
C_{2v}^{17}		$Aba2$	Aba	$Bba(2); \ B2ab; \ C2ca; \ Cc2a; \ Ab2a$,
C_{2v}^{18}		$Fmm2$	Fmm	$F2mm; \ Fm2m$
C_{2v}^{19}		$Fdd2$	Fdd	$F2dd; \ Fd2d$
C_{2v}^{20}		$Imm2$	Imm	$I2mm; \ Im2m$
C_{2v}^{21}		$Iba2$	Iba	$I2aa; \ Ib2a$
C_{2v}^{22}		$Ima2$	Ima	$Ibm(2); \ I2ma; \ I2am; \ Ib2m; \ Im2a$
$D_2^1,$	V^1	$P222$	$P222$	
$D_2^2,$	V^2	$P222_1$	$P222_1$	$P2_122; \ P22_12$
$D_2^3,$	V^3	$P2_12_12$	$P2_12_12$	$P22_12_1: \ P2_122_1$
$D_2^4,$	V^4	$P2_12_12_1$	$P2_12_12_1$	
$D_2^5,$	V^5	$C222_1$	$C222_1$	$A2_122; \ B22_12$

TABLE 7 — *Continued*

Schoenflies symbol	Complete Mauguin symbol	International symbol	International symbol for other orientations
D_2^6, V^6	$C222$	$C222$	$A222$; $B222$
D_2^7, V^7	$F222$	$F222$	
D_2^8, V^8	$I222$	$I222$	
D_2^9, V^9	$I2_12_12_1$	$I2_12_12_1$	
D_{2h}^1, V_h^1	$P\,2/m\,2/m\,2/m$	$Pmmm$	
D_{2h}^2, V_h^2	$P\,2/n\,2/n\,2/n$	$Pnnn$	
D_{2h}^3, V_h^3	$P\,2/c\,2/c\,2/m$	$Pccm$	$Pbmb$; $Pmaa$
D_{2h}^4, V_h^4	$P\,2/b\,2/a\,2/n$	$Pban$	$Pcna$; $Pncb$
D_{2h}^5, V_h^5	$P\,2_1/m\,2/m\,2/a$	$Pmma$	$Pmmb$; $Pmam$; $Pmcm$; $Pbmm$; $Pcmm$
D_{2h}^6, V_h^6	$P\,2/n\,2_1/n\,2/a$	$Pnna$	$Pnnb$; $Pnan$; $Pncn$; $Pbnn$; $Pcnn$
D_{2h}^7, V_h^7	$P\,2/m\,2/n\,2_1/a$	$Pmna$	$Pnmb$; $Pman$; $Pncm$; $Pbmn$; $Pcnm$
D_{2h}^8, V_h^8	$P\,2_1/c\,2/c\,2/a$	$Pcca$	$Pccb$; $Pbab$; $Pbcb$; $Pbaa$; $Pcaa$
D_{2h}^9, V_h^9	$P\,2_1/b\,2_1/a\,2/m$	$Pbam$	$Pmcb$
D_{2h}^{10}, V_h^{10}	$P\,2_1/c\,2_1/c\,2/n$	$Pccn$	$Pbnb$; $Pnaa$
D_{2h}^{11}, V_h^{11}	$P\,2/b\,2_1/c\,2_1/m$	$Pbcm$	$Pbma$; $Pcam$; $Pcmb$; $Pmab$; $Pmca$
D_{2h}^{12}, V_h^{12}	$P\,2_1/n\,2_1/n\,2/m$	$Pnnm$	$Pmnn$; $Pmnn$
D_{2h}^{13}, V_h^{13}	$P\,2_1/m\,2_1/m\,2/n$	$Pmmn$	$Pmnm$; $Pnmm$
D_{2h}^{14}, V_h^{14}	$P\,2_1/b\,2/c\,2_1/n$	$Pbcn$	$Pbna$; $Pcan$; $Pcnb$; $Pnab$; $Pnca$
D_{2h}^{15}, V_h^{15}	$P\,2_1/b\,2_1/c\,2_1/a$	$Pbca$	$Pcab$
D_{2h}^{16}, V_h^{16}	$P\,2_1/n\,2_1/m\,2_1/a$	$Pnma$	$Pnam$; $Pbnm$; $Pcmn$; $Pmnb$; $Pmcn$
D_{2h}^{17}, V_h^{17}	$C\,2/m\,2/c\,2_1/m$	$Cmcm$	$Ccmm$; $Amma$; $Amam$; $Bmmb$; $Bbmm$
D_{2h}^{18}, V_h^{18}	$C\,2/m\,2/c\,2_1/a$	$Cmca$	$Ccma$; $Abma$; $Abam$; $Bmab$; $Bbam$
D_{2h}^{19}, V_h^{19}	$C\,2/m\,2/m\,2/m$	$Cmmm$	$Ammm$; $Bmmm$
D_{2h}^{20}, V_h^{20}	$C\,2/c\,2/c\,2/m$	$Cccm$	$Amaa$; $Bbmb$
D_{2h}^{21}, V_h^{21}	$C\,2/m\,2/m\,2/a$	$Cmma$	$Abmm$; $Bmam$
D_{2h}^{22}, V_h^{22}	$C\,2/c\,2/c\,2/a$	$Ccca$	$Abaa$; $Bbab$
D_{2h}^{23}, V_h^{23}	$F\,2/m\,2/m\,2/m$	$Fmmm$	
D_{2h}^{24}, V_h^{24}	$F\,2/d\,2/d\,2/d$	$Fddd$	
D_{2h}^{25}, V_h^{25}	$I\,2/m\,2/m\,2/m$	$Immm$	
D_{2h}^{26}, V_h^{26}	$I\,2/b\,2/a\,2/m$	$Ibam$	$Ibma$; $Imaa$
D_{2h}^{27}, V_h^{27}	$I\,2_1/b\,2_1/c\,2_1/a$	$Ibca$	
D_{2h}^{28}, V_h^{28}	$I\,2_1/m\,2_1/m\,2_1/a$	$Imma$	$Imam$; $Ibmm$
S_4^1	$P\bar{4}$	$P\bar{4}$	$C\bar{4}$
S_4^2	$I\bar{4}$	$I\bar{4}$	$F\bar{4}$
D_{2d}^1, V_d^1	$P\bar{4}2m$	$P\bar{4}2m$	$C\bar{4}m2$
D_{2d}^2, V_d^2	$P\bar{4}2c$	$P\bar{4}2c$	$C\bar{4}c2$
D_{2d}^3, V_d^3	$P\bar{4}2_1m$	$P\bar{4}2_1m$	$C\bar{4}m2_1$
D_{2d}^4, V_d^4	$P\bar{4}2_1c$	$P\bar{4}2_1c$	$C\bar{4}c2_1$
D_{2d}^5, V_d^5	$C\bar{4}2m$	$C\bar{4}2m$	$P\bar{4}m2$
D_{2d}^6, V_d^6	$C\bar{4}2c$	$C\bar{4}2c$	$P\bar{4}c2$
D_{2d}^7, V_d^7	$C\bar{4}2b$	$C\bar{4}2b$	$P\bar{4}b2$
D_{2d}^8, V_d^8	$C\bar{4}2n$	$C\bar{4}2n$	$P\bar{4}n2$
D_{2d}^9, V_d^9	$F\bar{4}2m$	$F\bar{4}2m$	$I\bar{4}m2$
D_{2d}^{10}, V_d^{10}	$F\bar{4}2c$	$F\bar{4}2c$	$I\bar{4}c2$

TABLE 7 — *Continued*

Schoenflies symbol	Complete Mauguin symbol	International symbol	International symbol for other orientations
D_{2d}^{11}, V_d^{11}	$I\bar{4}2m$	$I\bar{4}2m$	$F\bar{4}m2$
D_{2d}^{12}, V_d^{12}	$I\bar{4}2d$	$I\bar{4}2d$	$F\bar{4}d2$
C_4^1	$P4$	$P4$	$C4$
C_4^2	$P4_1$	$P4_1$	$C4_1$
C_4^3	$P4_2$	$P4_2$	$C4_2$
C_4^4	$P4_3$	$P4_3$	$C4_3$
C_4^5	$I4$	$I4$	$F4$
C_4^6	$I4_1$	$I4_1$	$F4_1$
C_{4h}^1	$P4/m$	$P4/m$	$C4/m$
C_{4h}^2	$P4_2/m$	$P4_2/m$	$C4_2/m$
C_{4h}^3	$P4/n$	$P4/n$	$C4/a$
C_{4h}^4	$P4_2/n$	$P4_2/n$	$C4_2/a$
C_{4h}^5	$I4/m$	$I4/m$	$F4/m$
C_{4h}^6	$I4_1/a$	$I4_1/a$	$F4_1/d$
C_{4v}^1	$P4mm$	$P4mm$	$C4mm$
C_{4v}^2	$P4bm$	$P4bm$	$C4mb$
C_{4v}^3	$P4_2cm$	$P4cm$	$C4mc$
C_{4v}^4	$P4_2nm$	$P4nm$	$C4mn$
C_{4v}^5	$P4cc$	$P4cc$	$C4cc$
C_{4v}^6	$P4nc$	$P4nc$	$C4cn$
C_{4v}^7	$P4_2mc$	$P4mc$	$C4cm$
C_{4v}^8	$P4_2bc$	$P4bc$	$C4cb$
C_{4v}^9	$I4mm$	$I4mm$	$F4mm$
C_{4v}^{10}	$I4cm$	$I4cm$	$F4mc$
C_{4v}^{11}	$I4_1md$	$I4md$	$F4dm$
C_{4v}^{12}	$I4_1cd$	$I4cd$	$F4dc$
D_4^1	$P422$	$P42$	$C422$
D_4^2	$P42_12$	$P42_1$	$C422_1$
D_4^3	$P4_122$	$P4_12$	$C4_122$
D_4^4	$P4_12_12$	$P4_12_1$	$C4_122_1$
D_4^5	$P4_222$	$P4_22$	$C4_222$
D_4^6	$P4_22_12$	$P4_22_1$	$C4_222_1$
D_4^7	$P4_322$	$P4_32$	$C4_322$
D_4^8	$P4_32_12$	$P4_32_1$	$C4_322_1$
D_4^9	$I422$	$I42$	$F42$
D_4^{10}	$I4_122$	$I4_12$	$F4_12$
D_{4h}^1	$P\,4/m\,2/m\,2/m$	$P4/mmm$	$C4/mmm$
D_{4h}^2	$P\,4/m\,2/c\,2/c$	$P4/mcc$	$C4/mcc$
D_{4h}^3	$P\,4/n\,2/b\,2/m$	$P4/nbm$	$C4/amb$
D_{4h}^4	$P\,4/n\,2/n\,2/c$	$P4/nnc$	$C4/acn$
D_{4h}^5	$P\,4/m\,2_1/b\,2/m$	$P4/mbm$	$C4/mmb$
D_{4h}^6	$P\,4/m\,2_1/n\,2/c$	$P4/mnc$	$C4/mcn$
D_{4h}^7	$P\,4/n\,2_1/m\,2/m$	$P4/nmm$	$C4/amm$
D_{4h}^8	$P\,4/n\,2_1/c\,2/c$	$P4/ncc$	$C4/acc$

TABLE 7 — *Continued*

Schoenflies symbol	Complete Mauguin symbol		International symbol	International symbol for other orientations
D_{4h}^9	$P\,4_2/m\,2/m\,2/c$		$P4/mmc$	$C4/mcm$
D_{4h}^{10}	$P\,4_2/m\,2/c\,2/m$		$P4/mcm$	$C4/mmc$
D_{4h}^{11}	$P\,4_2/n\,2/b\,2/c$		$P4/nbc$	$C4/abc$
D_{4h}^{12}	$P\,4_2/n\,2/n\,2/m$		$P4/nnm$	$C4/amn$
D_{h4}^{13}	$P\,4_2/m\,2_1/b\,2/c$		$P4/mbc$	$C4/mcb$
D_{4h}^{14}	$P\,4_2/m\,2_1/n\,2/m$		$P4/mnm$	$C4/mmn$
D_{4h}^{15}	$P\,4_2/n\,2_1/m\,2/c$		$P4/nmc$	$C4/acm$
D_{4h}^{16}	$P\,4_2/n\,2_1/c\,2/m$		$P4/ncm$	$C4/amc$
D_{4h}^{17}	$I\,4/m\,2/m\,2/m$		$I4/mmm$	$F4/mmm$
D_{4h}^{18}	$I\,4/m\,2/c\,2/m$		$I4/mcm$	$F4/mmc$
D_{4h}^{19}	$I\,4_1/a\,2/m\,2/d$		$I4/amd$	$F4/ddm$
D_{4h}^{20}	$I\,4_1/a\,2/c\,2/d$		$I4/acd$	$F4/ddc$
C_3^1	$C3$		$C3$	$H3$
C_3^2	$C3_1$		$C3_1$	$H3_1$
C_3^3	$C3_2$		$C3_2$	$H3_2$
C_3^4	$R3$		$R3$	
$C_{3i}^1,\ S_6^1$	$C\bar{3}$		$C\bar{3}$	$H\bar{3}$
$C_{3i}^2,\ S_6^2$	$R\bar{3}$		$R\bar{3}$	
C_{3v}^1	$C3m1$		$C3m$	$H31m$
C_{3v}^2	$H3m1$	$C31m$	$H3m$	$C31m$
C_{3v}^3	$C3c1$		$C3c$	$H31c$
C_{3v}^4	$H3c1$	$C31c$	$H3c$	$C31c$
C_{3v}^5	$R3m$		$R3m$	
C_{3v}^6	$R3c$		$R3c$	
D_3^1	$H321$	$C312$	$H32$	$C312$
D_3^2	$C321$		$C32$	$H312$
D_3^3	$H3_121$	$C3_112$	$H3_12$	$H3_112$
D_3^4	$C3_121$		$C3_12$	$H3_112$
D_3^5	$H3_221$	$C3_212$	$H3_22$	$C3_212$
D_3^6	$C3_221$		$C3_22$	$H3_212$
D_3^7	$R32$		$R32$	
D_{3d}^1	$H\bar{3}2m1$	$C\bar{3}12m$	$H\bar{3}m$	$C\bar{3}1m$
D_{3d}^2	$H\bar{3}2c1$	$C\bar{3}12c$	$H\bar{3}c$	$C\bar{3}1c$
D_{3d}^3	$C\bar{3}2m1$		$C\bar{3}m$	$H\bar{3}1m$
D_{3d}^4	$C\bar{3}2c1$		$C\bar{3}c$	$H\bar{3}1c$
D_{3d}^5	$R\bar{3}2m$		$R\bar{3}m$	
D_{3d}^6	$R\bar{3}2c$		$R\bar{3}c$	
C_{3h}^1	$C\bar{6}$		$C\bar{6}$	$H\bar{6}$
D_{3h}^1	$C\bar{6}m2$		$C\bar{6}m2$	$H\bar{6}2m$
D_{3h}^2	$C\bar{6}c2$		$C\bar{6}c2$	$H\bar{6}2c$
D_{3h}^3	$H\bar{6}m2$	$C\bar{6}2m$	$H\bar{6}m2$	$C\bar{6}2m$
D_{3h}^4	$H\bar{6}c2$	$C\bar{6}2c$	$H\bar{6}c2$	$C\bar{6}2c$

TABLE 7 — *Continued*

Schoenflies symbol	Complete Mauguin symbol	International symbol	International symbol for other orientations
C_6^1	$C6$	$C6$	$H6$
C_6^2	$C6_1$	$C6_1$	$H6_1$
C_6^3	$C6_5$	$C6_5$	$H6_5$
C_6^4	$C6_2$	$C6_2$	$H6_2$
C_6^5	$C6_4$	$C6_4$	$H6_4$
C_6^6	$C6_3$	$C6_3$	$H6_3$
C_{6h}^1	$C6/m$	$C6/m$	$H6/m$
C_{6h}^2	$C6_3/m$	$C6_3/m$	$H6_3/m$
C_{6v}^1	$C6mm$	$C6mm$	$H6mm$
C_{6v}^2	$C6cc$	$C6cc$	$H6cc$
C_{6v}^3	$C6_3cm$	$C6cm$	$H6mc$
C_{6v}^4	$C6_3mc$	$C6mc$	$H6mc$
D_6^1	$C622$	$C62$	$H62$
D_6^2	$C6_122$	$C6_12$	$H6_12$
D_6^3	$C6_522$	$C6_52$	$H6_52$
D_6^4	$C6_222$	$C6_22$	$H6_22$
D_6^5	$C6_422$	$C6_42$	$H6_42$
D_6^6	$C6_322$	$C6_32$	$H6_32$
D_{6h}^1	$C\,6/m\,2/m\,2/m$	$C6/mmm$	$H6/mmm$
D_{6h}^2	$C\,6/m\,2/c\,2/c$	$C6/mcc$	$H6/mcc$
D_{6h}^3	$C\,6_3/m\,2/c\,2/m$	$C6/mcm$	$H6/mmc$
D_{6h}^4	$C\,6_3/m\,2/m\,2/c$	$C6/mmc$	$H6/mcm$
T^1	$P23$	$P23$	
T^2	$F23$	$F23$	
T^3	$I23$	$I23$	
T^4	$P2_13$	$P2_13$	
T^5	$I2_13$	$I2_13$	
T_h^1	$P\,2/m\,\overline{3}$	$Pm3$	
T_h^2	$P\,2/n\,\overline{3}$	$Pn3$	
T_h^3	$F\,2/m\,\overline{3}$	$Fm3$	
T_h^4	$F\,2/d\,\overline{3}$	$Fd3$	
T_h^5	$I\,2/m\,\overline{3}$	$Im3$	
T_h^6	$P\,2_1/a\,\overline{3}$	$Pa3$	
T_h^7	$I\,2/a\,\overline{3}$	$Ia3$	
T_d^1	$P\overline{4}3m$	$P\overline{4}3m$	
T_d^2	$F\overline{4}3m$	$F\overline{4}3m$	
T_d^3	$I\overline{4}3m$	$I\overline{4}3m$	
T_d^4	$P\overline{4}3n$	$P\overline{4}3n$	
T_d^5	$F\overline{4}3c$	$F\overline{4}3c$	
T_d^6	$I\overline{4}3d$	$I\overline{4}3d$	
O^1	$P432$	$P43$	
O^2	$P4_232$	$P4_23$	
O^3	$F432$	$F43$	
O^4	$F4_132$	$F4_13$	

TABLE 7 — *Continued*

Schoenflies symbol	Complete Mauguin symbol	International symbol	International symbol for other orientations
O^5	$I432$	$I43$	
O^6	$P4_332$	$P4_33$	
O^7	$P4_132$	$P4_13$	
O^8	$I4_132$	$I4_13$	
O_h^1	$P\ 4/m\ \bar{3}\ 2/m$	$Pm3m$	
O_h^2	$P\ 4/n\ \bar{3}\ 2/n$	$Pn3n$	
O_h^3	$P\ 4_2/m\ \bar{3}\ 2/n$	$Pm3n$	
O_h^4	$P\ 4_2/n\ \bar{3}\ 2/m$	$Pn3m$	
O_h^5	$F\ 4/m\ \bar{3}\ 2/m$	$Fm3m$	
O_h^6	$F\ 4/m\ \bar{3}\ 2/c$	$Fm3c$	
O_h^7	$F\ 4_1/d\ \bar{3}\ 2/m$	$Fd3m$	
O_h^8	$F\ 4_1/d\ \bar{3}\ 2/c$	$Fd3c$	
O_h^9	$I\ 4/m\ \bar{3}\ 2/m$	$Im3m$	
O_h^{10}	$I\ 4_1/a\ \bar{3}\ 2/d$	$Ia3d$	

$P2_1/a$, because the screw axis, 2_1, would have extinguished $0k0$ when k was odd, in addition to the extinctions listed. The decision between Pa and $P2/a$ depends entirely upon being able to determine by some other means whether the crystal class of the crystal contains a two-fold axis or not. If not, then the space group is Pa; if so, then the space group is $P2/a$.

Suppose that no extinctions whatever are shown by a monoclinic crystal. Then it is definitely known that it contains no screw axes or glide planes, so the space group cannot be $P2_1$, $P2_1/m$, Pa, $P2/a$, or $P2_1/a$. But from extinction information it cannot be discovered whether or not the space group contains a twofold rotation axis or a reflection plane or both, because neither of these involve translation components, so the space group may be either $P2$, Pm, or $P2/m$. A decision between these can be made only if it can be determined by some other means whether the crystal class contains a twofold axis, a reflection plane, or both.

In general, then, it may be said that (except for mutual interferences as provided below in precaution 2) extinctions give information toward identifying the space groups by:

(*a*) providing for the definite identification of the space lattice type, and by

(*b*) providing for the definite identification of any glide planes and screw axes in the space group, but

(*c*) are incapable of giving any information as to the presence or absence of rotation axes or reflection planes. In order to decide whether non-extinction implies the presence of either of these two or not, it is necessary to know the crystal class. If the crystal class includes one or the other of these, then non-extinction implies the presence of the rotation axis or the reflection plane in the space group.

2. The extinction of the more general classes of spectra automatically involves the extinction of some less general classes of spectra, whether the latter is caused by a corresponding symmetry element or not. For example, if a crystal is based upon a body-centered lattice, I, the class of general spectra, hkl, is extinguished when $h + k + l$ is odd. For the specialized class of hkl reflections $hk0$, $l = 0$, and this reduces to an extinction of $hk0$ when $h + k$ is odd; and when both $k = 0$ and $l = 0$, this reduces to an extinction of $h00$ when h is odd. The general extinction of hkl when $h + k + l$ is odd thus automatically implies the extinction of $hk0$ when $h + k$ is odd and $h00$ when h is odd (as well as certain other special cases). *Extinctions in the more general classes of spectra thus imply automatic extinctions in less general classes of spectra which are special cases, and automatically screen any characteristic extinctions in such classes due to symmetry elements.* When this occurs it destroys the usefulness of this special class of spectra in detecting a possible symmetry element in the space group. For example, the space groups $I222$ and $I2_12_12_1$ cannot be differentiated because the extinction caused by the screw axis, 2_1, is already present in the extinction caused by the body-centered lattice, I, and therefore whether 2_1 is present or absent cannot be determined.

LITERATURE

Orienting work

Paul Niggli. *Geometrische Kristallographie des Diskontinuums.* (Gebrüder Borntraeger, Leipzig, 1919.) Especially 463–503.

Complete general extinction tables

W. T. Astbury and Kathleen Yardley. Tabulated data for the examination of the 230 space-groups by homogeneous x-rays. *Phil. Trans. London (A)*, **224** (1924), 221–257, and plates 5–24.

Hermann Mark. *Die Verwendung der Röntgenstrahlen in Chemie und Technik. Handbuch der angewandtenphysikalischen Chemie.* Vol. XIV. (Johann Ambrosius Barth, Leipzig, 1926.) 387–394.

C. Hermann. Zur systematischen Strukturtheorie. I. Eine neue Raumgruppensymbolik. *Z. Krist. (A)*, **68** (1928), 257–287.

Arthur Schleede and Erich Schneider. *Röntgenspektroskopie und Kristallstrukturanalyse.* Vol. II. (Walter de Gruyter & Co., Berlin and Leipzig, 1929.) Pages 237–240.

Internationale Tabellen zur Bestimmung von Kristallstrukturen. I. (Gebrüder Borntraeger, Berlin, 1935.)

J. D. H. DONNAY and DAVID HARKER. Nouvelles tables d'extinctions pour les 230 groupes de recouvrements cristallographiques. *Le Naturaliste Canadien,* **67** (1940), 33–69.

Special extinction tables

RALPH W. G. WYCKOFF. The determination of the space-group of a cubic crystal. *Am. J. Sci.,* **4** (1922), 175–187.

RALPH W. G. WYCKOFF. Orthorhombic space-group criteria and their application to aragonite. *Am. J. Sci.,* **9** (1925), 145–175, also *Z. Krist.* (*A*), **61** (1925), 425–451.

RALPH W. G. WYCKOFF and HERBERT E. MERWIN. The space group of diopside [$CaMg(SiO_3)_2$]. [Contains criteria for monoclinic space groups.] *Am. J. Sci.,* **9** (1925), 379–394.

RALPH W. G. WYCKOFF. Kriterien für hexagonale Raumgruppen und die Kristallstruktur von β-Quarz. *Z. Krist.* (*A*), **63** (1926), 507–537.

RALPH W. G. WYCKOFF and STERLING B. HENDRICKS. Die Kristallstruktur von Zirkon und die Kriterien für spezielle Lagen in tetragonalen Raumgruppen. *Z. Krist.* (*A*), **66** (1927), 73–102.

K. HERMANN and M. BURAK. Röntgenographische Untersuchung des ortho- und meta-Nitranilins. *Z. Krist.* (*A*), **67** (1928), 189–225. [Contains extinction rules for orthorhombic space-groups.]

K. HERMANN. Röntgenographische Auslöschungstabellen. *Z. Krist.* (*A*), **68** (1928), 288–298.

E. BRANDENBERGER and P. NIGGLI. Die systematische Darstellung der kristallstrukturell wichtigen Auswahlregeln. *Z. Krist.* (*A*), **68** (1928), 301–329.

E. BRANDENBERGER. Systematische Darstellung der kristallstrukturell wichtigen Auswahlregeln trikliner, monokliner und rhombischer Raumsysteme. *Z. Krist.* (*A*), **68** (1928), 330–362.

E. BRANDENBERGER. Systematische Darstellung der kristallstrukturell wichtigen Auswahlregeln tetragonaler Raumsysteme. *Z. Krist.* (*A*), **71** (1929), 452–500.

International space group symbols

C. HERMANN. Zur systematischen Strukturtheorie. I. Eine neue Raumgruppensymbolik. *Z. Krist.* (*A*), **68** (1928), 257–287.

Ch. MAUGUIN. Sur le symbolisme des groupes de répétition ou de symétrie des assemblages cristallins. *Z. Krist.* (*A*), **76** (1931), 542–558.

C. HERMANN. Bemerkung zu der vorstehenden Arbeit von Ch. Mauguin. *Z. Krist.* (*A*), **76** (1931), 559–561.

CHAPTER 5

THE ROTATING-CRYSTAL METHOD, PRELIMINARY ACCOUNT

Some of the theoretical aspects of x-ray diffraction have already been considered. In this chapter some of the preliminary aspects of one of the simplest practical applications of x-ray diffraction will be discussed. In Chapter 3 (pages 45–47) it was shown that, in general, no diffracted rays develop if a parallel beam of x-rays impinges upon a crystal unless the crystal occupies a specialized orientation with respect to the x-ray beam. This specialization of orientation can be assured

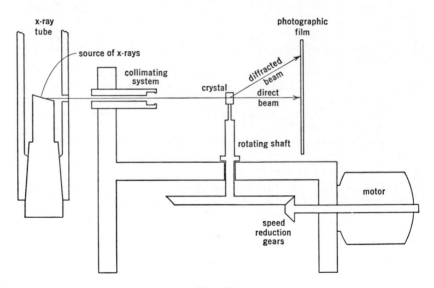

FIG. 50.

if the crystal is rotated about some axis, for, at the instant the crystal passes through a specialized orientation, the conditions for diffraction are satisfied and diffraction occurs. This method of producing diffraction of x-rays by crystals is the basis of the *rotating-crystal method*.

A diagram of the apparatus used in the rotating-crystal method is shown in Fig. 50. The x-rays are generated in the x-ray tube at the left of the rotating crystal apparatus, and they enter a system (essentially a tube) which permits only a beam of substantially parallel x-rays to reach the crystal. The crystal is mounted on a shaft which, for the sake of simplicity in interpreting the results, is arranged at right angles to the impinging beam. The shaft is slowly rotated at a uniform angular rate by a small motor. The diffracted radiation which is developed when the crystal reaches the several specialized orientations is recorded on a photographic plate or film. Two photographic arrangements are in common use, either a flat plate placed behind the crystal and at right angles to the direct x-ray beam, or a film rolled in the form of a cylinder and made coaxial with the rotation axis. The cylindrical film recording is preferable for a number of reasons which will become plain as the discussion continues. To anticipate a bit, it may be said that the chief advantage of the cylindrical film over the flat plate is that it is capable of recording a great range of values of the glancing angle, θ, and consequently gives a great wealth of reflections. The flat plate can record only a very limited range of θ values. The greater range of θ leads, ultimately, to superior accuracy in the refinement of cell constants, to a greater wealth of data for the identification of the correct space group, and finally, to greater accuracy in the determination of the positions of the atoms in the cell. In addition to this, it is easier to interpret cylindrical film records than flat plate records.

THE EXPERIMENTAL DETERMINATION OF CRYSTAL IDENTITY PERIODS

In Chapter 3, it was shown that the locus of diffraction directions for a single periodic line of atoms is a nest of coaxial cones. For a three-dimensional pattern of atoms, there is a possible nest of such cones coaxial with each possible rational lattice line. In order to account for all possible diffraction directions, it is unnecessary to consider cones about all such lines, however, for the possible directions of all diffraction beams are fixed when three such rational lattice lines are chosen. Under ordinary circumstances, these would be the three directions selected as crystallographic reference axes.

Suppose the crystal to be rotated in the x-ray beam about one of its lattice rows, i.e., about some rational line, which may or may not be a crystallographic reference axis. Since this row is the rotation axis, it maintains a constant angle with the incoming x-ray beam. Whatever diffraction develops, it is necessarily confined to directions along the generators of the cones coaxial with this axis, and these cones have fixed positions during the rotation, because the angle between the x-ray beam and the lattice row is constant.   These cones intersect a coaxial cylindrical film, Fig. 51, in a series of circles.   When the film is flattened out, the circles appear as straight lines.   These are known as *layer lines*.   If a flat plate recording is used, Fig. 52, the cones intersect the plate in a series of hyperbolas.   Layer lines constitute the loci of

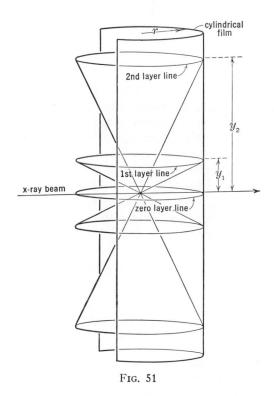

Fig. 51

all recorded diffraction spots provided that the crystal is rotated about a rational axis.   Figures 53 and 54 show, respectively, rotating crystal photographs recorded on cylindrical film and flat plate.

Since the cone angles are functions of the identity period of the row which is associated with them, the identity period of the lattice row which is used as rotation axis can be computed from an experimentally determined cone angle.   This angle is easily evaluated from the geometry of the apparatus and the locations of the layer lines.   The situation is shown diagrammatically in Fig. 55.

The angle $\bar{\nu}$ is the semi-opening angle of the cone as used in Chapter 3.   Its complement, ν, is customarily used in the discussion of rotating-crystal photographs.

In Fig. 55, the angle v, for the first layer cone, is fixed by

$$\sin v_1 = \frac{\lambda}{t}, \tag{1}$$

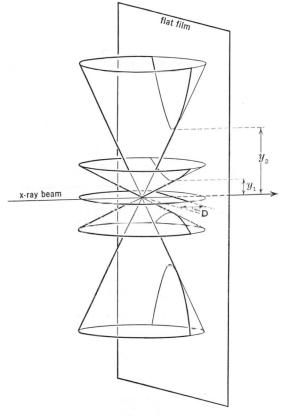

flat film

x-ray beam

y_2

y_1

D

FIG. 52.

and for the nth layer line,

$$\sin v_n = \frac{n\lambda}{t}. \tag{2}$$

The first layer line records at a height y_1 given by

$$\tan v_1 = \frac{y_1}{r}, \tag{3}$$

where r is the radius of the cylindrical film.   The nth layer line simi-

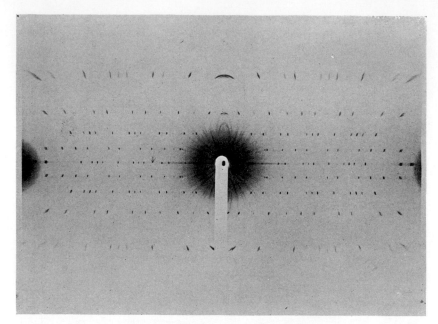

FIG. 53. Rotating-crystal photograph taken with cylindrical camera. (NaSbO$_3$·3H$_2$O, tetragonal; c-axis rotation; Cu$K\alpha$ radiation from gas x-ray tube, filtered through nickel foil.)

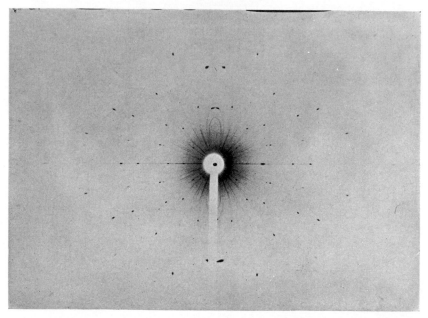

FIG. 54. Rotating-crystal photograph taken with flat-film camera. (Same data as Fig. 53.)

larly records at a height y_n, such that

$$\tan \nu_n = \frac{y_n}{r}. \tag{4}$$

Eliminating ν from (1) and (3) gives

$$t = \frac{\lambda}{\sin \tan^{-1} (y_1/r)}. \tag{5}$$

The corresponding relation obtained from (2) and (4) for the nth layer line is

$$t = \frac{n\lambda}{\sin \tan^{-1} (y_n/r)}. \tag{6}$$

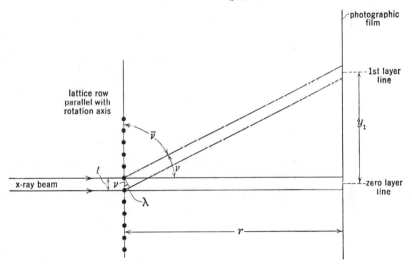

FIG. 55.

Any identity period of any crystal may thus be calculated from data obtained by rotating the crystal in an x-ray beam about the rational axis along which the identity period is required. For this purpose the layer lines must be recorded on a cylindrical film, not a flat plate. The reason for this is that, for a flat plate, relation (6) holds only for the nose of the hyperbola, Fig. 52; for a reflection to occur here would be fortuitous, and therefore no measurement of y can be made. Other spots not at the nose of the hyperbola cannot be used because the height of a layer line, y_n, varies with the x film coordinate of the spot for flat plate recording. This is a complication compared with cylindrical recording in which all spots on a given layer line have the same y film coordinate.   This is another reason why a cylindrical film is easier to interpret than a flat plate.

The most convenient and accurate simple way of determining y_n is to measure the doubled y_n distance from a spot on the upper, $+n$, layer line to the corresponding spot exactly under it on the lower, $-n$, layer line.   The measurement may be made with an ordinary steel scale, but it is easier and more accurate to spread the points of a micrometer caliper between these two spots.   Additional suggestions for measuring y are given on pages 186–187.

By rotating a crystal in an x-ray beam about each of its conventional crystallographic axes and measuring the identity periods along these axes, the lengths of the cell edges along these directions may be measured.   This ordinarily determines the unit cell.   However, if the axes of the cell are not symmetry-fixed (as they are, for example, in the normal class of the orthorhombic system, where they correspond with the twofold axes) then the original rational directions chosen for axes may prove to define an undesirable cell.   After studying the lengths of these identity periods as determined by the rotating-crystal method, it may prove desirable to adopt an alternative cell with different axes.

DETERMINATION OF SPACE GROUP

Introduction.   The determination of the space group of a crystal requires the identification of all reflections.   When the observed reflections are listed, data are at hand for a study of possible extinguished classes.   When these become known, the choice of space groups for the crystal is limited as outlined in the last section of Chapter 4.

Now, the determination of the unit cell of a crystal is a straightforward and simple matter.   The assignment of indices to the separate reflections, on the other hand, is a matter of considerable labor and difficulty if done in a straightforward manner.   The present chapter discusses this straightforward approach.   It should be said at the outset, however, that this approach is not the one to be recommended.   In the next chapter, a powerful tool will be developed, which, though not permitting a straightforward approach to the problem, simplifies it immensely.

The determination of the glancing angle, θ.   It has already been shown that, when a beam of x-rays strikes a single crystal, diffraction takes place only for certain definite orientations of the crystal such that the Bragg condition,

$$n\lambda = 2d \sin \theta, \tag{7}$$

is satisfied.   When the orientation of any plane in the crystal is such
that the glancing angle, θ, becomes equal to $\sin^{-1} \dfrac{n\lambda}{2d}$, then that plane
reflects x-rays.   From Fig. 56, it is evident that this is equivalent to
saying that, whenever the orientation of the plane becomes such that

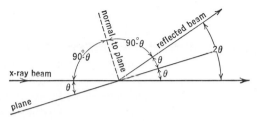

FIG. 56.

the angle between its normal and the x-ray beam equals $90° - \theta$
(i. e., $90° - \sin^{-1} \dfrac{n\lambda}{2d}$), then that plane reflects x-rays.   From Fig. 57,

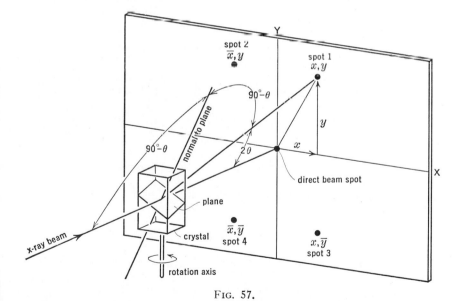

FIG. 57.

it is easy to see that, if the crystal is rotated about some axis, in general the angle between the x-ray beam and the normal to a given plane
varies continuously during the rotation of the crystal.   As this

angle instantaneously passes through the appropriate value of $90° - \sin^{-1}\dfrac{n\lambda}{2d}$ for each plane, a diffracted beam is developed, the direction of each diffracted beam depending on the slope of the plane relative to the rotation axis, and on its interplaner spacing, d. Each plane, in general, produces four spots on the photographic plate because its normal passes through four distinct positions each making an angle of $90° - \theta$ with the incoming beam: Fig. 57 shows the plane reflecting when the normal is directed right and up. A reflection also takes place when the normal is in a symmetrical position on the left side of the beam, thus producing spot 2. The plane also reflects when the negative end of the normal makes an angle of $90° - \theta$ with the beam, thus producing the lower spots, 3, and 4, distributed right and left of the primary beam. If the photographic plate surface (or film cylinder axis) is arranged parallel with the rotation axis, and if the x-ray beam strikes the crystal normal to its rotation axis as shown in Fig. 57, then these four spots have the coordinates x,y; $\bar{x},y$; $x,\bar{y}$; $\bar{x},\bar{y}$, on the photographic plate or film. If the reflecting plane under consideration has a specialized slope with respect to the rotation axis, some of these pairs of spots coalesce. The only important case is that in which the plane is parallel with the rotation axis, when spots $1 + 3$ and $2 + 4$ coalesce, the two resulting spots having film coordinates $x,0$; and $\bar{x},0$.

Assignment of indices. For every photographically recorded spot, the Bragg glancing angle, θ can be calculated, if desired, from measurements made on the film or plate. Thus, in Figs. 58 and 59, the angles 2θ, Υ, and χ, are related by the spherical trigonometric relation,

$$\cos 2\theta = \cos \Upsilon \cos \chi. \tag{8}$$

In the cylindrical camera, Fig. 58, the angles Υ and χ are determined as follows:

$$\Upsilon = \frac{x}{r} \text{ radians,} \tag{9}$$

$$\tan \chi = \frac{y}{r}. \tag{10}$$

Substituting these in (8) gives

$$\cos 2\theta = \cos\left(\frac{x}{r}\right) \cos\left(\tan^{-1}\frac{y}{r}\right). \tag{11}$$

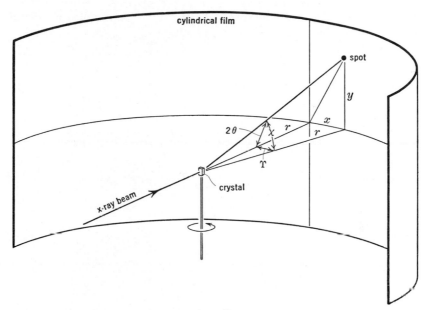

cylindrical film

spot

y

2θ

x

r

r

x

r

Υ

crystal

x-ray beam

Fɪɢ. 58.

Utilizing the value of θ given by (11) in the Bragg relation, (7), gives

$$d = \frac{n\lambda}{2 \sin \theta}$$

$$= \frac{n\lambda}{2 \sin \left[\frac{1}{2} \cos^{-1} \left\{ \cos \left(\frac{x}{r} \right) \cos \left(\tan^{-1} \frac{y}{r} \right) \right\} \right]}, \tag{12}$$

the interplanar spacing of the plane producing the spot.

For flat plates, Fig. 59, the corresponding relations are

$$\tan 2\theta = \frac{s}{D}. \tag{13}$$

Substituting this in (7) gives

$$d = \frac{n\lambda}{2 \sin \theta}$$

$$= \frac{n\lambda}{2 \sin \left(\frac{1}{2} \tan^{-1} \frac{s}{D} \right)}. \tag{14}$$

This relation may also be expressed directly in terms of film coordinates, x and y, by noting that

$$s = \sqrt{x^2 + y^2}. \tag{15}$$

Making this substitution, (14) becomes

$$d = \frac{n\lambda}{2 \sin\left(\frac{1}{2} \tan^{-1} \dfrac{\sqrt{x^2 + y^2}}{D}\right)}. \tag{16}$$

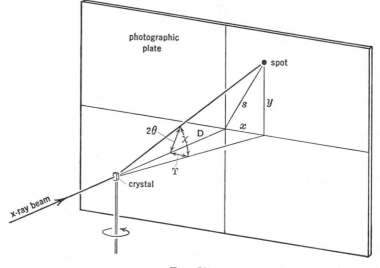

FIG. 59.

By measuring the coordinates of the several spots recorded on the photographic plate or film, and then substituting these measured values in the pertinent relations (12), (14), or (16), a list of observed interplanar spacings can be prepared for all planes reflecting within the recording range of the apparatus. *Such data can be obtained regardless of the orientation of the crystal on the rotation axis.*

Now, these same interplanar spacings can also be arrived at in another way, for they are functions of the axial translations, a, b, and c, and the indices of the reflecting plane, (hkl). These functions are simple for isometric crystals, but, because of the increasing generality of the axial coordinate systems, they become increasingly complicated for the other crystals in the order: tetragonal, hexagonal, (rhombohedral) orthorhombic, monoclinic, triclinic. The interplanar spacing formulas for the seven cases are:

Isometric system:

$$d = \frac{a}{\sqrt{h^2 + k^2 + l^2}}. \qquad (17)$$

Tetragonal system:

$$d = \frac{1}{\sqrt{\dfrac{h^2}{a^2} + \dfrac{k^2}{a^2} + \dfrac{l^2}{c^2}}}. \qquad (18)$$

Orthorhombic system:

$$d = \frac{1}{\sqrt{\dfrac{h^2}{a^2} + \dfrac{k^2}{b^2} + \dfrac{l^2}{c^2}}}. \qquad (19)$$

Hexagonal system, hexagonal indexing:

$$d = \frac{1}{\sqrt{\dfrac{4}{3a^2}\left(h^2 + k^2 + hk\right) + \dfrac{l^2}{c^2}}}. \qquad (20)$$

Hexagonal system, rhombohedral indexing:

$$d = \frac{1}{\dfrac{1}{a}\sqrt{\dfrac{(h^2 + k^2 + l^2)\sin^2 \alpha + 2(hk + hl + kl)(\cos^2 \alpha - \cos \alpha)}{1 + 2\cos^3 \alpha - 3\cos^2 \alpha}}}. \qquad (21)$$

Monoclinic system:

$$d = \frac{1}{\sqrt{\dfrac{\dfrac{h^2}{a^2} + \dfrac{l^2}{c^2} - \dfrac{2hl}{ac}\cos \beta}{\sin^2 \beta} + \dfrac{k^2}{b^2}}}. \qquad (22)$$

Triclinic system:

$$d = \frac{1}{\sqrt{\dfrac{\begin{vmatrix} \dfrac{h}{a} & \cos \gamma & \cos \beta \\ \dfrac{k}{b} & 1 & \cos \alpha \\ \dfrac{l}{c} & \cos \alpha & 1 \end{vmatrix} + \begin{vmatrix} 1 & \dfrac{h}{a} & \cos \beta \\ \cos \gamma & \dfrac{k}{b} & \cos \alpha \\ \cos \beta & \dfrac{l}{c} & 1 \end{vmatrix} + \begin{vmatrix} 1 & \cos \gamma & \dfrac{h}{a} \\ \cos \gamma & 1 & \dfrac{k}{b} \\ \cos \beta & \cos \alpha & \dfrac{l}{c} \end{vmatrix}}{\begin{vmatrix} 1 & \cos \gamma & \cos \beta \\ \cos \gamma & 1 & \cos \alpha \\ \cos \beta & \cos \alpha & 1 \end{vmatrix}}}}$$

$$(23)$$

It follows that, if one knows in advance the lengths and interlinear angles of the three axial vectors, a, b, and c, he can prepare a list of expected interplanar spacings, $d_{(hkl)}$. This list of calculated and indexed spacings can then be compared with the list of observed, but unindexed, spacings, and the indices of the individual spacings in the latter may thus be identified.

The list of observed interplanar spacings will display, in general, extinctions, or missing spectra of certain classes, as discussed fully in Chapter 4. A knowledge of these fixes the space group of the crystal within the limits discussed in Chapter 4.

Not only is indexing of reflection of importance for the purpose of fixing the space group of the crystal, but, if one expects to continue the investigation of the crystal structure further, and fix the positions of the atoms in the cell, then the identification of the indices of each reflection which appears on the film is of the greatest importance. This is because the relative intensities of reflections of definite index are functions of the positions of the atoms. To fix these positions, it is necessary to know the index associated with each reflection intensity.

It may be pointed out that, if the listing procedure just discussed were a practical one, the calculation of the spacings, d, could be highly simplified by the use of charts. Thus the variation of sin θ over either a flat plate or cylindrical film is easily prepared in the form of a chart of the same shape as the plate or film in use. With the aid of this template, the sin θ value of each spot could be read by inspection, and the value of d for each reflection observed could thus be quickly calculated with the aid of a slide rule. The preparation of a list of expected interplanar spacings from the known unit cell of the crystal, however, involves tedious calculations. These become so complex with the more general crystal systems that the procedure becomes intolerable. Fortunately, the indexing of planes may be accomplished by a very simple and powerful graphical method due to Bernal and based upon the important crystallographic concept of the *reciprocal lattice*, which is discussed in the following chapters.

Procedure when the orientation of the crystal is unknown. It should be emphasized again, in passing, that a list of observed interplanar spacings can be prepared from any film record of reflections according to relations (12), (14), or (16), regardless of the crystal orientation on the rotation axis. If, therefore, one has a fragment of a given crystal of unknown orientation, it may be attached to the rotation axis of the apparatus without regard to orientation, and a diffraction record obtained which will provide data for preparing a list of interplanar spacings. If nothing further is known about the crystallographic characteristics of the crystal, very little use can be made of this list.   If, however, it is known that the crystal is isometric, then reference to equation (17) shows that the interplanar spacing, d, for a given plane, (hkl), depends only on the cell edge, a.   The entire list of spacings for one isometric crystal is thus the same as that of any

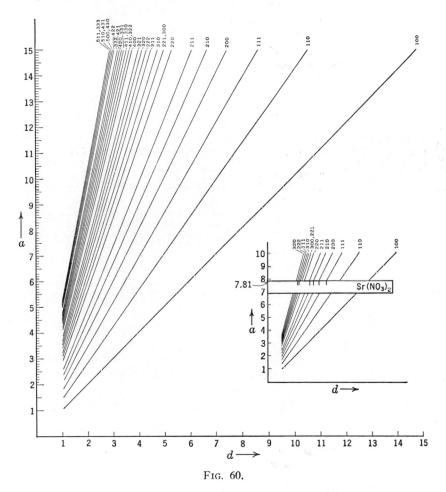

FIG. 60.

other isometric crystal, except that the entire list is multiplied by a different factor, a, in each case.   In other words, the spacings of isometric crystals differ from one another only in scale.   It is thus possible to seek the one variable, namely the scale or cell edge, a, and, when this is found, the reflections may be indexed.   The solution for a is easily carried out by means of the chart shown in Fig. 60.

If it is known that the crystal is tetragonal, or that it is hexagonal, then not only the scale, but the additional variable, the axial ratio, $\frac{c}{a}$, must be sought. There are graphical methods for accomplishing simultaneous determination of the scale and axial ratio for crystals of the tetragonal and hexagonal systems. The solution is possible because it can be carried out on the surface of a sheet of paper, which has two dimensions, and just two unknowns are involved in tetragonal and hexagonal crystals. A solution cannot be obtained in this manner for orthorhombic, monoclinic, or triclinic crystals because too many variables are involved.

LITERATURE

Original work

H. SEEMANN. Vollständige Spektraldiagramme von Kristallen. *Physik. Z.*, **20** (1919), 169–175.

On " Fiber diagrams," but containing a development of theory applicable to interpretation of ordinary rotating crystal photographs

R. O. HERZOG, WILLI JANCKE, and M. POLANYI. Röntgenspektrographische Beobachtungen an Zellulose. II. *Z. Physik*, **3** (1920), 343–348.

M. POLANYI. Faserstruktur im Röntgenlichte. *Naturwissenschaften*, **9** (1921), 337–340.

M. POLANYI. Das Röntgen-Faserdiagramm. *Z. Physik*, **7** (1921), 149–180.

K. WEISSENBERG. "Spiralfaser" und "Ringfaser" im Röntgendiagram. *Z. Physik*, **8** (1922), 20–31.

M. POLANYI and K. WEISSENBERG. Das Röntgen-Faserdiagramm. *Z. Physik*, **9** (1922), 123–130.

E. SCHIEBOLD. Bemerkungen zur Arbeit. Das Röntgenfaserdiagramm von P. Polanyi. *Z. Physik*, **9** (1922), 180–183.

M. POLANYI and K. WEISSENBERG. Das Röntgen-Faserdiagramm. *Z. Physik*, **10** (1922), 44–53.

Historical

M. POLANYI, E. SCHIEBOLD, and K. WEISSENBERG. Über die Entwickelung des Drehkristallverfahrens. *Z. Physik*, **23** (1924), 337–340.

General works

E. SCHIEBOLD. Über graphische Auswertung von Röntgenphotogrammen. *Z. Physik*, **28** (1924), especially 360–364.

E. SCHIEBOLD. Die Drehkristallmethode. *Fortschr. Mineral.*, etc., **11** (1927), 111–280.

Miscellaneous contributions

A. E. VAN ARKEL. Unikristallijn Wolfraam. *Physica*, **3** (1923), 76–87.

GILBERG GREENWOOD. Rotating crystal x-ray photographs. *Mineralog. Mag.*, **21** (1927), 258–271.

The deduction of interplanar spacing formulas

SURAIN SINGH SIDHU. The calculation of interplanar spacings of crystal systems by vectors. *Indian J. Phys.*, **11** (1937), 349–357.

CHAPTER 6

THE RECIPROCAL LATTICE

Since x-ray diffraction by a crystal may be thought of as reflection by sets of parallel planes in the crystal, these crystallographic planes take on a great importance. When planes of several slopes are to be dealt with in the same problem, difficulty arises in visualizing the several relative slopes of these two-dimensional surfaces. Now the slope of a plane is completely fixed either by the geometry of the plane itself or by the geometry of the normal to the plane. The normal has one less dimension than the plane and consequently it affords an easier means of thinking of the slope of a plane, especially when many planes are being considered at the same time. It is largely for this reason that, in many crystallographic problems, one thinks of the normals to a set of planes rather than of the planes themselves.

A number of conventional ways of simplifying crystallographic problems still further have been developed by projecting these normals in certain ways on certain surfaces. Among such simplifying devices are the spherical projection, the gnomonic projection, and the stereographic projection of classical crystallography. There will be no occasion to use any of these projections, for the problem of x-ray diffraction by crystals requires the introduction of one more variable than these projections can furnish.

A device for tabulating both the slopes and the interplanar spacings of the planes of a crystal lattice is provided by a concept known as the reciprocal lattice. An early treatment of the reciprocal lattice under the title " polar lattice " was given by Bravais, and later by Mallard. A vector analysis treatment of the reciprocal lattice was first given by Ewald, and crystallographers are indebted to Bernal for an intensive development of the application of the reciprocal lattice to x-ray diffraction problems. It can be fairly said that the reciprocal lattice provides one of the most important tools available in the study of the diffraction of x-rays by crystals. The properties which adapt it to this purpose will be developed in due course. In this chapter some of the fundamental geometrical properties of the reciprocal lattice will be studied. The chapter is divided into two parts. The first part discusses the reciprocal lattice by simple geometrical reasoning. The second part is a simple vector treatment of the subject.

GEOMETRICAL DISCUSSION OF THE RECIPROCAL LATTICE

Introduction. Given a " direct " crystal lattice, a second imaginary lattice, " reciprocal " to it, can be built up from it as follows: To every plane in the direct lattice, construct a normal; then limit the length of each normal so that it equals the reciprocal of the interplanar spacing of its plane (the interesting geometry which results from these rules will be developed presently). Place a point at the end of each limited normal. The collection of such points thus represents

(1) a collection of the slopes of the direct lattice planes, in the form of the directions of normals, and

(2) a collection of the interplanar spacings of the direct lattice, in the form of reciprocal spacings.

It will now be shown that this collection of points is a lattice array and that it therefore constitutes a new lattice, the *reciprocal lattice*.

In order to encourage a real understanding of the reciprocal lattice, the geometrical properties of this device will be developed in two dimensions, which can be completely represented in the plane of the paper. The geometry may be subsequently generalized and all the fundamental properties carried over to three dimensions.

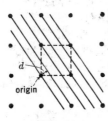

FIG. 61.

The square lattice. The basic relations between the direct and the reciprocal lattices may be easily grasped in the special case of the square lattice. Figure 61 shows a portion of a square lattice with all lines of index (hl), taken as (32) for the purpose of illustration, drawn through the central unit cell. The interlinear spacings, d, of all these lines are, of course, the same. The central unit cell is redrawn on an enlarged scale in Fig. 62A. Its edges have the absolute length, a. One (hl) line passes through the origin. The next one, at distance, d, crosses the cell edges at intercepts having absolute lengths $\dfrac{a}{h}$ and $\dfrac{a}{l}$. According to the Pythagorean theorem, the length of intercepted segment of the line (hl) is

$$AC = \sqrt{\left(\frac{a}{h}\right)^2 + \left(\frac{a}{l}\right)^2} = a\sqrt{\frac{1}{h^2} + \frac{1}{l^2}}. \qquad (1)$$

Now, in the similar triangles AOB and ACO,

$$\frac{OB}{OA} = \frac{CO}{CA}. \qquad (2)$$

Substituting the values given by (1) and indicated in Fig. 62*A* for these terms, (2) becomes

$$\frac{d}{a/h} = \frac{a/l}{a\sqrt{\dfrac{1}{h^2} + \dfrac{1}{l^2}}} \, . \tag{3}$$

Upon rearrangement, this gives the following value of the interlinear spacing:

$$d_{(hl)} = \frac{a}{hl\sqrt{\dfrac{1}{h^2} + \dfrac{1}{l^2}}}$$

$$= \frac{a}{\sqrt{h^2 + l^2}} \, . \tag{4}$$

According to the rules given above, the reciprocal lattice is built up by plotting lengths, σ, proportional to the reciprocals of d, one for each possible line (hl), i. e., by plotting

$$\sigma_{hl} = C\frac{1}{d_{(hl)}}, \tag{5}$$

where C is a proportionality constant. Each of the lengths must be properly directed normal to its line, in other words, directed parallel with the vector, d, of Fig. 62*A*. The plotting of such directed

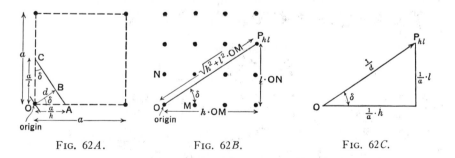

Fig. 62*A*.   Fig. 62*B*.   Fig. 62*C*.

lengths is illustrated in Fig. 62*B*. For convenience, the proportionality constant in this plotting is taken as unity, so (5) reduces to

$$\sigma_{hl} = \frac{1}{d_{(hl)}} \, . \tag{6}$$

The exact length of $d_{(hl)}$ to plot for any value of h and l is given by (4). Substituting this value in (6) gives

$$\sigma_{hl} = \frac{1}{a}\sqrt{h^2 + l^2}. \tag{7}$$

To plot the σ corresponding with the particular line illustrated in Fig. 61 and Fig. 62A, namely the line (32), one simply substitutes these h and l values in (7), thus:

$$\sigma_{hl} = \frac{1}{a}\sqrt{3^2 + 2^2}$$

$$= \frac{1}{a}\sqrt{13}. \tag{8}$$

When a corresponding plot is made for each possible value† of h and l, an array of points such as shown in Fig. 62B is produced.

The collection of points will now be examined for lattice properties. When l is zero, the general index (hl) reduces to $(h0)$. The vector, d, Fig. 62A, is normal to the vertical line represented by this index, and according to (7), the corresponding point on the reciprocal representation, Fig. 62B, is at a distance $\frac{1}{a}\cdot h$ from the origin. For example, when $h = 1$, the reciprocal lattice point is at a distance $\frac{1}{a}$, i. e., at M, Fig. 62B. When the index h takes on the integral values 2, 3, 4, $\cdots$, the corresponding reciprocal lattice point appears at dis-

† When all combinations of all possible values of h and l are formed, certain sets of combinations such as

$$(10), (20), (30) \cdots ,$$
$$(11), (22), (33) \cdots ,$$
$$(12), (24), (36) \ldots , \text{etc.}$$

appear to represent lines which have the same slope and which would, in classical crystallography, correspond with the same line. Thus, in classical crystallography the indices (200), (300), (400) $\cdots$, etc., are all identical with one another and would normally be represented by the indices (100). In lattice crystallography, however, the index (200) would imply a plane with intercepts $\frac{1}{2}$, ∞, ∞, and therefore a plane of slope identical with (100) but located at half its distance from the origin. The index (200) thus indicates a fictitious lattice plane of half the spacing of the real plane (100). For the development of the reciprocal lattice to be used in x-ray diffraction work, such indices are included. They give rise to reciprocal lattice points corresponding with fictitious interplanar spacings, but such lattice points are useful and in harmony with the system to be developed.

tances of $\frac{1}{a} \cdot 2$, $\frac{1}{a} \cdot 3$, $\frac{1}{a} \cdot 4 \cdots$, that is, at distances of 2, 3, 4 $\cdots$ times the distance of the first point, M, from the origin.

Similarly, when h is zero, the general index (hl) reduces to $(0l)$. The vector d, for this horizontal line, points at right angles to the first, and the corresponding reciprocal lattice point is at a distance of $\frac{1}{a} \cdot l$ from the origin. For example, when $l = 1$, the reciprocal lattice point is at a distance $\frac{1}{a}$, i. e., at N. When the index l takes on the integral values 2, 3, 4 $\cdots$, the corresponding reciprocal lattice point appears at distances of $\frac{1}{a} \cdot 2$, $\frac{1}{a} \cdot 3$, $\frac{1}{a} \cdot 4 \cdots$, i. e., at distances of 2, 3, 4, $\cdots$, times the distance of the first point, N, from the origin.

Now consider any line of more general index, (hl). Its normal vector d, Fig. 62A, makes an angle δ with the horizontal. From Fig. 62A, it is evident that

$$\tan \delta = \frac{a/h}{a/l} = \frac{l}{h},\tag{9}$$

and

$$\cos \delta = \frac{d}{a/h} = \frac{h/a}{1/d}.\tag{10}$$

The reciprocal point is located along a line in this direction at a distance of

$$\sigma_{hl} = \frac{1}{d_{(hl)}} = \frac{1}{a}\sqrt{h^2 + l^2},$$

which may be rewritten

$$\sigma_{hl} = \frac{1}{d_{(hl)}} = \sqrt{\left(\frac{1}{a}\cdot h\right)^2 + \left(\frac{1}{a}\cdot l\right)^2}.\tag{11}$$

This relation is evidently an expression of the Pythagorean theorem with quantities illustrated in Fig. 62C. Thus, to reach reciprocal lattice point P_{hl}, corresponding with direct lattice line (hl), one goes h reciprocal lattice vectors, of unit length $\frac{1}{a}$, along the horizontal axis, and then l reciprocal lattice vectors, of unit length $\frac{1}{a}$, along the vertical axis, thus attaining a point at a distance of $\frac{1}{a}\sqrt{h^2 + l^2}$ from the origin, and located on a line making an angle $\tan^{-1}\delta = \dfrac{l}{h}$ (or $\cos^{-1}\delta = \dfrac{h/a}{1/d}$) with the horizontal. In short, to reach a point P_{hl}, one goes h units right and l units up. Since h and l are integers, it is evident that all points† $[[hl]]^*$ fall on the intersections of a square grid, or ruled coordinate system, whose unit cell has an edge length of $\dfrac{1}{a}$. The coefficient $\dfrac{1}{a}$ appears as a term in any reciprocal lattice distance and is thus a constant characteristic of the particular lattice.

Accordingly, the reciprocal system of points is a lattice array. It is interesting to observe why this is: Since the direct lattice may have only rational lines, the indices of these lines may include only the integral numbers h and l. Therefore the only permissible points of the reciprocal set are those for which the horizontal and vertical distance components h and l (Fig. 62C) are restricted to integral numbers. It is this integral aspect which places the points of the reciprocal set on lattice positions. In other words, because the direct system of points has lattice properties, the reciprocal system of points also has lattice properties. It follows, of course, that the reciprocity is mutual, i. e., that the direct lattice is also the reciprocal of the reciprocal lattice.

The rectangular lattice. Instead of proceeding directly to the most general two-dimensional case of the parallelogram lattice, it will be easier to examine first the somewhat generalized case of the rectangular lattice. Here the specialized square cell of edge length a becomes generalized into a rectangular cell of edge lengths a and c, Fig. 63A. The intercepted line segment, formerly given by (1), now becomes

$$A C = \sqrt{\left(\frac{a}{h}\right)^2 + \left(\frac{c}{l}\right)^2} \; ; \tag{12}$$

and the interlinear spacing, derived from the similarity of triangles AOB and ACO, is developed from the proportion

$$\frac{d_{(hl)}}{a/h} = \frac{c/l}{\sqrt{\left(\frac{a}{h}\right)^2 + \left(\frac{c}{l}\right)^2}}. \tag{13}$$

† An asterisk following a crystallographic symbol signifies that the geometrical element symbolized occurs in the reciprocal lattice, not in the direct lattice. Thus, in this instance, $[[hl]]^*$ indicates a point on the reciprocal lattice having the same indices as the line (hl) in the direct plane-lattice. Further use is made of this convention on page 119 and beyond.

Thus,
$$d_{(hl)} = \frac{ac/hl}{\sqrt{\left(\dfrac{a}{h}\right)^2 + \left(\dfrac{c}{l}\right)^2}}, \tag{14}$$

and,
$$\sigma_{hl} = \frac{1}{d_{(hl)}} = \frac{hl}{ac}\sqrt{\frac{a^2}{h^2} + \frac{c^2}{l^2}} = \sqrt{\left(\frac{l}{c}\right)^2 + \left(\frac{h}{a}\right)^2}. \tag{15}$$

This interlinear spacing vector is directed from the origin at an angle from the horizontal (Fig. 63A) equal to

$$\delta = \cos^{-1}\frac{d}{a/h} = \cos^{-1}\frac{h/a}{1/d}. \tag{16}$$

With the elementary background of reciprocal lattice theory developed in the last section, equations (15) and (16) may now be interpreted as follows: For any line (hl) in the direct lattice there is a definite interlinear spacing, d, having vector properties, i.e., both length and direction. To this spacing there corresponds a point in

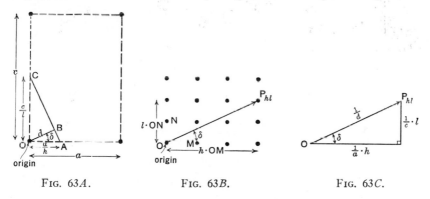

Fig. 63A.   Fig. 63B.   Fig. 63C.

so-called *reciprocal space* (meaning the space in which the reciprocal lattice is imagined to exist) at a distance, $\sigma_{hl} = \dfrac{1}{d_{(hl)}}$. This point is at the end of a vector making an angle with the horizontal of magnitude $\cos^{-1}\dfrac{h/a}{1/d}$. The length of the vector is given by (15). The graphical interpretation of (15) may be obtained from a right triangle whose lengths are as shown in Fig. 63C. In this diagram, the vector is shown broken down into components $\dfrac{1}{a}\cdot h$ and $\dfrac{1}{c}\cdot l$. Thus the reciprocals $\dfrac{1}{a}$ and $\dfrac{1}{c}$ are the magnitudes of the unit vectors in the a and

c directions, respectively. The symbols h and l stand for any integers; when they are multiplied by the reciprocals, $\dfrac{1}{a}$ and $\dfrac{1}{c}$, of the respective lengths of the axes of the direct unit cell, they become vector components in reciprocal space. To reach the point P_{hl}, one goes h of the $\dfrac{1}{a}$ units in the a direction, then l of the $\dfrac{1}{c}$ units in the c direction. Since all combinations of h and l appear in the original indices (hl), all integral coordinate points hl are occupied by corresponding

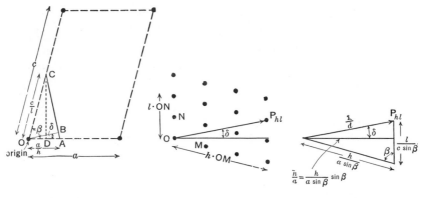

FIG. 64A.   FIG. 64B.   FIG. 64C.

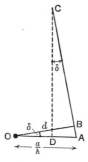

FIG. 64D.

reciprocal points. This system of points thus constitutes a lattice " reciprocal to " the original direct lattice.

The axial ratio of this reciprocal lattice is $\dfrac{1/c}{1/a} = \dfrac{a}{c}$. Since the unit vectors are normal to each other and parallel to the direct unit cell edges, it is apparent that the reciprocal to a rectangular lattice is another parallel rectangular lattice having an axial ratio which is the reciprocal of the axial ratio $\dfrac{c}{a}$ of the direct lattice. This is equivalent to saying that the reciprocal to a rectangular lattice is a similar lattice whose orientation is rotated 90° to that of the direct lattice.

The parallelogram lattice. The general plane lattice case may now be easily grasped. A cell of the parallelogram lattice is shown in Fig. 64A. An important part of the region in the vicinity of the acute axial angle, β, is removed for convenience to Fig. 64D. The segment AC, in this case, is given by

$$AC = \sqrt{\left(\frac{a}{h}\right)^2 + \left(\frac{c}{l}\right)^2 - 2\frac{a}{h}\cdot\frac{c}{l}\cos\beta}. \tag{17}$$

In triangles AOB and ACD, Fig. 64D,

$$\frac{OB}{OA} = \frac{CD}{CA}. \tag{18}$$

Making direct substitution, this is equivalent to

$$\frac{d_{(hl)}}{\frac{a}{h}} = \frac{\frac{c}{l}\sin\beta}{\sqrt{\left(\frac{a}{h}\right)^2 + \left(\frac{c}{l}\right)^2 - 2\frac{a}{h}\cdot\frac{c}{l}\cos\beta}}, \tag{19}$$

from which

$$d_{(hl)} = \frac{\frac{a}{h}\cdot\frac{c}{l}\sin\beta}{\sqrt{\left(\frac{a}{h}\right)^2 + \left(\frac{c}{l}\right)^2 - 2\frac{a}{h}\cdot\frac{c}{l}\cos\beta}}, \tag{20}$$

and the reciprocal vector length,

$$\sigma_{hl} = \frac{1}{d_{(hl)}} = \frac{hl}{ac\sin\beta}\sqrt{\left(\frac{a}{h}\right)^2 + \left(\frac{c}{l}\right)^2 - 2\frac{a}{h}\cdot\frac{c}{l}\cos\beta}$$

$$= \sqrt{\left(\frac{l}{c\sin\beta}\right)^2 + \left(\frac{h}{a\sin\beta}\right)^2 - 2\frac{h}{a\sin\beta}\cdot\frac{l}{c\sin\beta}\cos\beta}. \tag{21}$$

This vector, from Fig. 64A and 64D, is directed at an angle from the horizontal,

$$\delta = \cos^{-1}\frac{d}{a/h} = \frac{h/a}{1/d}. \tag{22}$$

Equation (21) is in the form of the modified Pythagorean relation for scalene triangles, and the graphical interpretation of it is given in Fig. 64C. To reach point P_{hl} one goes a distance $\dfrac{h}{a\sin\beta} = \dfrac{1}{a\sin\beta}\cdot h$ along an axis normal to the direct lattice lines $(h0)$, then a distance $\dfrac{l}{c\sin\beta} = \dfrac{1}{c\sin\beta}\cdot l$ parallel with the vertical axis (normal to the direct lattice lines $(0l)$. Here the coefficients of h and l, namely, $\dfrac{1}{a\sin\beta}$ and $\dfrac{1}{c\sin\beta}$, are, as in the rectangular lattice case, constants for the lattice.

They represent the lengths of unit vectors in reciprocal space. For, let (hl) reduce to the special index case (10), then equation (21) reduces to $\dfrac{1}{a \sin \beta}$, which gives the length of this unit vector. It is directed at an angle from the horizontal given by (22), which reduces to $\delta = \cos^{-1} (\sin \beta) = 90° - \beta$. Similarly, the length of the other unit vector is obtained by letting (hl) take the special value (01). In this case, the length of this vector reduces to $\dfrac{1}{c \sin \beta}$, and its direction from the horizontal, (22), reduces to $\delta = \cos^{-1} 0 = 90°$. This consideration locates lattice points M and N, respectively.

For the location of any point $[[hl]]*$, corresponding to the spacing of the lines in the direct lattice of index (hl), one proceeds, according to (21) and (22), in the direction from O toward M for a distance of h units of length OM, then proceeds vertically, parallel with ON, for l units of length ON. If all values of h and l are taken, points are defined which fall on the intersection of a parallelogram grid. The points are thus in lattice array as in the foregoing square and rectangular cases.

The axial ratio of the lattice reciprocal to a parallelogram lattice is

$$\frac{ON}{OM} = \frac{1/c \sin \beta}{1/a \sin \beta} = \frac{a}{c}. \tag{23}$$

Since the unit vectors are normal to the direct cell edges, it follows that the lattice reciprocal to a parallelogram lattice of axial ratio $\dfrac{c}{a}$ and interaxial angle β is a similar parallelogram lattice with axes normal to the direct axes, and thus mutually inclined at an angle $180° - \beta$ to one another. This is equivalent to saying that the lattice reciprocal to a parallelogram lattice is a similar parallelogram lattice having an orientation rotated 90° from the direct lattice.

Since the parallelogram lattice is the general case of plane lattices, it follows that the reciprocal to any plane lattice is a similar plane lattice whose orientation is rotated 90° from that of the direct lattice. This statement, of course, is accurate only when applied to the *shapes* of plane lattices. When the labeling of axes is important, it is necessary to return to the more cumbersome statement that the lattice reciprocal to any plane lattice is a similar lattice whose axial ratio is the reciprocal of that of the direct lattice and whose interaxial angle is the supplement of that of the direct lattice.

Reciprocal space lattices. A sufficient background in reciprocal plane lattices has now been developed to permit an easy introduction to reciprocal space lattices. To construct a lattice reciprocal to a given space lattice, one takes each interplanar spacing, $d_{(hkl)}$, and plots in reciprocal space, a vector, σ_{hkl}, parallel to it and in length proportional to the spacing reciprocal,

$$\sigma_{hkl} = C\frac{1}{d_{(hkl)}}, \tag{24}$$

where C is a proportionality constant, conveniently allowed to equal unity for most discussions. The lattice consists of the array of points, one at the end of each such vector, (24). This array can be represented by the collection

$$\mathbf{K}\sigma_{hkl} = \mathbf{K}C\frac{1}{d_{(hkl)}}. \tag{25}$$

Before proceeding any farther, it is first necessary to prove that this system of points is really a lattice array. This can be done with the aid of the relations developed for reciprocal plane lattices. In order to apply these relations to space lattices, however, the following two lemmas are first needed:

Lemma 1. The projection of a space lattice on a plane normal to any lattice row is a plane lattice. The reason for this is that each space lattice row projects as a single plane lattice point. This relation is illustrated in Fig. 65 for projections on two different planes.

Lemma 2. In a plane lattice formed by projection of a space lattice on a plane normal to a lattice row (according to lemma 1), all interlinear spacing vectors, d, are identically equal to the corresponding interplanar spacing vectors, d, in the space lattice. This is illustrated in Fig. 65 for the spacing $d_{(010)}$, which projects accurately as d in two projections. The general proposition is true because the spacing vectors of all planes common to the lattice row are perpendicular to the lattice row and therefore parallel to the plane of projection.

There is now sufficient geometrical equipment at hand to prove that the system of points defined by the ends of the vectors of (25) is a space lattice array. Let the points in Fig. 65 represent the direct space lattice. Choose, from this, any primitive unit cell, and project the space lattice onto planes perpendicular to two of its cell edges (in the illustration, the space lattice is projected onto planes perpendicular to a and c). Each projection is a plane lattice, according to lemma 1. To each of these plane lattices, there is a reciprocal plane lattice. By virtue of lemma 2, every spacing vector, d, in each direct plane lattice is also a spacing vector, d, in the space lattice. Therefore, each point in the reciprocal plane lattices (vector, $\sigma = \dfrac{1}{d}$) is also a point in the reciprocal to the space lattice, defined by (25). The two reciprocal plane lattices illustrated have in common the vector, $\sigma_{010} = \dfrac{1}{d_{(010)}}$; hence they intersect and join along this line.

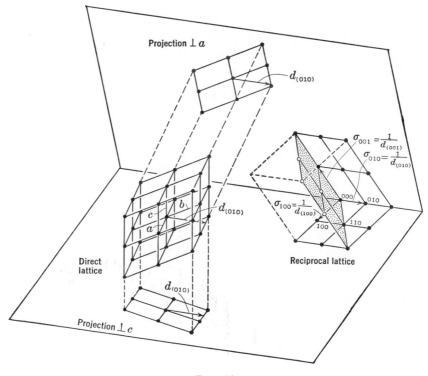

Fig. 65.

The two intersecting plane reciprocal lattices shown in Fig. 65 are, accordingly, identical with two particular planes of the array defined by (25). These two planes contain the three vectors σ_{100}, σ_{010}, and σ_{001}. These three vectors can be taken to define a (reciprocal) space lattice. It is now necessary to prove that every point of the array (25) corresponds with a point of this particular space lattice:

In the direct space lattice, every plane, (hkl), can be accounted for by examining the planes in a zone (lattice row), one zone at a time.

Consequently, every spacing vector, $d_{(hkl)}$, can be accounted for by examining projections perpendicular to each lattice row, one at a time.   Consider *any* such lattice row (not *a* or *c*, which have already been considered).   According to lemma 1, the entire space lattice projects along this row as a plane lattice (not illustrated).   The reciprocal to this plane lattice is indicated by shading in Fig. 65, and its lattice points are distinguished as open circles.   This plane lattice has points in common with the other planes it intersects, as proved by the following reasoning:

Every direct lattice plane is contained in numerous zones.   The point corresponding with it in reciprocal space, therefore, is common to each of the plane reciprocal lattices perpendicular to each of these zones.   Thus, in Fig. 65, the reciprocal lattice point, 110, is in the shaded plane as well as in the plane perpendicular to *c*.   Returning to the direct lattice, each two zones have a plane in common, and hence, in reciprocal space, each two intersecting plane reciprocal lattices have a point in common.   They also have the origin point in common, and these two points establish a row.   Hence every two intersecting plane lattices have an entire row in common.   In Fig. 65, the shaded plane lattice has a row in common with each of the other two plane lattices.   Indeed, these two rows define, or establish, the character of the plane lattice in the shaded plane, and this established character is such as to make it a plane of the space lattice defined by the translations σ_{100}, σ_{010}, and σ_{001}.   This is generally true for any reciprocal plane lattice perpendicular to any row in the direct space lattice. Thus, all points of the system represented by (25) fall in the lattice array defined by the three unit translations.

It is customary to designate the direct lattice cell edges and angles by the unstarred letters a, b, c, α, β, γ, and to designate the corresponding reciprocal cell elements by the starred elements a^*, b^*, c^*, α^*, β^*, γ^*.   The direct linear elements and reciprocal linear elements are evidently the reciprocals of one another's pinacoidal spacings,

$$a^* = \frac{1}{d_{(100)}}, \tag{26}$$

$$b^* = \frac{1}{d_{(010)}}, \tag{27}$$

$$c^* = \frac{1}{d_{(001)}}, \tag{28}$$

and,

$$a = \frac{1}{d^*_{(100)}}, \tag{29}$$

$$b = \frac{1}{d^*_{(010)}}, \tag{30}$$

$$c = \frac{1}{d^*_{(001)}}. \tag{31}$$

If indices, hkl, are attached to the points in the two plane reciprocal lattices of the unshaded planes in Fig. 65, the labels of all the other points follow. It is then evident that for any plane, (hkl), in the direct lattice, there is a point, hkl, in the reciprocal lattice. This point is attained by starting at the reciprocal lattice origin and proceeding h units of length a^* in the a^* direction, then k units of length b^* in the b^* direction, then l units of length c^* in the c^* direction. The two statements,

$$\sigma_{hkl} = \frac{1}{d_{(hkl)}} \tag{32}$$

and

$$\sigma_{hkl} = h\vec{a^*} + k\vec{b^*} + l\vec{c^*}, \tag{33}$$

are thus two different ways of defining the point hkl in reciprocal space, and the terms of the right member of (33) represent the vector components of the right member of (32) along the vector axes a^*, b^*, and c^*.

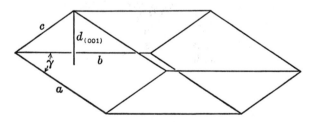

Fig. 66.

Fundamental dimensional relations. According to relations (26)–(28), the lengths of the reciprocal lattice cell axes are equal to the reciprocals of the pinacoid spacings of the direct lattice cell. These expressions may be further evaluated with the aid of Fig. 66, which represents a general triclinic cell. The volume of this cell is

$$V = \text{base} \times \text{altitude},$$

$$V = (a\, b \sin \gamma) \times d_{(001)}. \tag{34}$$

Transposing,

$$d_{(001)} = \frac{V}{a\, b \sin \gamma}. \tag{35}$$

This supplies the value of the $d_{(001)}$ in (28). Substituting this in (28) and also substituting similar relations for the spacings of the other pinacoids gives the following fundamental dimensional relations:

$$a^* = \frac{b\, c \sin \alpha}{V}, \tag{36}$$

$$b^* = \frac{a\, c \sin \beta}{V}, \tag{37}$$

$$c^* = \frac{a\, b \sin \gamma}{V}. \tag{38}$$

It can be shown (Chapter 18) that the volume, V, of a triclinic cell, in terms of its own dimensions a, b, c and α, β, γ, is

$$V = a\, b\, c\, \sqrt{1 - \cos^2 \alpha - \cos^2 \beta - \cos^2 \gamma + 2 \cos \alpha \cos \beta \cos \gamma} \tag{39}$$

$$= a\, b\, c\, \sqrt{\sin^2 \alpha + \sin^2 \beta + \sin^2 \gamma - 2 + 2 \cos \alpha \cos \beta \cos \gamma}. \tag{39A}$$

Similar relations for the direct cell in terms of reciprocal cell dimensions, and corresponding with (29)–(31), are

$$a = \frac{b^* c^* \sin \alpha^*}{V^*}, \tag{40}$$

$$b = \frac{a^* c^* \sin \beta^*}{V^*}, \tag{41}$$

$$c = \frac{a^* b^* \sin \gamma^*}{V^*}, \tag{42}$$

where V^* is the volume of the reciprocal lattice cell.

These relations may be regarded as fundamental ones. In a subsequent chapter, there will be occasion to develop some further relations between the cells of reciprocal lattices.

Finally, it should be pointed out that the dimensional relations given in this chapter appear complicated because they are the general relations of triclinic cells. These relations become highly simplified in other cases, and are reasonably simple even for the monoclinic system.

VECTOR DEVELOPMENT OF THE RECIPROCAL LATTICE

The development of the reciprocal lattice by vector methods is comparatively compact. The main object is to prove that, if, to each plane, (hkl), in the direct space lattice, a normal is drawn of length

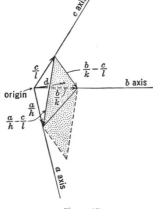

$$\sigma_{hkl} = \frac{1}{d_{(hkl)}}, \tag{43}$$

the sum total of the points at the ends of these vectors is a lattice array.

Fig. 67.

To prove this, consider Fig. 67. In this diagram, the first plane from the origin of a system (hkl) is shown. This first plane intercepts the $\mathbf{a}$, $\mathbf{b}$, and $\mathbf{c}$ lattice axes at the ends of the vectors $\frac{\mathbf{a}}{h}$, $\frac{\mathbf{b}}{k}$, $\frac{\mathbf{c}}{l}$. The two vectors outlining the shaded area of the plane are $\frac{\mathbf{a}}{h} - \frac{\mathbf{c}}{l}$ and $\frac{\mathbf{b}}{k} - \frac{\mathbf{c}}{l}$. The area of the parallelogram defined by these last two vectors is

$$\mathbf{Area}_{\square \frac{a}{h}, \frac{b}{k}, \frac{c}{l}} = \left(\frac{\mathbf{a}}{h} - \frac{\mathbf{c}}{l}\right) \times \left(\frac{\mathbf{b}}{k} - \frac{\mathbf{c}}{l}\right)$$

$$= \frac{1}{hk}(\mathbf{a} \times \mathbf{b}) - \frac{1}{hl}(\mathbf{a} \times \mathbf{c}) - \frac{1}{lk}(\mathbf{c} \times \mathbf{b}) + 0$$

$$= \frac{l}{hkl}(\mathbf{a} \times \mathbf{b}) + \frac{k}{hkl}(\mathbf{c} \times \mathbf{a}) + \frac{h}{hkl}(\mathbf{b} \times \mathbf{c})$$

$$= \frac{1}{hkl}[h(\mathbf{b} \times \mathbf{c}) + k(\mathbf{c} \times \mathbf{a}) + l(\mathbf{a} \times \mathbf{b})]. \tag{44}$$

Now, the plane just considered is merely the first of an infinite series of planes, equally spaced from the origin, Fig. 68. The plane of this series which makes the first rational intercepts with the particular axes chosen is the $\sqrt{rst \cdot hkl}$th plane, according to the discussion on pages 9–10. The parallelogram defined by the three lattice point intercepts, A, B, C, is evidently the base of a possible primitive cell, whose altitude is $d_{(hkl)}$ provided that A, B, and C are prime to one another in pairs. If this is not the case, then the cell is multiply primitive. For, suppose that A and B both contain the highest common factor t. Then the line AB not only has lattice points at A and B, but also other lattice points between them. Counting only one corner of the cell, there are t lattice points along the line AB within

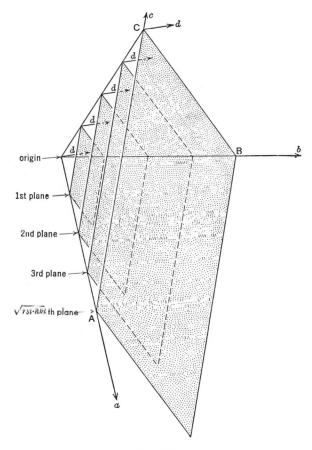

FIG. 68.

the cell, and therefore, according to page 6, the cell is t-fold primitive. Similarly, if the pairs of intercept numbers, B and C, C and A, A and B, contain the highest common factors r, s, and t, respectively, then the cell is rst-fold primitive. If V is the volume of a primitive cell, the volume of the cell, built upon parallelogram ABC and the spacing $d_{(hkl)}$, is

$$\text{Area}_{\square ABC} \cdot d_{(hkl)} = rst \cdot V, \qquad (45)$$

which, by rearrangement, gives

$$\frac{1}{d_{(hkl)}} = \frac{\text{Area}_{\square ABC}}{rst \cdot V}. \tag{46}$$

The volume, V, of course, is the same for any primitive cell of the lattice, and consequently is a constant for the lattice. In order to evaluate $\text{Area}_{\square ABC}$, note that this is the $\sqrt{rst \cdot hkl}$th similar equally spaced parallelogram, Fig. 68. Its edges are therefore $\sqrt{rst \cdot hkl}$ times as large as those of the first parallelogram, and its area is therefore $rst \cdot hkl$ times as large as that of the first parallelogram. The area of the latter was given by (44). The area of $\square_{ABC}$ is, therefore,

$$\text{Area}_{\square ABC} = rst \cdot hkl \ \text{Area}_{\square \frac{a}{h}, \frac{b}{k}, \frac{c}{l}}$$

$$= rst\{h(\mathbf{b} \times \mathbf{c}) + k(\mathbf{c} \times \mathbf{a}) + l(\mathbf{a} \times \mathbf{b})\}. \tag{47}$$

Note, from the derivation of (44), that (47) represents a vector normal to the plane of $\square_{ABC}$, and therefore a vector parallel with $d_{(hkl)}$, and parallel with σ_{hkl} of (43). Let the unit vector normal to the area of $\square_{ABC}$ be represented by $\mathbf{n}$. Then equations (43) and (46) can be rewritten in vector notation as follows:

from (43),

$$\sigma_{hkl} = \frac{1}{d_{(hkl)}} \mathbf{n}; \tag{48}$$

from (46),

$$\frac{1}{d_{(hkl)}} \mathbf{n} = \frac{\text{Area}_{\square ABC}}{rst \cdot V}. \tag{49}$$

Substituting the value of the area from (47) in (49) gives

$$\frac{1}{d_{(hkl)}} \mathbf{n} = \frac{h(\mathbf{b} \times \mathbf{c}) + k(\mathbf{c} \times \mathbf{a}) + l(\mathbf{a} \times \mathbf{b})}{V}; \tag{50}$$

and, finally, combining (48) and (50) yields

$$\sigma_{hkl} = \frac{h(\mathbf{b} \times \mathbf{c}) + k(\mathbf{c} \times \mathbf{a}) + l(\mathbf{a} \times \mathbf{b})}{V}. \tag{51}$$

Equation (51) is a vector expression for the location of a single reciprocal lattice point corresponding to plane (hkl). The entire reciprocal lattice is the collection of all such points for all values of h, k, and l, and can be abbreviated

$$\mathbf{K}\sigma_{hkl} = \frac{\mathbf{K}\{h(\mathbf{b} \times \mathbf{c}) + k(\mathbf{c} \times \mathbf{a}) + l(\mathbf{a} \times \mathbf{b})\}}{V}. \tag{52}$$

Note that, for any given lattice, only h, k, and l change from plane to plane, and that V, $(b \times c)$, $(c \times a)$, and $(a \times b)$ are constants for any given lattice.   Note, further, that

$$\frac{b \times c}{V}$$ is a vector $\perp$ plane of b and c, i. e., $\perp$ (100);

$$\frac{c \times a}{V}$$ is a vector $\perp$ plane of c and a, i. e., $\perp$ (010);

$$\frac{a \times b}{V}$$ is a vector $\perp$ plane of a and b, i. e., $\perp$ (001).

For abbreviations, call these vectors

$$\left. \begin{aligned} \frac{b \times c}{V} &= a^*, \\ \frac{c \times a}{V} &= b^*, \\ \frac{a \times b}{V} &= c^*. \end{aligned} \right\} \tag{53}$$

Then (51) and (52) can be rewritten:

reciprocal lattice point: $\sigma_{hkl} = ha^* + kb^* + lc^*,$ (54)

entire reciprocal lattice: $\mathbf{K}\sigma_{hkl} = \mathbf{K}\{ha^* + kb^* + lc^*\}.$ (55)

These expressions can now be easily interpreted.   In order to reach the reciprocal lattice point corresponding with plane (hkl), one proceeds,

$$h \text{ units of magnitude } |a^*| = \frac{|b \times c|}{V} \text{ in direction } \perp (100),$$

then,

$$k \quad `` \quad `` \quad `` \quad |b^*| = \frac{|c \times a|}{V} \quad `` \quad `` \quad \perp (010),$$

then,

$$l \quad `` \quad `` \quad `` \quad |c^*| = \frac{|a \times b|}{V} \quad `` \quad `` \quad \perp (001).$$

By taking all values of h, k, and l, one fills all integral points of a coordinate system based upon the unit vectors a^*, b^*, and c^*.   The sum total of such points, (52), therefore constitutes a lattice array.

The lengths of the unit vectors, (53), may be easily evaluated as follows:

$$|\mathbf{a}^*| = \frac{|\mathbf{b} \times \mathbf{c}|}{V} = \frac{b\,c\,\sin\alpha}{V}, \tag{56}$$

$$|\mathbf{b}^*| = \frac{|\mathbf{c} \times \mathbf{a}|}{V} = \frac{a\,c\,\sin\beta}{V}, \tag{57}$$

$$|\mathbf{c}^*| = \frac{|\mathbf{a} \times \mathbf{b}|}{V} = \frac{a\,b\,\sin\gamma}{V}. \tag{58}$$

Relations (56), (57), and (58) are the same as (36), (37), and (38).

An interesting and useful relation can be derived by expanding the volume, V, of (53). The unit reciprocal vectors then become

$$\mathbf{a}^* = \frac{\mathbf{b} \times \mathbf{c}}{\mathbf{a} \cdot \mathbf{b} \times \mathbf{c}}, \tag{59}$$

$$\mathbf{b}^* = \frac{\mathbf{c} \times \mathbf{a}}{\mathbf{a} \cdot \mathbf{b} \times \mathbf{c}}, \tag{60}$$

$$\mathbf{c}^* = \frac{\mathbf{a} \times \mathbf{b}}{\mathbf{a} \cdot \mathbf{b} \times \mathbf{c}}. \tag{61}$$

By multiplying both members of (59) by $\mathbf{a}$, of (60) by $\mathbf{b}$, and (61) by $\mathbf{c}$, there results

$$\mathbf{a} \cdot \mathbf{a}^* = \frac{\mathbf{a} \cdot \mathbf{b} \times \mathbf{c}}{\mathbf{a} \cdot \mathbf{b} \times \mathbf{c}} = 1, \tag{62}$$

$$\mathbf{b} \cdot \mathbf{b}^* = \frac{\mathbf{a} \cdot \mathbf{b} \times \mathbf{c}}{\mathbf{a} \cdot \mathbf{b} \times \mathbf{c}} = 1, \tag{63}$$

$$\mathbf{c} \cdot \mathbf{c}^* = \frac{\mathbf{a} \cdot \mathbf{b} \times \mathbf{c}}{\mathbf{a} \cdot \mathbf{b} \times \mathbf{c}} = 1. \tag{64}$$

In this important sense, the direct and starred unit vectors are reciprocals.

This relation may now be used to prove another interesting result: *The reciprocal of the reciprocal lattice is the direct lattice.* To prove this, first form the reciprocal vector $(\mathbf{a}^*)^*$ according to the usual rules indicated by (53):

$$(\mathbf{a}^*)^* = \frac{\mathbf{b}^* \times \mathbf{c}^*}{V^*}$$

$$= \frac{\mathbf{b}^* \times \mathbf{c}^*}{\mathbf{a}^* \times \mathbf{b}^* \cdot \mathbf{c}^*}. \tag{65}$$

Multiplying the right member by $\mathbf{a \cdot a^*}$, which, according to (62), is unity:

$$(\mathbf{a}^*)^* = \mathbf{a \cdot a}^* \frac{\mathbf{b}^* \times \mathbf{c}^*}{\mathbf{a}^* \times \mathbf{b}^* \cdot \mathbf{c}^*}$$

$$= \mathbf{a} \, \frac{\mathbf{a}^* \cdot \mathbf{b}^* \times \mathbf{c}^*}{\mathbf{a}^* \times \mathbf{b}^* \cdot \mathbf{c}^*}$$

$$= \mathbf{a}. \tag{66}$$

Similarly,

$$(\mathbf{b}^*)^* = \mathbf{b}, \tag{67}$$

and,

$$(\mathbf{c}^*)^* = \mathbf{c}. \tag{68}$$

By combining (65) and (66) it follows that

$$\mathbf{a} = \frac{\mathbf{b}^* \times \mathbf{c}^*}{V^*}, \tag{69}$$

$$\mathbf{b} = \frac{\mathbf{c}^* \times \mathbf{a}^*}{V^*}, \tag{70}$$

and

$$\mathbf{c} = \frac{\mathbf{a}^* \times \mathbf{b}^*}{V^*}. \tag{71}$$

LITERATURE

Early form of reciprocal lattice theory

A. BRAVAIS. Abhandlung über die Systeme von regelmässig auf einer Ebene oder in Raum vertheilten Punkten (1848). (Translated by C. and E. Blasius, and appearing as No. 90 of Ostwald's *Klassiker der exakten Wissenschaften;* Wilhelm Engelmann, Leipzig, 1897.) Pages 112–139.

ERNEST MALLARD. *Traité de cristallographie géométrique et physique.* (Dunod, Paris, 1879.) Vol. I, 26–32, 305–309.

Recent treatments

P. P. EWALD. Das " reziproke Gitter " in der Strukturtheorie. *Z. Krist. (A),* **56** (1921), 148–150.

J. D. BERNAL. On the interpretation of x-ray, single crystal, rotation photographs. *Proc. Roy. Soc. London (A),* **113** (1926), especially 118–120.

M. J. BUERGER. The application of plane groups to the interpretation of Weissenberg photographs. *Z. Krist. (A),* **91** (1935), especially 276–283.

CHAPTER 7

GEOMETRICAL INTERPRETATION OF BRAGG'S LAW: THE APPLICATION OF THE RECIPROCAL LATTICE TO THE SOLUTION OF X-RAY DIFFRACTION PROBLEMS

The reciprocal lattice provides a convenient and powerful method of dealing with problems arising in the diffraction of x-rays by crystals. The fundamental Bragg condition for the reflection of x-rays by a crystal plane, (hkl),

$$n\lambda = 2d \sin \theta, \tag{1}$$

may be rearranged in the following manner to give the glancing angle, θ, in terms of the variables involved:

$$\sin \theta_{(hkl)} = \frac{n\lambda}{2d_{(hkl)}}. \tag{2}$$

Equation (2) is still in the form of the original Bragg notation in which n and d refer to the order of reflection and spacing of a rational plane, (hkl), in the direct lattice. It will be recalled (page 9) that the nearest plane to the origin in the system designated (hkl) cuts the crystallographic axes at distances $\frac{1}{h}, \frac{1}{k}$, and $\frac{1}{l}$ from the origin. The notation $(nh\ nk\ nl)$ determines a plane which would cut the crystallographic axes at distances $\frac{1}{n} \cdot \frac{1}{h}, \frac{1}{n} \cdot \frac{1}{k}$, and $\frac{1}{n} \cdot \frac{1}{l}$ from the origin. The notation $(nh\ nk\ nl)$ thus represents a plane parallel with the lattice plane (hkl) but located at $\frac{1}{n}$ th of its distance, d, from the origin, that is,

$$d_{(nh\ nk\ nl)} = \frac{1}{n} d_{(hkl)}. \tag{3}$$

Since the nearest plane to the origin of slope $(nh\ nk\ nl)$ is (hkl), the plane $(nh\ nk\ nl)$ does not go through lattice points and therefore is fictitious.

Now, if there were a first-order reflection from the fictitious plane

(nh nk nl), its glancing angle from equation (2) would be

$$\sin \theta_{(nh\ nk\ nl)} = \frac{1\lambda}{2d_{(nh\ nk\ nl)}}. \tag{4}$$

But, according to (3), this spacing, $d_{(nh\ nk\ nl)}$, may be expressed in terms of $d_{(hkl)}$, so that (4) may be written

$$\sin \theta_{(nh\ nk\ nl)} = \frac{1\lambda}{2 \cdot \dfrac{1}{n} d_{(hkl)}}, \tag{5}$$

which simplifies to

$$\sin \theta_{(nh\ nk\ nl)} = \frac{n\lambda}{2d_{(hkl)}}. \tag{6}$$

Note that the glancing angle, θ, for the first order of reflection from the fictitious plane, (nh nk nl), is identical with the glancing angle, θ, for the nth order of the real, parallel plane (hkl). The idea of order of reflection, therefore, may be uniquely expressed in the indices of the reflection by multiplying the three indices of the reflecting plane through by the order of reflection. This index triple is the reflection index. *A collection of index triples, hkl, for all values of the numbers h, k, and l, therefore, uniquely expresses all the possible orders of reflection of all the possible real crystallographic planes. The highest common factor of the three numbers of the reflection index is the order of reflection, and the factored index triple is the index of the real reflecting plane.* Note that to each set of reflection indices there corresponds a reciprocal lattice point, with identical indices. This is because all values of h, k, and l were used in forming the reciprocal lattice (page 110), and not simply those true crystallographic triples in which the numbers were prime to one another. Thus, reciprocal lattice points having indices which are not prime to one another represent non-unit orders of reflection of the plane to which their distance vectors are normal.

Returning, now, to equation (2), it should be observed that this is expressed in notation referring to orders of reflection, n, from a plane whose index is the crystallographic triple (hkl). This may be expressed in notation referring to the reflection index, in which the order of reflection is already included, by eliminating the term n, thus:

$$\sin \theta_{hkl} = \frac{\lambda}{2d_{(hkl)}}. \tag{7}$$

Another way of writing (7), in which the reciprocal lattice vector $\dfrac{1}{d_{(hkl)}}$ occurs as a term, is

$$\sin \theta_{hkl} = \frac{\lambda \dfrac{1}{d_{(hkl)}}}{2}. \tag{8}$$

A direct geometrical interpretation of (8) is given in Fig. 69. Here θ is represented as the angle between the diameter of a circle of radius 1,

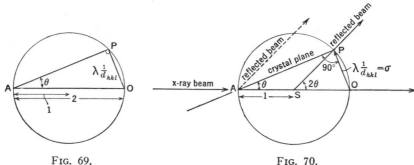

FIG. 69.   FIG. 70.

and the line to the end of the reciprocal lattice vector $\lambda \dfrac{1}{d_{(hkl)}}$.   This geometrical interpretation of (8) may appear, at first glance, to be trivial, but it is a vector diagram having some important properties, which are further developed in Fig. 70.   Note that:

(a) If AO is taken not only as a length, but also as the direction of the x-ray beam, then, since AP makes the angle θ with AO, AP is the slope of the reflecting crystal plane.

(b) OP is normal to the reflection plane, AP, and it is therefore the direction of the vector from the origin to the reciprocal lattice point representing this plane of slope AP which is responsible for the reflection.

(c) Since the length of OP was originally constructed equal to the reciprocal lattice vector, $\lambda \dfrac{1}{d_{(hkl)}}$, OP actually represents this reciprocal lattice vector in length (with proportionality constant, $C = \lambda$), as well as in direction with respect to the incoming x-ray beam.

(d) Since $\angle OSP = 2\angle OAP = 2\theta$, the vector from the center of the circle to the reciprocal lattice point, P_{hkl}, represents the direction of the x-ray reflection.

Now, the importance of this geometrical construction lies in the fact that it provides an easy solution of (8) or (2) which can be readily visualized. For any given experimental set-up, the direction of the x-ray beam is defined as AO; the wavelength, λ, is fixed (thus fixing

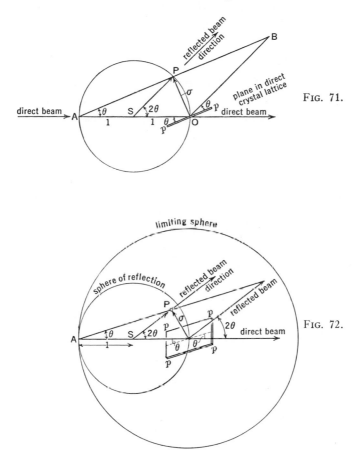

FIG. 71.

FIG. 72.

the constant $C = \lambda$, for which the entire reciprocal lattice is constructed); and the variables involved in the problem of crystal diffraction are the interplanar spacings, $d_{(hkl)}$, and the orientations of the planes (hkl). The solution of the problem of when, and in what direction, diffracted beams occur, is neatly provided by diagrams like Figs. 70 and 71. Diffraction occurs only when the orientation of the crystal is such that a reciprocal lattice point, P, comes to lie on the circumference of a circle, S, of unit radius. When this occurs, a diffracted beam develops in the direction SP.

It is important to realize that Figs. 70 and 71 have been purposely drawn in two dimensions for simplicity of demonstration. Actually, the more general space representation of these figures is given in Fig. 72. If any reciprocal lattice point comes to cut a sphere of radius 1, then (8) or (2) is satisfied and a diffracted beam develops in direction *SP*. This permits only reciprocal lattice points having $\sigma \leq 2$ to possibly reflect. The locus of such lattice points is therefore within a sphere, known as the *limiting sphere*, of radius 2.

LITERATURE

J. D. BERNAL. On the interpretation of x-ray, single crystal, rotation photographs. *Proc. Roy. Soc. London* (*A*), **113** (1926), especially 120–123.

CHAPTER 8

ROTATING-CRYSTAL PHOTOGRAPHS AND THEIR INTERPRETATION

The reciprocal lattice relations developed in the last chapter are of immense aid in the solution of problems arising in the rotating-crystal method. The crystal may be envisaged as a collection of reciprocal lattice points. If the crystal is rotated about any axis, the reciprocal lattice accompanies it, and those reciprocal lattice points within range pass through the sphere. Each time a reciprocal point cuts the

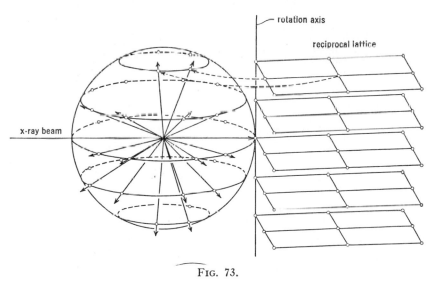

Fig. 73.

sphere, a diffracted beam is developed in the direction *SP*, Fig. 72. The sum total of all directions of reflection from a crystal with known reciprocal lattice, rotated about a given axis, can thus be visualized (Fig. 73) and, if necessary, calculated. In deriving reflection directions, each lattice point may be imagined rotated about the rotation axis from its initial position until it cuts the sphere, which occurs, in general, twice for each point. The sum total of the rays from the center of the sphere to the cutting points on the sphere's surface represents the sum total of diffraction directions for a complete rotation of the crystal.   Of course, the most efficient way of recording these rays would be on a spherical film, but since this would be an impractical kind of film, the rays are customarily recorded on a cylindrical film whose axis is parallel with the rotation axis, or else on a plane plate placed normal to the x-ray beam.

The reciprocal lattice points which actually reflect during a complete rotation of the crystal consist of all those which have passed through the sphere of reflection.   These points occupy the tore, Fig. 74A, swept out of the rotating reciprocal lattice space by the stationary sphere of reflection.   The general case in which the rotation axis is

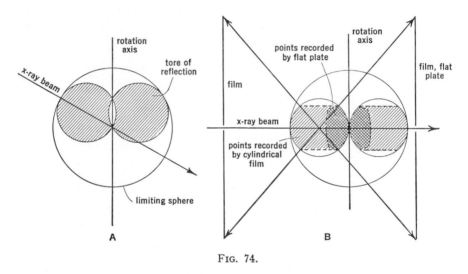

FIG. 74.

inclined to the direction of the incident x-ray beam is shown in Fig. 74A.   Although this case is of very little practical importance in the simple rotating-crystal technique, it will be found to be of importance in the Weissenberg method, developed in later chapters.   In the ordinary rotating-crystal technique, the x-ray beam is customarily made normal to the axis of rotation (*normal-beam method*), because this brings about a simplicity of interpretation of the films.   The tore of reflection for this procedure is shown in Fig. 74B.

All the reciprocal lattice points contained within the tore of reflection actually produce reflections, but not all of them reach the film. This is illustrated in Fig. 74B.   For a complete cylindrical film, only the lattice points in a wide equatorial region of the tore give rise to reflections which can be intercepted by the film.   For a flat plate, only lattice points within the very central section of the tore have reflection directions which can be intercepted by the plate. The advantage of recording reflections with a cylindrical film rather than with a plate is obvious.

CYLINDRICAL COORDINATES

The direction of a reflected ray is determined by the space position of a point with respect to the axis of crystal rotation. A very practical way of specifying the location of a reciprocal lattice point is therefore by specifying its *cylindrical coordinates* with respect to the rotation

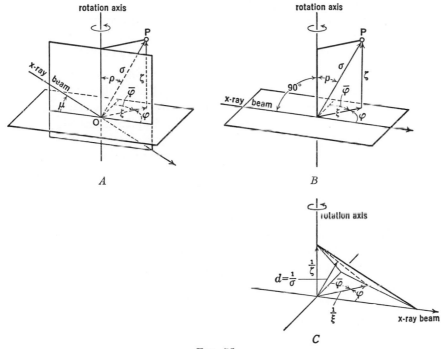

Fig. 75.

axis. For this purpose, the position vector, σ, is resolved into components parallel with, and perpendicular to, the rotation axis (Figs. 75A and B). These two coordinates are designated: ζ (zeta), parallel with the rotation axis; and ξ (xi), perpendicular to the rotation axis, and they lie in a plane containing the rotation axis and the original vector, σ. This plane makes an angle, φ, with the plane containing the direct x-ray beam. The three quantities, φ, ξ, and ζ, completely fix the point P and are its cylindrical coordinates with respect to the rotation axis and with respect to the direct beam as origins.

In vector notation,

$$\sigma = \xi + \zeta. \tag{1}$$

The absolute lengths of these vectors are related by the equation

$$\sigma^2 = \xi^2 + \zeta^2. \tag{2}$$

The reciprocals to these vectors in the direct lattice, shown in Fig. 75C, have the lengths $\dfrac{1}{\sigma}$, $\dfrac{1}{\xi}$, and $\dfrac{1}{\zeta}$. The first quantity is evidently $d_{(nh\ nk\ nl)}$, and the last quantity is of special importance in determining the identity period along the rotation axis as will appear later.

THE INDEXING OF ROTATING-CRYSTAL PHOTOGRAPHS

Given a reciprocal lattice, it is possible, as pointed out in the first section of this chapter, to find each of the reflection directions of the crystal by rotating each reciprocal lattice point about the rotation axis until it cuts the sphere of reflection. The line from center of the sphere to this intersection point is a direction of reflection due to this lattice point, and the reflection is assigned the index of the lattice point. Conversely, given a reflection direction, a point on the sphere is determined and, with it, two of the three cylindrical coordinates of the reciprocal lattice point, namely ζ and ξ. The third coordinate, φ, is undetermined; it represents the amount of rotation from the initial position of the crystal until the point touches the sphere. In spite of this unknown coordinate, it is still possible to swing the point about the rotation axis from its position on the sphere and seek a point on the reciprocal lattice with the same ζ and ξ coordinates. It is possible that more than one point will be found to have these coordinates by pure coincidence, but, if the crystal is of high symmetry and has short identity periods (i.e., large reciprocal identity periods), it is possible to locate uniquely the reciprocal lattice point corresponding with the reflection point. By carrying out this operation point by point, the reflection points on the sphere can be identified with reciprocal lattice points and consequently indexed.

Now, the reflections are actually recorded on a cylindrical or flat film, so that the pattern of spots on the film is not the same as the pattern of reflection points on the sphere, but the former is a projection of the latter. Thus, every spot on the film corresponds with a unique spot on the sphere and, therefore, with a unique ζ, ξ coordinate pair. If the transformation from film coordinates, x, y, to sphere coordinates, ζ, ξ, is known, then it is possible to seek a reciprocal lattice point corresponding with each film spot and thus index the reflections from the crystal. This may be done in either of two ways:

(a) The film coordinates, x and y (Fig. 57), of each spot may be measured, and, the appropriate transformation being known, the cylindrical coordinates ζ and ξ of the corresponding reciprocal lattice point can be calculated. These transformations are cumbersome, and the calculations involved in them are extremely tedious. Because of the labor involved in finding ζ and ξ, this method cannot be recommended.

(b) The positions on the film where reflections due to reciprocal lattice points of known ζ and ξ coordinates record can be determined through appropriate transformations. A chart can therefore be prepared showing the ζ and ξ coordinates of any spot on the film. In this way one set of calculations of transformations suffices once for all, and the chart and film need only be superposed to permit reading the ζ and ξ coordinates of each film spot directly. This procedure is highly recommended and is followed in this book.

In either case, when the ζ and ξ coordinates of each spot on the film have been determined, the reciprocal lattice point having the same coordinates may be sought graphically, and thus each spot on the film can be assigned the appropriate reciprocal lattice (i. e., reflection) index. We now turn to the development of the transformation formulas.

TRIGONOMETRIC RELATIONS EXPRESSED IN RECIPROCAL LATTICE CYLINDRICAL COORDINATES

It is customary to express the direction of the reflected ray in terms of an azimuth angle, Υ (upsilon), normal to the plane containing the reflection and the rotation axis, and an inclination angle, χ (chi), in the plane of the rotation axis (Figs. 58 and 59). Trigonometric functions of these, as well as other angles, may be easily expressed in terms of the coordinates of the reciprocal lattice point P. From Fig. 76A and B it is evident that

$$\sin \chi = \zeta. \tag{3}$$

In oblique triangle SON, by the law of cosines,

$$\xi^2 = 1^2 + \left(\sqrt{1 - \zeta^2}\right)^2 - 2 \cdot 1 \cdot \sqrt{1 - \zeta^2} \cos \Upsilon, \tag{4}$$

which reduces to

$$\xi^2 = 2 - \zeta^2 - 2\sqrt{1 - \zeta^2} \cos \Upsilon, \tag{5}$$

from which,

$$\cos \Upsilon = \frac{2 - \zeta^2 - \xi^2}{2\sqrt{1 - \zeta^2}}. \tag{6}$$

The slope, ρ, of the crystal plane giving rise to the reflection is given by

$$\tan \rho = \frac{\xi}{\zeta}. \tag{7}$$

All the above relations hold regardless of the direction of the x-ray beam with respect to the rotation axis.   In the usual case of the x-ray

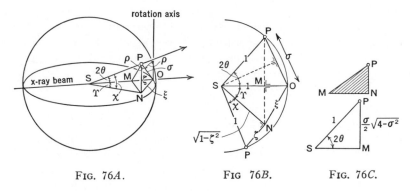

Fig. 76A.   Fig 76B.   Fig. 76C.

beam normal to the axis of crystal rotation, the glancing angle, θ, may be derived from triangle SOP, Fig. 76:

$$\sin \theta = \frac{\sigma}{2}, \tag{8}$$

and combining with (2),

$$\sin \theta = \frac{\sqrt{\zeta^2 + \xi^2}}{2}. \tag{9}$$

The expression for the total deviation angle, 2θ, is more complicated and may be derived from (8) with the aid of the trigonometric identity

$$\sin 2\theta = 2 \sin \theta \cos \theta, \tag{10}$$

from which it turns out that

$$\sin 2\theta = \frac{\sigma}{2} \sqrt{4 - \sigma^2}; \tag{11}$$

or, substituting for σ from (2),

$$\sin 2\theta = \tfrac{1}{2}\sqrt{(\zeta^2 + \xi^2)(4 - \zeta^2 - \xi^2)}. \tag{12}$$

This is evidently the length of PM, Fig. 76B.

The length SM is cos 2θ. This can be evaluated with the aid of the auxiliary diagram, Fig. 76C, as

$$\cos 2\theta = \sqrt{1 - \frac{\sigma^2}{4}(4 - \sigma^2)},$$

which reduces to

$$\cos 2\theta = \tfrac{1}{2}(2 - \sigma^2), \tag{13}$$

or, substituting for σ from (2),

$$\cos 2\theta = \tfrac{1}{2}(2 - \zeta^2 - \xi^2). \tag{14}$$

These trigonometric relations are summarized for convenience in Table 8. Several of them will prove to be of importance in the following developments.

TABLE 8

TRIGONOMETRIC FUNCTIONS OF IMPORTANT ANGLES, EXPRESSED IN TERMS OF CYLINDRICAL COORDINATES OF RECIPROCAL LATTICE POINTS

(X-ray beam perpendicular to crystal-rotation axis)

	θ		2θ		Υ		χ		ρ	
sin	$\dfrac{\sqrt{\zeta^2 + \xi^2}}{2}$	(9)	$\tfrac{1}{2}\sqrt{(\zeta^2 + \xi^2)(4 - \zeta^2 - \xi^2)}$	(12)			ζ	(3)		
cos			$\tfrac{1}{2}(2 - \zeta^2 - \xi^2)$	(14)	$\dfrac{2 - \zeta^2 - \xi^2}{2\sqrt{1 - \zeta^2}}$	(6)				
tan									$\dfrac{\xi}{\zeta}$	(7)

THE TRANSFORMATION FROM RECIPROCAL LATTICE COORDINATES TO FILM COORDINATES

Cylindrical film. The x, y coordinates of a cylindrical film are easily expressed in terms of the ξ, ζ reciprocal lattice coordinates with the aid of Fig. 77. Figure 77A shows a perspective view of the production of a reflection by a reciprocal lattice point whose cylindrical coordinates are ξ, ζ, and the recording of this reflection as a spot with film coordinates x, y on a cylindrical film. In Fig. 77B, some of the important geometrical features of the situation are shown laid out in plan to give true dimensional relations.

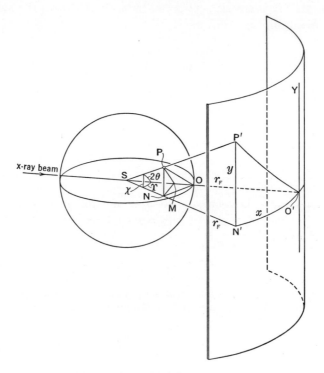

Fig. 77*A*.

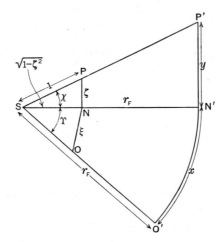

Fig. 77*B*.

The film coordinate x is derivable from the arc-angle proportion (lower part of Fig. 77B):

$$\frac{x}{2\pi r_F} = \frac{\Upsilon}{360°},$$ (15)

from which

$$x = \frac{2\pi r_F}{360°}\Upsilon.$$ (16)

The value of Υ in terms of reciprocal lattice coordinates has already been derived and is given by (6). Substituting this value in (16) gives

$$x = \frac{2\pi r_F}{360°}\cos^{-1}\left(\frac{2 - \zeta^2 - \xi^2}{2\sqrt{1 - \zeta^2}}\right).$$ (17)

The film coordinate y is easily obtained from the similar triangles $P'N'S$ and PNS (upper part of Fig. 77B):

$$\frac{y}{r_F} = \frac{\zeta}{\sqrt{1 - \zeta^2}},$$ (18)

from which

$$y = r_F\frac{\zeta}{\sqrt{1 - \zeta^2}}.$$ (19)

Flat plate. The x, y coordinates of a flat plate may be expressed in terms of the ξ, ζ reciprocal lattice coordinates with the aid of Fig. 78. This shows a perspective view of the production of a reflection by a reciprocal lattice point whose coordinates are ξ, ζ and the recording of this reflection as a spot with film coordinates, x, y, on a flat plate.

In Fig. 78,

$$\tan \Upsilon = \frac{x}{D},$$ (20)

or

$$x = D \tan \Upsilon.$$ (21)

Substituting, in this relation, the value of Υ given by (6), yields

$$x = D \tan \cos^{-1}\left(\frac{2 - \zeta^2 - \xi^2}{2\sqrt{1 - \zeta^2}}\right).$$ (22)

In the similar pyramids, $SMNP$ and $SO'N'P'$, the following proportion obtains:

$$\frac{y}{D} = \frac{\zeta}{\cos 2\theta},$$ (23)

from which

$$y = D \frac{\zeta}{\cos 2\theta}.$$ (24)

The term $\cos 2\theta$ has already been evaluated in (14). Substituting this value in (24) gives

$$y = D \frac{2\zeta}{2 - \zeta^2 - \xi^2}.$$ (25)

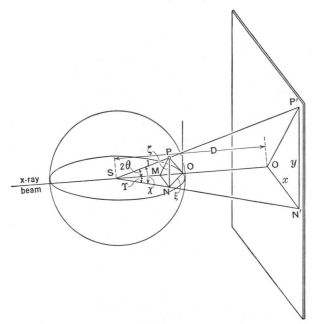

FIG. 78.

THE TRANSFORMATION FROM FILM COORDINATES TO RECIPROCAL LATTICE COORDINATES

From the value of ξ^2 given in (5),

$$\xi = \sqrt{2 - \zeta^2 - 2\sqrt{1 - \zeta^2} \cos \Upsilon}.$$ (26)

For cylindrical films, it has already been shown, (15), that

$$\Upsilon = \frac{360°}{2\pi r_F} x,$$ (27)

and for flat films, from Fig. 78,

$$\cos \Upsilon = \frac{D}{\sqrt{D^2 + x^2}}.$$ (28)

Substituting these values in (26) gives

For cylindrical films:

$$\xi = \sqrt{2 - \varsigma^2 - 2\sqrt{1 - \varsigma^2} \cos\left(\frac{360°}{2\pi r_F} x\right)}; \tag{29}$$

For flat plates:

$$\xi = \sqrt{2 - \varsigma^2 - 2\sqrt{1 - \varsigma^2} \frac{D}{\sqrt{D^2 + x^2}}}. \tag{30}$$

For any coordinate transformation, $x, y \rightarrow \xi, \varsigma$, the value of ς is one of the quantities which must be calculated anyway, and its value may be substituted from the appropriate expression to be derived below:

The transformations from x, y to ς take a much simpler form. The value of ς has already been given in (3) as

$$\varsigma = \sin \chi \tag{31}$$

For a cylindrical film, Fig. 77 B,

$$\sin \chi = \frac{y}{\sqrt{r_F^2 + y^2}}; \tag{32}$$

and for a flat plate, Fig. 78,

$$\sin \chi = \frac{y}{\sqrt{D^2 + x^2 + y^2}}. \tag{33}$$

Making the appropriate substitution of (32) or (33) in (31) gives

For cylindrical films:

$$\varsigma = \frac{y}{\sqrt{r_F^2 + y^2}}; \tag{34}$$

For flat plates:

$$\varsigma = \frac{y}{\sqrt{D^2 + x^2 + y^2}}. \tag{35}$$

For convenience, the several transformations for deriving cylindrical reciprocal lattice coordinates and film coordinates from one another are assembled in Table 9.

TABLE 9

Collected Transformations between Film Coordinates and
Reciprocal Lattice Coordinates

Coordinate sought	Cylindrical film		Flat plate	
$\xi =$	$\sqrt{2 - \zeta^2 - 2\sqrt{1 - \zeta^2}\cos\left(\dfrac{360°}{2\pi r_F}x\right)}$	(29)	$\sqrt{2 - \zeta^2 - 2\sqrt{1 - \zeta^2}\dfrac{D}{\sqrt{D^2 + x^2}}}$	(30)
$\zeta =$	$\dfrac{y}{\sqrt{r_F^2 + y^2}}$	(34)	$\dfrac{y}{\sqrt{D^2 + x^2 + y^2}}$	(35)
$x =$	$\left[\dfrac{2\pi r_F}{360°}\right]\cos^{-1}\left(\dfrac{2 - \zeta^2 - \xi^2}{2\sqrt{1 - \zeta^2}}\right)$	(17)	$D\tan\cos^{-1}\left(\dfrac{2 - \zeta^2 - \xi^2}{2\sqrt{1 - \zeta^2}}\right)$	(22)
$y =$	$r_F\dfrac{\zeta}{\sqrt{1 - \zeta^2}}$	(19)	$D\dfrac{2\zeta}{2 - \zeta^2 - \xi^2}$	(25)

**CHARTS FOR THE DETERMINATION OF THE ζ, ξ COORDINATES
DIRECTLY FROM THE FILM**

It has already been mentioned that the most convenient way of interpreting rotating-crystal films is to prepare a chart of the film showing the ζ and ξ coordinates of every point directly. Now that the transformations from reciprocal lattice coordinates to film coordinates have been developed, the means are available for the preparation of such charts. Given a pair of values ζ and ξ for a reciprocal lattice point, the position, x, y, of the corresponding reflection, as recorded on the film and due to this point, is given by formulas (17) and (19) for a cylindrical film, and formulas (22) and (25) for a flat plate.

In order to map out the area of the film in terms of reciprocal lattice coordinates, it is only necessary to assume decimal values of the pair ζ, ξ, such as .1, .1; .1, .2; .1, .3; etc., and plot the x, y film coordinates where such pairs of values would record with the aid of (17), (19), or (22), (25). Connecting all points for which $\zeta = .1$, for example, gives a line for which ζ has the constant value .1, etc.; and connecting all points for which $\xi = .3$, for example, gives a line for which ξ has the constant value .3, etc. Proceeding in this way, the entire area of the film may be blocked out in a ζ, ξ net from which the value of ζ and ξ may be read for any spot.

Such charts have been prepared by Bernal for both cylindrical films and flat plates. These are shown reduced for a cylindrical camera (Fig. 79) of 57.296-mm. diameter (1 mm. $x = 2°\,\Upsilon$, which is a particularly convenient dimension for all cylindrical camera work) and for a flat plate (Fig. 80) distant $D = 5$ cm. from the crystal (also a particularly convenient dimension). In practice, it is most convenient to

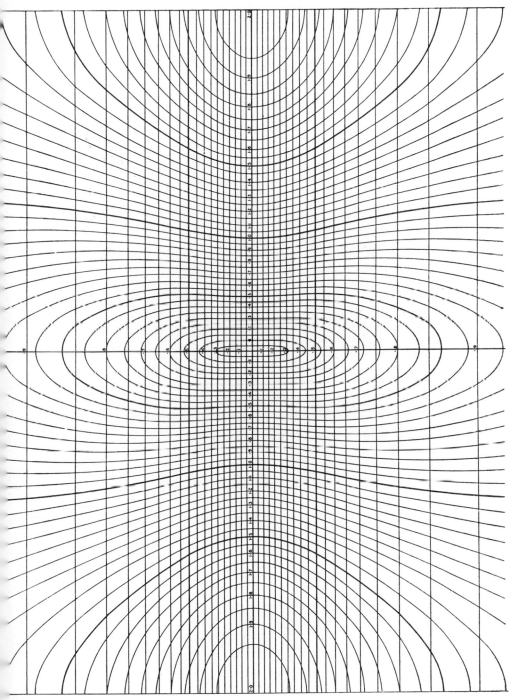

FIG. 79. Bernal chart for cylindrical film, camera diameter 57.3 mm.

145

have the chart photographed on a transparent film or plate. The x-ray film is placed over this film, Fig. 81, and viewed against an

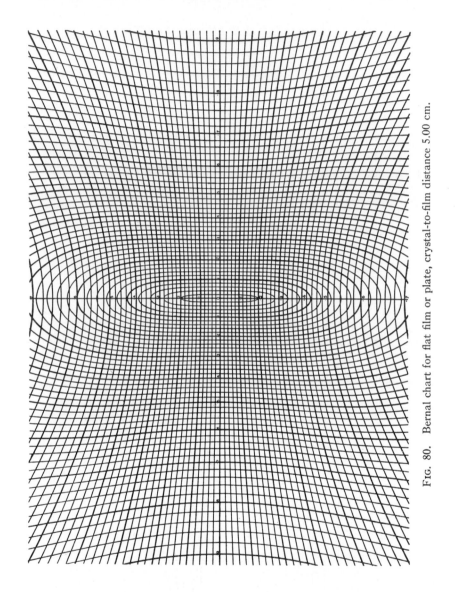

Fig. 80. Bernal chart for flat film or plate, crystal-to-film distance 5.00 cm.

illuminated opal glass background. The reciprocal lattice coordinates, ζ, ξ, may then be read for each diffraction spot with the greatest ease. Figure 82 shows a rotation photograph which bears a striking

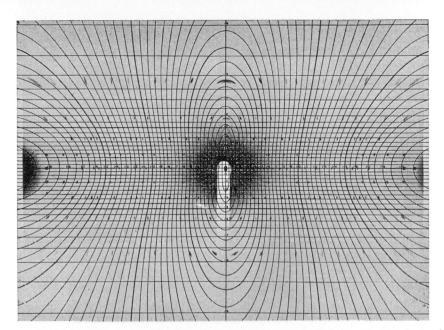

FIG. 81. Rotating-crystal photograph with Bernal chart overlaid. Note particularly the alignment of spots on the photograph along the horizontal lines of the chart. These lines of spots are the so-called *layer lines of the first kind*. (The rotating-crystal photograph is the same as that shown above in Fig. 53.)

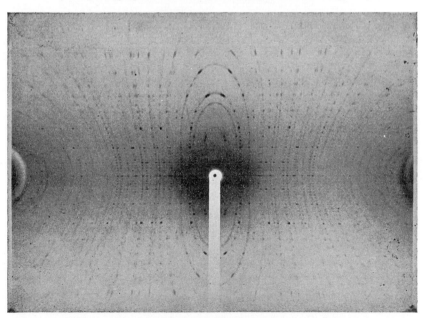

FIG. 82. Rotating-crystal photograph taken with cylindrical camera. Note the alignment of spots along curves of the same family as those of the Bernal chart in Fig. 81. These lines of spots are the so-called *layer lines of the second kind*. (Low-tridymite, SiO_2, orthorhombic; c-axis rotation; $CuK\alpha$ radiation from hot-cathode x-ray tube, filtered through nickel foil.)

147

resemblance to the Bernal cylindrical film chart. The crystal (low
tridymite rotated about its *c*-axis) is characterized by a very short

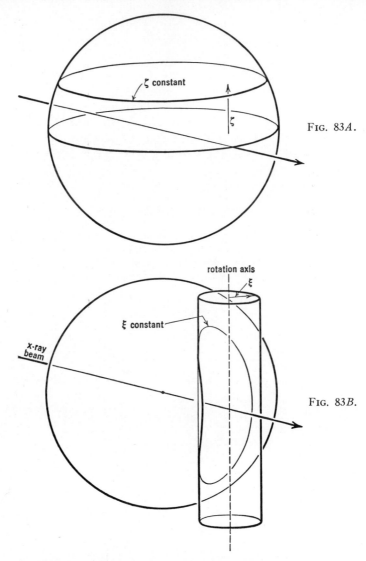

Fig. 83*A*.

Fig. 83*B*.

reciprocal lattice spacing along the rotation axis, and orthogonal axes.
The spots therefore form almost continuous lines of constant ξ values.
    The curves on the chart inherit important aspects of their shapes
from corresponding curves on the sphere of reflection.   Typical curves

on the sphere are illustrated in Fig. 83. All the reciprocal lattice points having the same value of the ζ coordinate occupy a plane normal to the rotation axis, Fig. 83A, and consequently the locus of their contacts with the sphere, on rotation, lie on the intersection of this plane with the sphere of reflection; this intersection is a circle. The locus of rays from the center of the sphere to this circle is, in general, a cone. This cone cuts a coaxial cylindrical film in a circle. When the film is opened out flat, these circles become straight lines. The constant-ζ curves on a cylindrical film are therefore straight lines parallel with the X-axis. The cone from the center of the sphere of reflection to the constant-ζ curve on the sphere intersects a flat plate in a hyperbola, so that the constant-ζ curves on a flat plate are a series of hyperbolas symmetrical with respect to the Y-axis.

All the reciprocal lattice points having the same value of the ξ coordinate lie on the surface of a cylinder whose axis is the rotation axis and whose radius is ξ, Fig. 83B. This cylinder intersects the sphere in a closed ovoid curve which represents the locus of points where reciprocal lattice points of the same ξ coordinate strike the sphere to produce reflections. These curves projected from the center of the sphere produce curves of rather similar shape, both on the cylindrical film and on the flat plate.

Figure 84 illustrates an interesting feature of rotating-crystal photographs

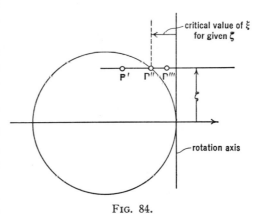

FIG. 84.

taken with the normal beam method. A reciprocal lattice point with the particular coordinate ζ rotated about the rotation axis will not touch the sphere unless it has a certain minimum value of the coordinate ξ. Thus, while P' and P'' touch the sphere and give rise to reflections, P''' has too small a ξ coordinate for its ζ coordinate, and so cannot touch the sphere; therefore it does not give rise to a reflection on rotation of the crystal. For this reason the Bernal charts show, for each value ζ, a lower limit of ξ coordinates in the coordinate net. The corresponding analytical explanation of this is furnished by the expression for the reflection angle, Υ (6), which enters the expressions for the

x coordinate of a spot on the film of given ζ, ξ coordinates:

$$\cos \Upsilon = \frac{2 - \zeta^2 - \xi^2}{2\sqrt{1 - \zeta^2}}. \tag{36}$$

Given a value of ζ, unless ξ has a certain minimum value, the numerator of the right member of the equation exceeds the denominator, and $\cos \Upsilon > 1$.   When this occurs, $\cos \Upsilon$ becomes meaningless and the reflection has no existence.

THE CURVES FOR CONSTANT ρ

One other kind of chart is of some importance in understanding the appearance of rotating-crystal photographs: Consider a crystal plane whose normal makes the angle ρ with the axis of rotation.   Along this normal lie all the reciprocal lattice points which represent the orders of

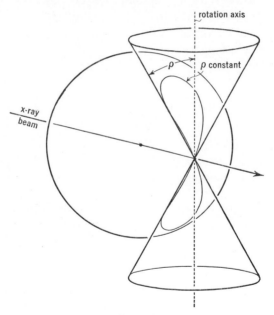

Fig. 85.

reflection, $nh\ nk\ hl$, of this plane.   As the crystal rotates, the plane normal and its lattice points sweep out the surface of a cone of half-angle ρ (see Fig. 85).   The intersection of this conical surface with the surface of the sphere of reflection is the locus of points where the line of lattice points normal to the plane (hkl) intersects the sphere.   In other words, this intersection represents the locus of reflection directions of the orders of reflection from the same plane.   The projection of these curves on a cylindrical film for the plane-slope coordinates ρ differing by 5° intervals is shown in Fig. 86.

The orders of reflection from the same plane lie along the curves of the chart, Fig. 86. The chart also has another function which is somewhat less useful, but nonetheless illuminating: The *characteristic x-radiation* from a given target metal is not strictly monochromatic. It consists chiefly of the $K\alpha$ radiation of the target material, but also a lesser amount of the somewhat shorter-wavelength $K\beta$ radiation, and small amounts of general radiation mainly of still shorter wavelength. Now, in Fig. 70 of the last chapter it will be noted that, in terms of the unit sphere of reflection convention, the distance, σ, of a reciprocal lattice point from the origin is $\lambda \dfrac{1}{d_{(hkl)}}$. If more than one wavelength radiation is present in the x-ray beam, then the reciprocal lattice distance, $\sigma = \lambda \dfrac{1}{d}$, becomes variable, although, for a given plane, (hkl), the lattice points still lie along the generator of the cone whose half-angle is ρ. Therefore, the locus of reflections from a plane rotated in polychromatic radiation is the sheaf of rays to the intersection of the cone of half-angle, ρ, with the surface of the unit sphere of reflection. Along the constant-ρ lines of the chart of Fig. 86 there thus occur:

    (a) the several orders of reflection from the plane (hkl);

    (b) the β radiation reflection spot for each α radiation reflection spot, hkl;

    (c) the general radiation reflection trail from the plane (hkl). This trail traces out a constant-ρ curve.

These features may be seen especially clearly in the rotating-crystal photograph, Fig. 87, which has been deliberately overexposed.

LAYER LINE PHOTOGRAPHS

No mention has been made so far of the orientation of the crystal with respect to the rotation axis, and the discussion has been completely general.

If the reciprocal lattice of a crystal is oriented so that one of its planes is normal to the axis of rotation, then evidently each of the points in one of the levels parallel with that plane will have the same ζ coordinate. As the crystal is rotated, each point of the level within range touches the sphere within the confines of the circumference of a circle whose elevation is ζ, and the reflections of all the points on the level thus lie on a constant-ζ curve on the photographic film, i.e., they lie along a straight line on the opened cylindrical film, or along a hyper-

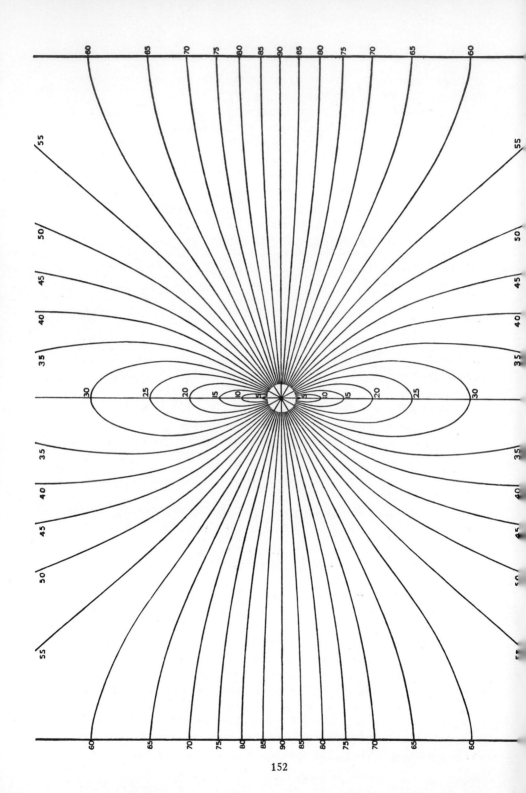

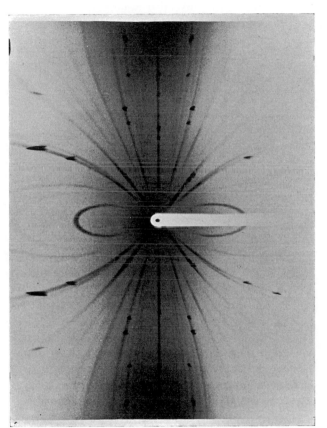

FIG. 86 (on page 152). Chart showing constant-ρ curves. Values of ρ are indicated by labels. (Camera diameter, 57.3 mm.)

FIG. 87 (on page 153). Rotating-crystal photograph made by rotating a large diamond crystal about its a-axis in unfiltered CuK radiation. Compare the shapes of the trails with the shapes of the curves in Fig. 86.

bola on the flat plate. It is evident that, since the reciprocal lattice is composed of levels, the sum total of all its reflections must lie along several such lines, known as *layer lines*. (These are specifically designated *layer lines of the first kind* by German authors. The so-called layer lines of the second kind, clearly seen in Fig. 82, lose their identities in axial-rotation photographs made with crystals belonging to inclined systems and, in any case, are ordinarily very non-obvious properties of a rotating-crystal photograph, especially for crystals having large axes.) Since the reciprocal lattice point 000 lies at the origin, the zero level of the reciprocal lattice has a zero ζ coordinate, and the nth level has a coordinate n times that of the first level:

$$\zeta_n = n\zeta_1. \tag{37}$$

Now, between the direct and reciprocal lattices, the relation mutually holds that an interplanar spacing in one lattice is the reciprocal of the identity period parallel with it in the other. The interplanar spacing ζ_1, in the reciprocal lattice, which is under discussion, is therefore the reciprocal of the identity period along the rotation axis in the direct lattice:

$$T_{\text{rotation axis}} = \lambda \cdot \frac{1}{\zeta_1}, \tag{38}$$

and, from (37),

$$T_{\text{rotation axis}} = \lambda \cdot \frac{n}{\zeta_n}. \tag{39}$$

Looking at the situation in another way, if a crystal is rotated about a rational crystallographic direction (i. e., about a translation direction), a reciprocal lattice net plane is normal to the rotation axis, and the net spacing, ζ_1, is given by the usual spacing : identity-period reciprocity as

$$\zeta_1 = \lambda \cdot \frac{1}{T_{\text{rotation axis}}}. \tag{40}$$

It is evident that the identity period, T, along the rotation axis can be calculated after measuring the layer spacing, ζ_1, of the reciprocal lattice. This is easily and conveniently done by placing the rotating crystal film against the Bernal chart, Fig. 81, reading the layer spacing directly, and, according to (38), dividing the wavelength used to make the film, by this value. This determination of the identity period, T, may be refined somewhat by measuring the ζ_n coordinates of all the layers present on the film, calculating a T for each one with the aid of (39), and averaging the several results.

It should be observed that the discussion of the rotation of a crystal about a rational direction and the determination of the identity period in that direction is the reciprocal lattice equivalent of the discussion in Chapter 5, pages 93–98.

DETERMINATION OF THE DIMENSIONS OF THE UNIT CELL

If one has a crystal whose symmetry is known, then the dimensions of the cell based upon the crystallographic axes can easily be determined by rotating the crystal about each of the three crystallographic axes.

Isometric Crystals. For isometric crystals, the three crystallographic axes are the same, and a single rotation photograph about any one of them gives a layer spacing, $\zeta_{\perp a}$, from which a can be calculated.

Tetragonal Crystals. For tetragonal crystals, the a_1- and a_2-axes are the same, but the c-axis has a different length. Two rotation photographs may be taken to measure the cell completely: one with the crystal rotated about an a-axis, and the other with the crystal rotated about the c-axis. From the $\zeta_{\perp a}$ and $\zeta_{\perp c}$ layer spacings, respectively, of these photographs, the cell dimensions can be determined. *Subsequent indexing of the reflections of one of these photographs (most conveniently, the c-axis rotation) may prove that the cell chosen is C-centered, or face-centered, in which case a smaller unit cell (primitive and body-centered, respectively) can be had by taking a new a-axis at 45° with the first.*

Hexagonal crystals. For hexagonal crystals, the a_1-, a_2-, and a_3-axes are the same, but the c-axis has a different length. Two rotation photographs may be taken to measure the cell completely: one with the crystal rotated about an a-axis, and the other with the crystal rotated about the c-axis. From the $\zeta_{\perp a}$ and $\zeta_{\perp c}$ layer spacings, respectively, of these photographs, the cell is determined. *Subsequent indexing of the reflections of one of these photographs (most conveniently, the c-axis rotation) may prove that the cell chosen is multiply primitive, in which case a smaller, primitive unit cell can be had by taking a new a-axis at 30° to the first.*

Trigonal crystals. Trigonal crystals are based upon either a hexagonal lattice or a rhombohedral lattice, and the correct lattice type is not ordinarily known in advance. In the investigation of trigonal crystals it is therefore best to treat them tentatively as hexagonal and refer them to a hexagonal coordinate system with axes A_1, A_2, A_3, and C, until the reflections are indexed. As in the case of hexagonal crystals, the cell is then determined with the aid of two rotation photographs, one for rotation about one of the A-axes and one for rotation about the C-axis.   From the $\zeta_{\perp A}$ and $\zeta_{\perp C}$ coordinates of the layer lines of these photographs, the lengths of the A- and C-axes are determined.   Now, if the crystal is based upon a rhombohedral lattice, then subsequent indexing of the reflections will show that the reflections not satisfying the rhombohedral condition (pages 68–70),

$$-H + K + L = \text{an integer divisible by 3,}$$

will be extinguished.   In this case, the true rhombohedral axial length, a, and interaxial angle, α, may be derived from the hexagonal axes A and C, measured on the rotation photographs, with the aid of the transformations (to be derived subsequently, pages 503–504):

$$a = \sqrt{3A^2 + C^2}, \tag{41}$$

$$\alpha = 2\sin^{-1}\frac{3A}{2\sqrt{3A^2 + C^2}}. \tag{42}$$

If it is known in advance from previous work that the crystal is based upon a rhombohedral lattice, then, since all three rhombohedral axes are the same, a single rotation photograph about the rhombohedral edge permits a measurement of the layer spacing $\zeta_{\perp a}$, from which the length of the axis, a, can be calculated.   Subsequent indexing of the reflections may prove

(a) that the cell based upon this edge is face-centered, in which case a smaller, primitive rhombohedral cell can be had by taking a new a-axis in the [110] direction; or

(b) that the cell based upon this edge is body-centered, in which case a smaller, primitive rhombohedral cell can be had by taking a new a-axis in the [11$\bar{1}$] direction.

Orthorhombic crystals.   All three axes, a, b, and c, of an orthorhombic crystal, are distinct, and to determine the cell dimensions three rotation photographs may be taken, one for rotation about each axis. From the $\zeta_{\perp a}$, $\zeta_{\perp b}$, and $\zeta_{\perp c}$ layer spacings, the lengths of the three axes a, b, and c may be calculated.   This procedure uniquely defines the smallest unit cell, and alternative units consistent with the symmetry do not come up for consideration.

Monoclinic crystals.   Rotation photographs for each of the three crystallographic axes give layer spacings, $\zeta_{\perp a}$, $\zeta_{\perp b}$, and $\zeta_{\perp c}$, for which the identity periods along the classical a-, b-, and c-axes can be calculated. The b-axis so determined is also uniquely the correct unit cell axis, but it may be possible to choose more suitable a- and c-axes. *Thus, if indexing proves that the cell is face-centered, a smaller cell can be chosen by a recombination of the measured a and c vectors to give either of two cells having half the volume of the original cell. One of these is body-centered; the other is base-centered. Of the two, the one should be chosen having the angle β nearer a right angle.*

Triclinic crystals. Rotation of a triclinic crystal about its three crystallographic axes gives three photographs from whose layer spacings, $\zeta_{\perp a}$, $\zeta_{\perp b}$, and $\zeta_{\perp c}$, the lengths of the three classical crystallographic axes may be calculated. *The cell so defined may be primitive, end-centered, body-centered, face-centered, or otherwise non-primitive, a situation which may be detected in indexing the reflections.* The selection of the simplest unit from the original cell dimensions is not easy from rotation photographs alone, and the discussion of the selection of the correct unit is deferred until after moving-film photographs are treated.

THE INDEXING OF ROTATING-CRYSTAL PHOTOGRAPHS

The indexing of a rotating-crystal photograph consists of the assignment of the three numbers h, k, and l to each reflection. This is equivalent to the assignment of the numbers h, k, and l to the point on the reciprocal lattice responsible for the recorded reflection on the film. The two cylindrical coordinates, ζ and ξ, of each reciprocal lattice point may be read directly from the x-ray photograph, and the problem of indexing is essentially one of transforming from the cylindrical, ζ, ξ, reciprocal lattice coordinates to the ha^*, kb^*, lc^* reciprocal lattice coordinates. In order to do this it is first necessary to know the unit vectors, a^*, b^*, and c^*; in other words, it is necessary to know the dimensions of the reciprocal unit cell. If the direct unit cell dimensions are known, the reciprocal unit cell dimensions may be calculated with the aid of the relations given on page 121, in which the proportionality constant, λ, is introduced:

$$a^* = \lambda \, \frac{bc \sin \alpha}{V}, \tag{43}$$

$$b^* = \lambda \, \frac{ac \sin \beta}{V}, \tag{44}$$

$$c^* = \lambda \, \frac{ab \sin \gamma}{V}, \tag{45}$$

where

$$V = abc\sqrt{1 - \cos^2 \alpha - \cos^2 \beta - \cos^2 \gamma + 2 \cos \alpha \cos \beta \cos \gamma}. \qquad (46)$$

(Relation (46) as well as the following three will be deduced in Chapter 18.)

The reciprocal cell angles can be computed from the direct cell angles with the aid of either set of the following relations:

$$\alpha^* = \cos^{-1}\left(\frac{\cos \beta \cos \gamma - \cos \alpha}{\sin \beta \sin \gamma}\right), \qquad (47)$$

$$\beta^* = \cos^{-1}\left(\frac{\cos \alpha \cos \gamma - \cos \beta}{\sin \alpha \sin \gamma}\right), \qquad (48)$$

$$\gamma^* = \cos^{-1}\left(\frac{\cos \alpha \cos \beta - \cos \gamma}{\sin \alpha \sin \beta}\right), \qquad (49)$$

or, alternatively,

$$\alpha^* = 2 \sin^{-1}\sqrt{\frac{\sin \sigma \sin (\sigma - \alpha)}{\sin \beta \sin \gamma}}, \qquad (50)$$

$$\beta^* = 2 \sin^{-1}\sqrt{\frac{\sin \sigma \sin (\sigma - \beta)}{\sin \alpha \sin \gamma}}, \qquad (51)$$

$$\gamma^* = 2 \sin^{-1}\sqrt{\frac{\sin \sigma \sin (\sigma - \gamma)}{\sin \alpha \sin \beta}}, \qquad (52)$$

where

$$\sigma = \frac{\alpha + \beta + \gamma}{2}.$$

These relations are deduced in Chapter 18. For all except triclinic crystals, they reduce to considerably simplified forms which are listed in Table 10.

Now, all the right members of the above relations are either the identity periods along the crystallographic axes or the angles between crystallographic axes. The former may be easily calculated from layer-line data of not more than three axial rotation photographs, as outlined in the last section. Usually the interaxial angles, in the cases where these differ from 90° and 120°, can be measured on the crystal used for the three rotation photographs, because if the crystal had a good enough surface development to permit orientation for the rotation photographs, its interaxial angles could be derived from measurements with a reflection goniometer. In most instances, the interaxial angles of the direct lattice can be taken from classical inter-axial angles of the crystals as listed in standard crystallographic or mineralogical references.

TABLE 10

DIMENSIONS OF SPECIAL RECIPROCAL LATTICE CELLS

	Monoclinic	Orthorhombic	Hexagonal	Rhombohedral	Tetragonal	Isometric
a^*	$\lambda \dfrac{1}{a \sin \beta}$	$\lambda \dfrac{1}{a}$	$\lambda \dfrac{1}{a}\dfrac{2}{\sqrt{3}}$	$\lambda \dfrac{1}{a}\dfrac{\sin \alpha}{\sqrt{1 - 3\cos^2 \alpha + 2\cos^3 \alpha}}$	$\lambda \dfrac{1}{a}$	$\lambda \dfrac{1}{a}$
b^*	$\lambda \dfrac{1}{b}$	$\lambda \dfrac{1}{b}$	"	"	"	"
c^*	$\lambda \dfrac{1}{c \sin \beta}$	$\lambda \dfrac{1}{c}$	$\lambda \dfrac{1}{c}$	"	$\lambda \dfrac{1}{c}$	"
α^*	90°	90°	90°	$\cos^{-1}\left(-\dfrac{\cos \alpha}{1 + \cos \alpha}\right)$	90°	90°
β^*	180° − β	90°	90°	"	90°	90°
γ^*	90°	90°	60°	"	90°	90°
V (direct cell)	$abc \sin \beta$	abc	$\dfrac{\sqrt{3}}{2}a^2c$	$a^3\sqrt{1 - 3\cos^2 \alpha + 2\cos^3 \alpha}$ $= a^3 \dfrac{(1 - \cos \alpha)^2}{\sin \alpha\,(1 + \cos 2\alpha)}$	a^2c	a^3

With the dimensions of the unit cell of the reciprocal lattice determined as above, two indexing procedures are available:

(*a*) The reciprocal lattice may be drawn and the [[*hkl*]]* point corresponding with the cylindrical coordinates ζ, ξ of each spot may be found graphically.

(*b*) The cylindrical coordinates, ζ, ξ, of each spot of the reciprocal lattice may be calculated and the list compared with the list of ζ, ξ coordinates for each spot on the film. In this way each spot on the film may be identified with a reciprocal lattice point of index [[*hkl*]]* and therefore receive the reflection index *hkl*.

The latter procedure is tedious, and time spent upon it is largely wasted. The simplest, easiest, and most rapid method is graphical indexing:

Graphical indexing. Suppose a crystal to be rotated about its *c*-axis. Then all the points on the zero level of the reciprocal lattice (Fig. 88*A*) have indices [[*hk*0]]*; all the points on the first level have indices [[*hk*1]]*; all the points on the second level have indices [[*hk*2]]*; and, in general, all the points on the *n*th level have indices [[*hkn*]]*. In a similar manner, if a crystal is rotated about any crystallographic axis, the index in the position corresponding to that axis is the number of the level on which the point occurs. Therefore one index of every spot on the rotation photograph is determined by the layer number of the spot. The cylindrical coordinate ζ thus determines this one index, *n*. The graphical determination of the other two coordinates, in this example *h* and *k* for the zero layer, is illustrated in Fig. 88*B*. The zero level of the reciprocal lattice is first graphically laid out, using the dimensions as calculated above. Any radial line, *OQ*, is drawn from the origin, and the ξ value of each spot on the zero layer of the photograph is plotted along it. Each of these radial distances of the reciprocal lattice points is then swung about the origin with the aid of a pair of dividers until its end is found to coincide with a reciprocal lattice point [[*hk*0]]*. When this spot is found, the original spot along the *OQ* axis is given the label, *hk*0, of the reciprocal lattice point, and the point itself should also be checked off as having been found to give rise to a reflection.

The *n*th levels may also be indexed with the same procedure, the only difference being that, if the reciprocal axis *OC**, Fig. 88*A*, is not at right angles to the levels, then the origin of the level coordinates, *h*, *k*, is offset from the rotation axis, which is the origin of the ξ coordinates.

Analytical indexing. Instead of graphically correlating a spot of cylindrical coordinates, ζ, ξ, with a reciprocal lattice point whose index is $[[hkn]]^*$, the cylindrical coordinates of each point within reflecting range may be calculated from the known reciprocal lattice constants and then compared with the observed ζ, ξ coordinates of the spots as

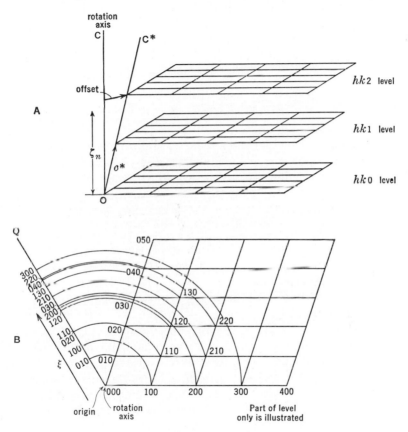

FIG. 88.

measured on the x-ray film. If the level net has its origin of coordinates coinciding with the rotation axis, as in the example given in Fig. 88B (this is true for all zero levels), then the radial vector, ξ, may be expressed as a sum of the reciprocal lattice vectors,

$$\boldsymbol{\xi} = h\mathbf{a}^* + k\mathbf{b}^*, \tag{53}$$

or, in absolute lengths,

$$\xi = \sqrt{h^2 a^{*2} + k^2 b^{*2} - 2hka^*b^* \cos \gamma^*}. \tag{54}$$

In the cases of n layers of:

(a) triclinic crystals rotated about any crystallographic axis,

(b) monoclinic crystals rotated about the a- or c-axis,

(c) rhombohedral crystals rotated about a crystallographic axis,

the origin of the nth level is offset from the rotation axis by a vector (Fig. 88A) whose length is $\zeta_n \tan (c \wedge c^*)$. The length of ξ then becomes

$$\xi = \left\{ \frac{l^2 c^{*2}}{\sin^2 \gamma^*} (\cos^2 \beta^* - 2 \cos^2 \alpha^* \cos^2 \beta^* \cos^2 \gamma^*) + h^2 a^{*2} + k^2 b^{*2} \right.$$
$$\left. + 2klb^*c^* \cos \alpha^* + 2lhc^*a^* \cos \beta^* + 2hka^*b^* \cos \gamma^* \right\}^{1/2} \quad (55)$$

These analytical methods are tedious, and any possible additional accuracy they might possess is completely wasted, because the position of a spot on an x-ray film can be determined with only limited accuracy.

INDETERMINATENESS IN INTERPRETING ROTATING-CRYSTAL PHOTOGRAPHS

Index indeterminateness. In three-dimensional space, three independent coordinates are required to specify, or locate, a point completely. A reciprocal lattice point is completely located by its three coordinates h, k, and l. It is also completely located by its three cylindrical coordinates ξ, ζ, and φ. The first two of these cylindrical coordinates are given by the location of the diffraction spot on the photographic film, but the angular coordinate, φ, of the reciprocal lattice point cannot be determined, because there is no record of the amount of rotation necessary to bring a reciprocal lattice point from its initial position to the position when it makes contact with the sphere of reflection and gives rise to a reflection. In other words, the coordinate φ is indeterminate in rotating-crystal photographs. This gives rise to uncertainty in two of the three indices in indexing layer-line photographs. Thus, if two reciprocal lattice points $[[hkn]]^*$ and $[[h'k'n]]^*$ on the same level have nearly the same distance from the rotation axis, they have nearly the same ξ coordinates, and it may be impossible to decide which one of these gives rise to a certain spot of this observed ξ measurement. (In the example given in Fig. 88B, 020 and 110 are such a pair of spots.) If it were possible to determine the φ coordinate of a reciprocal lattice point from the diffraction record it leaves on the film, then it would be possible to decide between these two alternatives, for, even if two lattice points $[[hkn]]^*$ and $[[h'k'n]]^*$

had the same ξ coordinate, they could not simultaneously have the same φ coordinate.

For cubic crystals, and for tetragonal, hexagonal, and trigonal crystals (referred to hexagonal axes) rotated about the c-axis, the symmetry of this crystallographic axis reduces some of the uncertainty in indexing. For cubic crystals rotated about a crystallographic axis and for tetragonal crystals rotated about the c-axis, (54) reduces to

$$\xi = a^*\sqrt{h^2 + k^2}. \tag{56}$$

In this case, the sequence of radial coordinates, ξ, for the different values of the indices h and k is the same for all crystals, except for the scale a^* of the reciprocal lattice level and except for possible omissions due to extinction. The rotating-crystal spot sequences are therefore analogous for all cubic crystals and for tetragonal crystals rotated about the c-axis. All spots on the same level for which $h^2 + k^2$ in (56) is the same, reflect to the same position and produce superposed records. The reflections which exactly superpose in this manner are as follows:

$h^2 + k^2$	index pair h k			
25	0 5	3 4	4 3	5 0
50	1 7	5 5		7 1
65	1 8	4 7	7 4	8 1
85	2 9	6 7	7 6	9 2
100	0, 10	6 8	8 6	10, 0
125	2, 11	5, 10	10, 5	11, 2
130	3, 11	7 9	9 7	11, 3
145	1, 12	8 9	9 8	12, 1
169	0, 13	5, 12	12, 5	13, 0
170	1, 13	7, 11	11, 7	13, 1
185	4, 13	8, 11	11, 8	13, 4
200	2, 14	10, 10		14, 2
.	.	.	.	.
.	.	.	.	.
.	.	.	.	.

Except for these exact superpositions, all other spots have different ξ coordinates, and are therefore resolved provided that the reciprocal a^*-axis is not too short (i.e., provided that the direct a-axis is not too long).

For hexagonal crystals and for trigonal crystals referred to hexagonal coordinates, both rotated about the c-axis, (53) reduces to

$$\xi = a^* \sqrt{h^2 + k^2 + hk}. \tag{57}$$

As in the case just discussed, the sequence of ξ values with different h, k values is fixed for any crystal except for the scale a^* of the reciprocal lattice level, and except for possible omissions due to extinctions. For this case, a list analogous to that given above could be prepared showing exact superpositions. Again, except for such exact superpositions, all other spots have different ξ values, sufficiently separated, provided that the a-axis is not too long, so that the indices of reciprocal lattice points within reflecting range can be identified.

For orthorhombic crystals rotated about a crystallographic axis, say c, for definiteness, (54) takes the form

$$\xi = \sqrt{h^2 a^{*2} + k^2 b^{*2}}. \tag{58}$$

The sequence of spots in this instance depends not only upon h and k, but also upon two incommensurable constants, a^* and b^*. The sequence of ξ values is therefore different for every crystal, and fortuitous overlaps of ξ values occur which are unpredictable. Orthorhombic crystals cannot therefore be unequivocally indexed by the ordinary rotation method. The same is true for monoclinic and triclinic crystals.

So far nothing has been said specifically about the possibility that certain critical reciprocal lattice points might be out of reflection range. This occurs, for example, for the lattice points along a rotation axis, for these can never touch the sphere (Fig. 84). Consequently $00l$ spectra of tetragonal, hexagonal, and trigonal crystals, which furnish important space group data, cannot appear on photographs taken for rotations about the c-axis. If rotations are made about the a-axis instead of the c-axis, then (54) reduces to a case similar to that of the rotation of an orthorhombic crystal about a crystallographic axis, namely:

$$\xi = \sqrt{h^2 a^{*2} + l^2 c^{*2}} \tag{59}$$

and the discussion given for the orthorhombic case applies here also. In other words, it is not possible to make an unequivocal and complete list of reflection indices of tetragonal, hexagonal, and trigonal crystals. *Therefore only the reflections of isometric crystals can be unequivocally and completely indexed (with the exceptions of the exact superpositions noted) by the rotating-crystal method.*

There are three methods of partially or completely rectifying the indeterminateness of indexing by methods involving rotating crystals:

(a) By use of the rotating-crystal method in which the crystal is purposely tipped out of orientation by a known direction and amount. This causes a tip in the reciprocal lattice levels, and different parts of the same level record at slightly different ζ levels, and the amount of departure of ζ from the mean value is a measure of the φ coordinate of the reciprocal lattice point. Even this method does not have sufficiently great power to resolve all pairs of spots. It has the disadvantages of requiring a separate set of films for indexing and for determining axial lengths, and of requiring a somewhat complicated interpretative procedure. This method is infrequently used.

(*b*) By the oscillating-crystal method, which is treated in Chapter 10. This method permits a determination of φ as between certain limits for each reflection.

(*c*) By moving-film methods. These methods, particularly the Weissenberg method, permit an exact determination of φ for each reflection; they are discussed at length in subsequent chapters.

Symmetry indeterminateness. The rotating-crystal method provides only meager or indirect evidences of symmetry. The chief kind of information supplied is through the reflection identity of photographs taken about different rotation axes. For example, if the three photographs about three orthogonal crystallographic axes are duplicates, the crystal may be assumed to be isometric; if two are duplicates while the third is different, then the crystal may be assumed to be tetragonal; if all three are different, the crystal may be either orthorhombic, monoclinic, or triclinic.

Information of this sort is meager, indirect, and usually known or suspected in advance. For this reason, the Laue method (discussed in substantially every book treating x-ray methods even slightly) is frequently used in conjunction with the rotating-crystal method for symmetry information. Complete symmetry information in an extremely useful form is provided by the method discussed in the last part of Chapter 10 and by the moving-film methods treated in subsequent chapters.

LITERATURE

J. D. BERNAL. On the interpretation of x-ray, single crystal, rotation photographs. *Proc. Roy. Soc. London* (*A*), 113 (1926), especially 123–141.

E. SCHIEBOLD. Die Drehkristallmethode. *Fortschr. Mineral.*, etc., 11 (1927), 111–280.

H. GEORGE. On the interpretation of x-ray crystal photographs.
    I. *Phil. Mag.* (7), 7 (1929), 373–384.
    II. *Phil. Mag.* (7), 8 (1929), 442–456.

LUCIEN J. B. LACOSTE. Reciprocal lattice projecting ruler and chart. *Rev. Sci. Instruments*, 3 (1932), 356–364.

CHAPTER 9

PRACTICAL ASPECTS OF ROTATING-CRYSTAL INVESTIGATIONS

DESIGN OF ROTATING-CRYSTAL APPARATUS

Pinhole system design. X-rays have a refractive index practically equal to unity and are consequently undeviated as they pass through matter. They cannot, therefore, be controlled with the aid of lenses like ordinary light. In order to obtain a beam of x-rays, one must be content to limit a portion of the radiation by means of two holes in some opaque material, ordinarily lead. In other words, one can only crudely control x-rays by defining a beam of roughly parallel rays by a pair of pinholes, as shown in Fig. 89. The x-ray beam

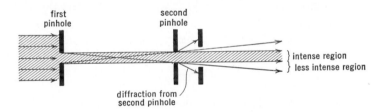

FIG. 89.

emerging from such a pinhole system consists of all the possible ray directions which can be drawn through the pinhole pair. The sum total of these create a beam having an intense central region surrounded by a region in which the intensity falls off toward the outer boundary.

In addition to the primary ray directions just mentioned, the emergent beam contains some secondary ray directions. These are due to diffraction of the primary x-ray beam by the crystals in the metal of the second pinhole. These diffraction halos, if not removed, would reach the film and produce a background. To prevent this from occurring, this relatively weak radiation is usually caught on the solid part of a third diaphragm whose aperture is so large that it passes the direct beam, but which is still small enough to intercept the diffracted radiation from the second pinhole. This requires that the third pinhole must have an aperture small enough to intercept rays

166

having direction 2θ corresponding with the largest unextinguished spacing, $d_{(hkl)}$ of the metal of the second pinhole. This third pinhole is shown in Fig. 89, and a simple design for an actual pinhole system is shown in Fig. 90.

Camera design. In order to permit x-rays to reach the photographic film and at the same time to prevent light from doing so, x-ray films are ordinarily protected by black paper. The simplest camera designs result by enclosing the entire film in a black paper envelope. The design of this envelope depends on the general camera design, and will be discussed subsequently.

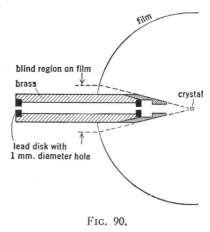

FIG. 90.

There are a number of schemes for holding the photographic film in the form of a cylinder. These may be divided into two general categories. In one of these the film is wrapped around the *outside* of a skeleton metal cylinder, and in the other it is forced against the *inside* of a hollow metal cylinder. The first design should be avoided

FIG. 91. Cylindrical camera for rotating-crystal apparatus.

because this arrangement leaves the film unprotected against stray x-radiation impinging on the film from the outside. In the second design, the film is completely protected from this stray external radiation by the metal cylinder itself. A camera of this design is shown in Fig. 91.

In a well-designed camera, the film covers a maximum Υ range, i.e., it occupies a maximum arc of the camera cylinder. At the same time, the film must lie tightly against the cylinder.. The last requirement cannot be satisfied by holding the film against the inside of a

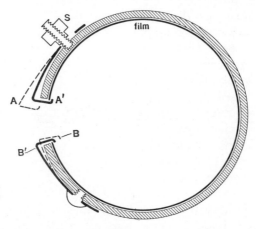

FIG. 92.

cylinder with the aid of spring rings or wedge rings, for these permit the film to sag away from the cylinder in the region between the rings. An excellent scheme for holding the film against the inside of the cylinder for its entire area is shown in Figs. 91 and 92. Figure 91 shows the outside of the camera; Fig. 92 shows a section across the cylinder axis of the camera. The free edges of the film are held by a pair of thin metal clips. With correct dimensioning, the clips push against these free edges of the film, and thus develop a centrifugal force which holds the entire film against the walls of the cylinder. To load the film in the cylinder, thumbscrew S, Fig. 92, is first loosened, and the upper spring shifted to position A. The film, enclosed in its paper envelope, is then curled into the shape of a cylinder and slipped down the camera tube. The lower spring clip is now pressed into position B, after which the film may be rotated clockwise until a free edge is under the clip. Release of this clip to position B' fixes one edge of the film firmly. The upper clip, which is free to move, is now hooked over the remaining free edge of the film and pulled back to position A', forcing the film outward against the cylinder. The upper spring is finally permanently fixed in position A' by turning thumbscrew S.

There are several ways of making satisfactory film envelopes. The simplest of these is illustrated in Figs. 93A and 93B. This envelope is cut from a single rectangular sheet of black photographic masking paper. Figure 93*A* shows the dimensions of the original rectangle, together with the positions of the folds for a camera of standard diameter, namely 57.3 mm. (see discussion in Chapter 12), and using a 5- by 7-inch photographic film trimmed to 5 by $6\frac{21}{32}$ inches. The rectangle is folded along the indicated lines. The resulting envelope is shown in Fig. 93*B*, curled into the form of a cylinder. This envelope cylinder, with its contained film, is slipped into the metal camera cylinder so that its $5\frac{1}{4}$ inch junction faces metal, and hence is light-sealed. The top opening is light-sealed by dropping down upon it a metal ring whose overlapping flange

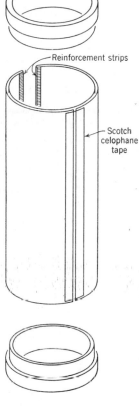

Reinforcement strips

Scotch celophane tape

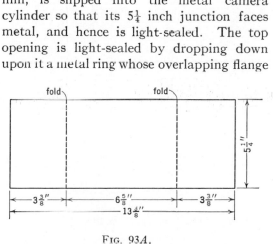

Fig. 93*A*.   Fig. 93*B*.

covers the opening in the paper. The lower opening of the envelope is closed by fitting under the flange of a corresponding metal ring which is permanently attached to the inside of the camera. This envelope, in the form described, often causes difficulty by becoming unwrapped while it is being manipulated within the camera. This can be prevented by permanently closing the junction with adhesive tape as shown in Fig. 93*B*. If the joining tape is of the transparent Scotch cellulose variety, the additional thickness is so slight that no appreciable distortion of the film from truly cylindrical form occurs. The life and manipulability of the envelope in the camera are improved by shellacking thin, narrow, metal strips to the inside free edges, as shown in Fig. 93*B* or by reinforcing the same region with Scotch cellulose tape.

A somewhat more elaborate envelope is shown in Figs. 94*A* and 94*B*. In this envelope, the junction is on the inside of the cylinder, and therefore must be inherently light-tight. It has the disadvantage of casting a shadow due to the doubled paper thickness at this point. The bottom of this envelope is pasted shut and so does not rely on a lower ring for light-tightness in this region. Because of the pasting operations involved, this envelope is difficult to make, and several tries are usually required to prepare a perfect, light-tight, smooth product. To make the envelope, a rectangle of black photographic masking paper having the dimensions indicated in

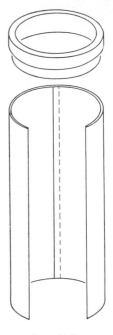

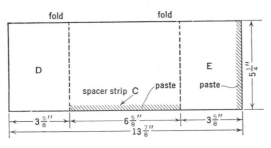

Fig. 94*A*.   Fig. 94*B*.

Fig. 94*A* is first cut out. Extra spacer strip *C* is next pasted onto what is to become the inside of the envelope. Paste is then applied to the two indicated edges, and the flaps *D* and *E* are folded in rapid succession. The envelope is then quickly wrapped about a cylinder of approximately the same diameter as the camera cylinder and held with rubber bands until dry. When removed from its form, the envelope has the appearance shown in Fig. 94*B*. The only overlap of the paper is located away from the camera cylinder so that nothing prevents the film from taking a truly cylindrical form against it.

The direct x-ray beam must be prevented from impinging on the film. This is arranged by catching the direct beam in a deep lead cup, Fig. 95, which lies back of the crystal but within the camera cylinder. Behind the cup there is, of course, a blind spot on the film in which no reflections can be recorded. This area is reduced to a minimum by making the cup small, and shaping it to wedge toward the crystal.

The direct radiation which is intercepted by the bottom of the cup is partly scattered back to the region near the point of entrance of the direct beam into the camera cylinder. The dimensions of the hole in

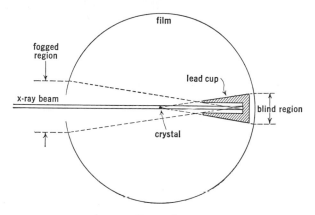

FIG. 95.

the cup should be arranged so that the rim of the cup prevents this radiation from reaching the film, as far as possible, and so that this back-scattered radiation is confined to the open region of the camera where the pinhole system enters.

CHOICE OF RADIATION

Introduction. In crystal-structure investigations it is desirable (as explained below) to be able to select, at will, the kind of x-radiation suitable to the particular investigation. Since the x-radiation emitted by an x-ray tube is characteristic of the target metal employed, the radiation may be changed by simply changing the target metal in the tube. To arrange for this, x-ray tubes used in crystal-diffraction investigations are ordinarily fitted with replaceable target inserts, and the insert with the metal appropriate for the investigation can thus be employed at will. An even more convenient mechanical arrangement is to have an x-ray tube with a rotatable target faced with different metals, Fig. 232. With this design, a different target material and therefore radiation may be had in a moment's time by simply rotating the target insert to the appropriate position.

The various radiations vary in wavelength and in penetration. The wavelength emitted by a metal decreases with increasing atomic number. Table 11 lists the important characteristic wavelengths of the elements whose radiations are of practical value in crystal structural investigations.

The total number of reflections developed by a crystal. The total number of reflections of different index developed by a crystal during a complete rotation is equal to the number of its reciprocal lattice points which have passed through the surface of the sphere of reflection. This is evidently equal to the number of reciprocal lattice points included in the tore of reflection. This number can be determined in the following manner: In a primitive lattice, there is one lattice point per unit cell volume. If the reciprocal cell is centered, it has N lattice points per reciprocal unit cell, where N is

2 for end-centered cells.

2 for body-centered cells.

3 for rhombohedral lattices referred to a hexagonal cell.

4 for face-centered cells.

$$\text{Number of cells per tore} = \frac{\text{Volume of tore}}{\text{Volume of cell}}. \qquad (1)$$

$$\text{Number of lattice points in the tore} = \frac{\text{Volume of tore}}{\text{Volume of cell}} N. \qquad (2)$$

The volume of the tore of reflection is $2\pi^2 r^3 = 2\pi^2 \cdot 1^3 = 2\pi^2$. The volume of the reciprocal lattice cell, V^*, is a function of $\left(\lambda\dfrac{1}{a}, \lambda\dfrac{1}{b}, \lambda\dfrac{1}{c}\right)$ and specifically equal to $\lambda^3 \cdot \dfrac{1}{V}$, where V is the volume of the direct lattice cell. Substituting this volume value in (2) gives

$$\text{Number of lattice points in the tore} = \frac{2\pi^2}{\lambda^3 \cdot \dfrac{1}{V}} N. \qquad (3)$$

The total number of reflections developed by the crystal is the same as this and can be put in the form

$$\text{Reflections}_{\text{total}} = 2\pi^2 \frac{V}{\lambda^3} N$$

$$= 19.74 \frac{V}{\lambda^3} N. \qquad (4)$$

It is of interest to see how these reflections are distributed on different layer lines. In the first place, layer lines can be produced up to a height equal to the radius of the sphere of reflection. If ζ_1 is the height of the first layer, then the total number of layers which can give rise to reflections is

$$\text{Number of layers} = \frac{\text{Radius of sphere}}{\zeta_1} = \frac{1}{\zeta_1}. \tag{5}$$

Since, from (40), Chapter 8,

$$\zeta = \lambda \frac{1}{T}, \tag{6}$$

where T is the translation period along the rotation axis, (5) may be written

$$\text{Number of layers} = \frac{1}{\lambda} T. \tag{7}$$

The number of reflections per layer is evidently equal to the number of reciprocal lattice points included in that level of the tore. If R is the radius of the circular section of the sphere of reflection at level height, ζ, then it can easily be shown that the area of the section of the tore at this level is $4\pi R$. It can also be easily shown (see pages 235 and 236) that $R = \sqrt{1 - \zeta^2}$, so that the area of the section of the tore at level ζ is $4\pi \sqrt{1 - \zeta^2}$. The number of reflections for this nth level is

$$\text{Number of reflections} = \frac{\text{Area of tore level}}{\text{Area of reciprocal cell base}},$$

$$\text{Reflections}_n = \frac{4\pi \sqrt{1 - \zeta^2}}{a^* \cdot b^* \cdot \sin \gamma^* \text{ (for example)}}. \tag{8}$$

Substituting the values given by (36) and (37), Chapter 6, for a^* and b^*, this becomes

$$\text{Reflections}_n = \frac{4\pi \sqrt{1 - \zeta^2}}{\left(\lambda \dfrac{bc \sin \alpha}{V} \right) \left(\lambda \dfrac{ac \sin \beta}{V} \right) \sin \gamma^*}$$

$$= 4\pi \sqrt{1 - \zeta^2} \frac{1}{\lambda^2} \frac{1}{c} \frac{V^2}{abc \sin \alpha \sin \beta \sin \gamma^*}. \tag{9}$$

It will subsequently be shown (Chapter 18, equation (50)) that

$$V = abc \sin \alpha \sin \beta \sin \gamma^*.$$

Hence, (9) reduces to

$$\text{Reflections}_n = 4\pi \sqrt{1 - \zeta^2} \frac{1}{\lambda^2} \frac{V}{c}$$

$$= 4\pi \sqrt{1 - \zeta^2} \frac{A}{\lambda^2}, \tag{10}$$

where A is the area of the base of the direct lattice cell.

This number of reflections must be multiplied by 2 if the reciprocal cell is base-centered or face-centered (but not if it is body-centered or rhombohedral) because each level of such cells has two lattice points per basal plane unit.

Another interesting bit of information is the number of orders of reflection produced by a given crystal plane reflecting to the zero layer line. If $d_{(hkl)}$ is the spacing of the crystal plane (hkl), the ξ value of its first-order reflection is

$$\xi_1 = \lambda \frac{1}{d_{(hkl)}}. \tag{11}$$

Reflections are produced by all reciprocal lattice points which pass through the sphere of reflection. The sphere has an equatorial diameter of 2, and therefore all reciprocal lattice points in the zero level having ξ values of 2 or less will reflect. The total number of orders which reflect will be the number of ξ's contained in 2, or

$$\text{Number of orders} = \frac{2}{\xi} = \frac{2}{\lambda \dfrac{1}{d_{(hkl)}}} = 2 \frac{d_{(hkl)}}{\lambda}. \tag{12}$$

In these discussions the crystal is assumed to be triclinic. If it is not, then the number of reflection directions is less than the number of reflections because (1) all faces of the same crystallographic form in any layer reflect to the same position on the film, and (2) extinctions may remove certain special reflections. Furthermore, as a practical matter of recording, even a cylindrical film does not intercept all possible reflections; those in a very small region having high ξ values are not recorded because the film does not extend into the region of the pinhole entrance; the film also misses reflections having reflections above a certain ζ value (Fig. 74B). (The last group can be recorded with the aid of a conical film, but the interpretation of films of such shape is more difficult than the interpretation of cylindrical films.)

It is desirable to choose a wavelength such that there will be a sufficient number of reflections to provide a true statistical picture of the systematic absences (extinctions) which determine the space group. According to a relation (4), the total number of possible reflections available from a given crystal varies inversely as the cube of the wavelength employed, so that the total number of data available for study decreases enormously with a small increase in wavelength. Different classes of the data are available for fixing various aspects of the space group study. The variation of these various classes with wavelength is indicated in the following table:

Information	Class of reflections	Formula number	Number of reflections available proportional to
Space lattice type	All hkl	(4)	$\dfrac{1}{\lambda^3}$
Symmetry plane type	$hk0$ type	(10)	$\dfrac{1}{\lambda^2}$
Symmetry axis type	$h00$ type	(12)	$\dfrac{1}{\lambda}$

It will be observed that, for a given crystal and wavelength, the number of data available for the fixing of the space lattice type are relatively numerous, while the number of data for the fixing of the symmetry axis type are relatively scarce. Specifically, the probabilities of having sufficient data for correctly fixing the lattice type, the symmetry plane types, and the symmetry axis types for a given crystal are inversely proportional to the cube, square, and first power, respectively, of the wavelength. For these reasons one might be tempted to use as short a wavelength as possible in order to be sure of sufficient data. Against this desirability, however, are the disadvantages that:

(1) much of this wealth of data is unnecessary, and it carries with it an enormous increase in the labor of indexing, inversely proportional to the cube of the wavelength;

(2) it is difficult to experimentally shield the shorter wavelengths (Mo, Cb, Zr); and, most important of all,

(3) the certainty in indexing the reflections, especially those of great ξ, decreases exceedingly rapidly with decreasing wavelength owing to the increasing density of possible lattice points in the region of the measured ξ coordinate.

In the initial stages of the investigation, one ordinarily does not know the identity periods involved, but the following rough reasoning may suggest an appropriate wavelength: The identity periods of simple crystals are in the vicinity of 6Å.   One ought to have at least three layer lines available for study.   Allowing an additional layer line to escape recording, and substituting these values in (7), gives

$$4 = \frac{1}{\lambda} \cdot 6,$$

from which λ turns out to be about 1.5 Å.   From Table 11, this is the wavelength of copper radiation which would thus be the appropriate radiation to try as a guess for a simple crystal, unless the chemical composition (discussed below) prohibits this radiation.

Specific absorption factor.   A much more important reason for choosing a particular kind of radiation is that the elements present in the chemical composition of the crystal have a tendency to absorb certain specific radiations very highly, and such radiations should be avoided. When this absorption occurs, not only are the diffracted beams enormously weakened, but part of the absorbed radiation is diffusely scattered, causing a fogged background on the film.   The resulting photograph consists of weak diffraction spots on a badly fogged background.

Every radiation has an element which highly absorbs and diffusely scatters it.   These data, together with the wavelengths of the radiations involved, are listed in Table 11.   This table is to be interpreted as follows: Any given target element emits radiation containing three components important for structural investigations.   These are known as the $K\alpha_1$, $K\alpha_2$, and $K\beta_1$ radiations. $K\alpha_1$ and $K\alpha_2$ have nearly the same wavelengths and thus give rise to diffraction spots which are overlapping and unseparated for low glancing angles, and just separated for the higher glancing angles.   The $K\beta_1$ wavelength is somewhat shorter than the $K\alpha$ wavelengths and always produces a diffraction spot which is resolved from the α spots.

The $K\beta$ radiation of an element is highly absorbed by a crystal containing the element one (or two) atomic numbers less than that of the radiating element, and the $K\alpha$ radiation is highly absorbed by, the element two (or three) atomic numbers less than that of the radiating element.   These wavelengths are also absorbed by the elements of immediately smaller atomic numbers than those just mentioned (indicated in the table by an arrow pointing to the elements of lower atomic number), but the absorption decreases with the difference in atomic number.   On the other hand, elements of immediately higher atomic number (in the direction of the tail of the arrows of the last columns of Table 11) are highly transparent to the same

TABLE 11

WAVELENGTHS AND ABSORPTIONS OF USEFUL X-RAY TARGET MATERIALS

Radiating element	Atomic No.	K_{α_2}	K_{α_1}	K_{β_1}	K_α	K_β
					Highly absorbed and badly scattered by	
		The elements above 22 give rise to such weakly penetrating radiations that they are easily absorbed, even by air.				
Ti	22	2.74681Å	2.74317Å	2.5090Å	Sc ↑	Sc ↑
V	23	2.50213	2.49835	2.2797	Sc ↑	Ti ↑
Cr	24	2.28891	2.28503	2.0806	Ti ↑	V ↑
Mn	25	2.10149	2.09751	1.90620	V ↑	Cr ↑
Fe	26	1.936012	1.932076	1.753013	Cr ↑	Mn ↑
Co	27	1.78919	1.78529	1.61744	Mn ↑	Fe ↑
Ni	28	1.65835	1.65450	1.49705	Fe ↑	Co ↑
Cu	29	1.541232	1.537395	1.38935	Co ↑	Ni ↑
Zn	30	1.43603	1.43217	1.29255	Ni ↑	Cu ↑
Ga	31					
Ge	32					
As	33					
Se	34	The elements in this range constitute poor material for x-ray targets because of their physical properties.				
Br	35					
Kr	36					
Ru	37					
Sr	38					
Y	39					
Zr	40	0.78851	0.78430	0.70028	Ru ↑	Y ↑
Cb	41	0.74889	0.74465	0.66438	Sr ↑	Zr ↑
Mo	42	0.712805	0.707831	0.630978	Y ↑	Cb ↑

The elements beyond 42 give rise to highly penetrating radiations which are difficult to shield; they also give rise to large quantities of general radiation in addition to the characteristic K_α and K_β.

radiation. It should be observed that, since the $K\alpha$ and $K\beta$ radiations are differentially absorbed by the same material, it is possible so to select a material that it is highly transparent to the $K\alpha$ radiation but practically opaque to $K\beta$ radiation.   For example, CuKβ radiation is highly absorbed by nickel, according to the last column of the table. Nickel is highly transparent to $CuK\alpha$ radiation, however, as it stands just below the most absorbing metal, Co (second last column).   Thus, nickel foil acts as a filter which passes $CuK\alpha$ but screens off $CuK\beta$ radiation.   All the elements in the last column of Table 11 except Sc can similarly be used as screens to pass only the $K\alpha$ wavelengths of the corresponding radiation in the first column.   Metal foils about 0.001 inch thick are appropriate filters, where practicable.   Filters considerably increase exposure times, however, for the $K\alpha$ is also considerably absorbed, although relatively much less so than the $K\beta$ radiation.

In choosing an appropriate radiation for investigating a crystal of given composition, one should make sure that the target material is such that no element in the chemical composition of the crystal stands in, or shortly above, the place of the greatest absorber in the $K\alpha$ column of Table 11.   The cleanest film (with regard to background scattering) results if no element stands in, or shortly above, the position of maximum absorber *in either the $K\alpha$ or $K\beta$ column*.

Ordinarily, the lighter non-metallic elements, such as O, S, Cl, etc., which form part of the composition of many common crystals, lie so far above the maximum absorber in Table 11 that they do not come up for consideration, and only the heavier metals need be considered in selecting an appropriate wavelength.   For a given metal present in the composition of the crystal, the optimum target material is the metal itself (or the metal of next lower atomic number, if $K\beta$ scattering is not objectionable).   The optimum condition can, of course, be realized only for the relatively few suitable target metals listed in Table 11.   Other target materials are unsuitable for the reasons mentioned in the corresponding places in Table 11.

CHOICE OF CRYSTAL SIZE

Elementary absorption theory.   With the aid of elementary absorption theory, the optimum size of a crystal to use in any given case may be predicted.   Let I represent the intensity of the beam of x-rays.   As the radiation passes through a small thickness, dt, of an absorbing medium, the intensity is decreased by an amount dI.   The decrease is proportional to the thickness, i.e.,

$$-dI \sim dt. \tag{13}$$

The absolute amount of decrease, dI, is also proportional to the original intensity, I, and so

$$-dI \sim I. \tag{14}$$

Combining (13) and (14) gives

$$-dI \sim I \, dt. \tag{15}$$

A constant of proportionality, μ (the linear absorption coefficient expressed in cm.$^{-1}$ units), being introduced, (15) takes the form

$$-dI = \mu \, I \, dt,$$

or

$$-\frac{dI}{I} = \mu \, dt. \tag{16}$$

Integration of this gives

$$-\log I_0 + \log I = -\mu t,$$

or

$$\log \left(\frac{I}{I_0}\right) = -\mu t. \tag{17}$$

This is usually expressed in the exponential forms:

$$\frac{I}{I_0} = e^{-\mu t} \tag{18}$$

or

$$I = I_0 e^{-\mu t}. \tag{19}$$

The meaning of (19) is that, after the radiation passes through a thickness, t, the value of the intensity, I, is equal to the initial intensity, I_0, times the factor $e^{-\mu t}$, where e is the base of the natural logarithms and μ is the linear absorption coefficient.

Optimum crystal size. As the crystal thickness, t, increases, the absorption of the x-rays cuts down the amount of radiation transmitted through the crystal according to (19). At the same time, the total amount deviated to a diffraction spot on the film increases with the thickness, and approximately as the volume of the crystal i.e., as t^2. Those reflections traveling through the longest path in the body of the crystal are thus cut down by absorption proportional to $I_0 e^{-\mu t}$, and built up proportional to t^2. This may be expressed in symbols as

$$I_{\text{longest path}} = I_l \sim I_0 e^{-\mu t} t^2, \tag{20}$$

or

$$I_l = K \, I_0 e^{-\mu t} t^2, \tag{21}$$

where K is a proportionality constant.

This net intensity is a maximum when the first derivative is zero. The first derivative with respect to thickness t is

$$\frac{dI_l}{dt} = K\,I_0\,(e^{-\mu t} \cdot 2t - t^2\mu e^{-\mu t})$$

$$= K\,I_0 e^{-\mu t}\, t\,(2 - t\mu).\tag{22}$$

A maximum is attained when the last term on the right of (22) is zero,

$$2 - t\mu = 0,$$

under which conditions,

$$t = \frac{2}{\mu}.\tag{23}$$

So far as the longest diffraction path in the crystal is concerned, therefore, the amount of diffracted radiation increases with thickness up to a thickness $\frac{2}{\mu}$, after which the intensity starts to decrease. The optimum cross-sectional diameter of a crystal for diffraction work in which the entire crystal is immersed in the x-ray beam is thus roughly $\frac{2}{\mu}$. If a thicker crystal is used, the reflections traversing the longest internal paths are reduced, but it should be noted that the total amount of diffracted radiation continues to increase for reflections which traverse somewhat shorter paths, as well as those which are direct reflections from the external surfaces and which do not need to be transmitted through the body of the crystal. The intensities of the last class of reflections increase indefinitely with thickness of crystal. The thickness $\frac{2}{\mu}$, however, gives those reflections which are most reduced by absorption their maximum chance of recording. Use of a *much thicker* crystal might suppress this class of reflection altogether.

It should be noted that, if other assumptions are made about the dependence of building-up of intensity on thickness, the results will come out about the same order of magnitude. Thus, if a thin slip of crystal is used, not surrounded by the x-ray beam, the building-up of intensity is proportional to t, not t^2. The optimum thickness then comes out $\frac{1}{\mu}$, not $\frac{2}{\mu}$. These results are so close together that it is obvious that any approximation error made in the assumption regarding the dependence of the building-up of intensity on thickness will not vitiate the conclusion that the optimum thickness is of the order of $\dfrac{2}{\mu}$.

Calculation of linear absorption coefficients. The linear absorption coefficient can be calculated from a knowledge of the chemical composition of the crystal, its density, and a table of *mass absorption coefficients*, which are functions of both chemical composition and the wavelength of the x-rays employed in the investigation. The mass absorption coefficients of the elements for certain important wavelengths are collected in table form in Volume 2 of *International Tabellen zur Bestimmung von Kristallstrukturen*, pages 577–578. The advantage of using the mass absorption coefficient $\dfrac{\mu}{\rho}$, over using the linear absorption coefficients μ, is that the former is a constant for an element and does not depend on state of aggregation, chemical combination, etc. The linear absorption coefficient for use in (23) is simply calculated from the mass absorption coefficient by the relation

$$\mu = d\left(\frac{\mu}{\rho}\right), \tag{24}$$

where d = the density of the crystal. For a chemical compound, or mixture, $ABC\cdots$, the linear absorption coefficient is calculated from the individual mass absorption coefficients of the elements by the relation

$$\mu = d\sum p\left(\frac{\mu}{\rho}\right)$$

$$= d\left\{p_A\left(\frac{\mu}{\rho}\right)_A + p_B\left(\frac{\mu}{\rho}\right)_B + p_C\left(\frac{\mu}{\rho}\right)_C \cdots\right\}, \tag{25}$$

where p = the proportion of the element present in the compound. This is, of course, $\left(\dfrac{\%}{100}\right)_A$, $\left(\dfrac{\%}{100}\right)_B$, $\left(\dfrac{\%}{100}\right)_C$, etc. For minerals, the weight per cent is given for each species under " chemical composition " in *Dana's Textbook of Mineralogy*.

As a concrete illustration, this computation for the crystal marcasite, FeS_2, is as follows: The analysis and density of marcasite, given by *Dana's Textbook*, are:

$$
\begin{array}{ll}
\text{Fe} \quad 46.6\% & \\
\text{S} \quad \underline{53.4\%} & \qquad d = 4.9 \\
\phantom{\text{S} \quad} 100.0\% &
\end{array}
$$

The computation of (25) may be tabulated:

	p	$\dfrac{\mu}{\rho}$	$p\left(\dfrac{\mu}{\rho}\right)$
Fe	$\dfrac{46.6}{100}$	38.3 cm.²/gm.	17.85 cm.²/gm.
S	$\dfrac{53.4}{100}$	10.03	5.36

$$\Sigma p\left(\frac{\mu}{\rho}\right) = 23.21 \text{ cm.}^2/\text{gm.}$$

$$\mu = d\Sigma p\left(\frac{\mu}{\rho}\right) = 4.9 \times 23.21 = 114 \text{ cm.}^{-1}.$$

The optimum thickness for rotation photographs of marcasite crystals, according to (23), is

$$t = \frac{2}{\mu}$$

$$= \frac{2}{114}$$

$$= .019 \text{ cm.}$$

SELECTING, MOUNTING, AND ADJUSTING THE CRYSTAL

Since the cross section of the intense region of the x-ray beam is ordinarily about 1 mm. in diameter, it is desirable to select a crystal which does not exceed this diameter, even if the optimum thickness as calculated from the absorption coefficient is greater than this. The crystal selected for the investigation should be one whose surface morphology is sufficiently well developed to permit the crystal to be oriented for rotation about the desired axis or axes. The crystal faces should be smooth and bright so that the axis of the identity period can be adjusted to the rotation axis (as described beyond) with the aid of a reflecting goniometer.

The crystal is to be attached to an adjusting device for final exact orientation purposes. In order to make the attachment, the crystal is first stuck to a mounting pin, Fig. 96. This is made as follows: a short brass pin is cut of $\frac{3}{32}$-inch brass rod, and the cutting burrs rounded off with a file. The pin, held with pliers, is heated in a Bunsen flame, and a dab of hot piceïn wax is melted onto one end. Into this piceïn lump is thrust a short length of hollow glass fiber made by pulling out a piece of glass tubing. After this the wax should be gently reheated to cause it to flow about and adhere to the glass fiber. The finished mounting pin appears as in Fig. 96A. The end of the glass fiber is next wet with shellac, " Duco " household cement (thinned with amyl acetate), or glue. (The less there is of this the better, since it diffusely scatters x-rays and causes fogging of the film.) While the manipulator watches the process with the aid of a binocular microscope, the sticky end of the glass fiber is touched to the appropriate face of the crystal. If this has been done properly, the crystal

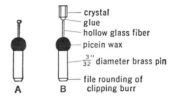

FIG. 96.

is now stuck to the fiber in almost correct orientation. The orientation may be improved by judiciously pushing against the body of the crystal with a mounted needle. The lower end of the brass pin is now set into the pinhole of the adjusting crystal holder (Fig. 97B). It is best to make a rough adjustment working under the binocular; in this way the crystal can be watched while the pin is rotated in its hole until the orientation of the crystal faces is correct with respect to the adjusting arcs of the adjusting crystal holder (see below). The set screw is finally tightened on the pin, and the whole put away until the shellac or other adhesive becomes hard. It is inadvisable to use the crystal for at least a day's time because the crystal invariably tips out of orientation as the result of the flow of the adhesive during this period.

A simple device for adjusting the crystal axis to parallelism and coincidence with the rotation axis is shown in Fig. 97A. A more elaborate device is shown in Fig. 97B. Two arcs, at right angles to each other and centered at the crystal, permit the crystal axis to be tipped in any direction, and two translation sledges at right angles to each other permit the crystal axis to be moved parallel to itself. These four motions are controlled by four screws. The entire device has a base which may be screwed onto a reflecting goniometer for the purpose of adjusting the crystal optically. After this has been accomplished, the adjusting device bearing the crystal may be removed bodily and screwed onto the rotation axis of the rotating crystal apparatus, without loss of adjustment.

If the investigator is not accustomed to adjusting crystals on the reflecting goniometer, the following advice should be observed: Mount the crystal so that one of the reflecting faces to be used in the adjustment is parallel with one of the arc adjusting screws. A simple illustration of this is shown in Fig. 97*C*, which is a diagrammatic top view of the crystal on the adjusting holder. Before the adjusting device is attached to the reflecting goniometer, and when the crystal pin is being given a preliminary setting under the binocular, some face, say 1, should be made approximately parallel to an adjusting screw, say *I*.

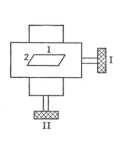

FIG. 97*A*. Simple device for adjusting crystals. This is based on a ball-and-socket joint for angular adjustment, and on a sliding plane for translation adjustment.

FIG. 97*B*. More elaborate device for adjusting crystals. This is based on two arc motions for angular adjustment, and on two sledge motions for translation adjustment.

FIG. 97*C*.

When the holder is placed on the goniometer, a reflection should be first obtained from this face 1, which should be brought parallel with the rotation axis by bringing this reflection to the cross hair of the telescope *with the aid of the other screw II*, which is at right angles to face 1. Next, a reflection should be obtained from face 2, and its reflection should be adjusted to the cross hair *with the aid of screw I*. The reason for this is that this second adjustment, namely face 2 with screw *I*, does not disturb the first adjustment of face 1, because 1 is parallel with screw and arc *I*. Actually, since 1 has been set parallel to *I* only approximately by eye, screw *I* will disturb the adjustment of face 1 a trifle, so the entire adjusting sequence should be repeated once or twice more until both adjustments are perfect. Unless this simple strategy is resorted to, the investigator is likely to waste a great deal of time in adjusting the crystal. There are other systematic ways of adjusting it, but if the student is unfamiliar with crystal adjustment he is advised to learn this method first.

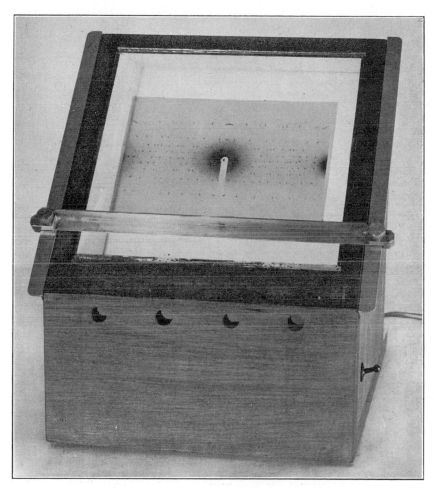

Fig. 98. Viewing stand for examining x-ray photographs.

EXAMINING THE FILM

The best background for examining an x-ray film is illuminated opal glass. An efficient device for producing this background and for holding and examining the film is shown in Fig. 98. Figure 99 shows the cross section of this device. The following points in the design should be noticed in duplicating the frame:

The film itself is laid against a piece of clear glass, and the opal glass held some distance behind this. The reason for this is twofold: first, any imperfection or specks in the opal glass are not directly against the film, and parallax prevents mistaking them for x-ray reflection spots on the film; and second, the air space between the two pieces of glass acts as a heat insulator. The lamps themselves (two 100-watt bulbs or a fluorescent tube) are placed outside the frame enclosure so that the heat is easily dissipated by convection. The lamps of a well-designed frame may be left burning for hours without heating the frame to an uncomfortable temperature. The film can

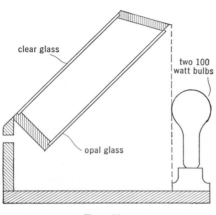

clear glass

two 100
watt bulbs

opal glass

FIG. 99.

be adjusted to height with the aid of an adjustable film rest. This permits the use of special measuring frames, mentioned in later chapters, for precision measurements and for Weissenberg photographs.

The ξ, ζ coordinates of the spots on the rotation photograph are read by placing a transparent film copy of Fig. 79 back of the x-ray film. The most convenient method to follow is to adjust the x-ray film to a chart film, and then to fasten the two together with the aid of paper clips. The pair is then placed on the viewing stand and coordinates read without further adjustment.

For measurements of the coordinate, y, of a layer line, a steel millimeter scale may be laid directly against the x-ray film from $-y$ to $+y$ and the double distance measured. A much more accurate method is to use a vernier caliper. For this purpose the author uses one designated No. 12, Glogau & Co., Chicago. This has a range of 120 mm. and a vernier reading directly to 0.1 mm. The distances from spot to spot, or from layer line to layer line, are conveniently measured with the pointed jaws of this caliper.

In order to obtain even more accurate values of the layer spacing, the following procedure should be used: Before taking the rotating-crystal photograph, the *end* of the crystal is adjusted to the *center* of the x-ray beam. After exposure and development, the photograph is placed in the film measuring device shown in Fig. 229. The movable frame of this device has a fine hair line, which should be aligned with the parts of the layer lines representing the same end of the crystal which was adjusted to the center of the x-ray beam. A scale and vernier permit making accurate readings of the y film coordinates of these positions, and these lead to accurate values of $2y_n$, from which accurate values of t can be computed with the aid of (6), Chapter 5.

LITERATURE

Rotating-crystal apparatus

J. D. BERNAL. A universal x-ray photogoniometer.
  I. *J. Sci. Instruments*, **4** (1927), 273–284.
  II. *J. Sci. Instruments*, **5** (1928), 241–250, 281–290.
  III. *J. Sci. Instruments*, **6** (1929), 314–318, 343–353.

HAROLD P. KLUG. An improved device and method for adjusting the cylindrical camera of the Bernal universal x-ray photogoniometer and similar instruments. *Rev. Sci. Instruments*, **3** (1931), 439.

RALPH W. G. WYCKOFF. *The structure of crystals.* Second edition. (Chemical Catalog Company, New York, 1931.) Pages 106–111.

ROBERT B. HULL and VICTOR HICKS. A universal x-ray photogoniometer. *Z. Krist. (A)*, **96** (1937), 311–321.

Preparation of specimens

H. KERSTEN and W. LANGE. A method for preparing crystals for rotation photographs. *Rev. Sci. Instruments*, **3** (1932), 790–791.

X-ray wavelengths

MANNE SIEGBAHN. *Spektroskopie der Röntgenstrahlen.* (Julius Springer, Berlin, 1931.) Especially pp. 171–179.

Table of mass absorption coefficients

Internationale Tabellen zur Bestimmung von Kristallstrukturen. (Gebrüder Borntraeger, Berlin, 1935.) Vol. 2, pp. 577–578.

CHAPTER 10

OSCILLATING–CRYSTAL PHOTOGRAPHS AND THEIR INTERPRETATION

In the preceding chapter, attention was called to the fact that the lack of certainty in indexing rotating-crystal photographs was due to the impossibility of determining, from this type of photograph, the third cylindrical coordinate, ω, of the reciprocal lattice point producing the reflection.   It is possible, however, to reduce the amount of uncertainty by limiting the number of points allowed to reflect for each photograph.   This is done by limiting the crystal rotation so that only the reciprocal lattice points between certain selected limits are permitted to come in contact with the sphere of reflection to produce spots on the film.   This amounts to the predetermination, between limits, of the value of the ω coordinate of each film spot.   The uncertainty in the indexing of films made by this method, over that in which ω may have all possible values, is consequently much reduced.   The reduction of the range of rotation is accomplished by causing the crystal to rotate back and forth, i.e., to *oscillate*, about the axis of rotation through a predetermined angle.   There are several mechanisms for accomplishing this oscillation or limited rotation.

MECHANISMS FOR OSCILLATING THE CRYSTAL ABOUT THE ROTATION AXIS

The customary way of causing a crystal to oscillate about the rotation axis is by means of a cam-and-follower mechanism.   Such a device is illustrated in Fig. 100A.   A heart-shaped cam is rotated by a motor; a follower arm, held against the cam and attached to the crystal-rotation shaft, causes the crystal to undergo alternate clockwise and counterclockwise rotation through a limited angle fixed by the dimensions of the cam-and-follower mechanism.   This mechanism has several disadvantages:

1. It is difficult to change from oscillation to complete rotation without reassembling the entire driving machinery in a new way.

2. Only a single definite oscillation range can be had with a given cam.

3. The making of a cam is a hand-plotting, hand-cutting, and hand-finishing job, so the cam cannot give a truly uniform speed of rotation to the crystal.   The perfection of the uniformity of rotation is dependent on the care taken in the above-mentioned hand processes.

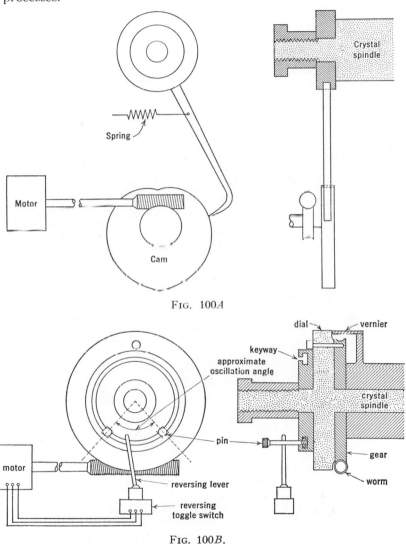

FIG. 100A

FIG. 100B.

A much neater and more flexible arrangement for causing the crystal to oscillate back and forth about the rotation axis is illustrated in Fig. 100B.   In this mechanism, a motor is permanently geared to

FIG. 101*A* (above). Oscillating-crystal apparatus in operating position.

FIG. 101*B* (below). Oscillating-crystal apparatus in position for attaching crystal. Note that the whole upper part of the instrument has been translated to the left, thus permitting one to look down the pinhole system at the crystal. The camera and the direct-beam stop (seen standing at the left) have been removed.

190

the crystal-rotation axis shaft. The gearing is preferably of the worm-and-pinion type, because this imparts an exceedingly uniform rotation speed to the crystal. After rotation has proceeded to a predetermined limit, an arm, attached to the crystal-rotation axis, throws a switch and reverses the motor. The motor then rotates the crystal shaft in the reverse direction until a second arm throws the switch, causing the motor to reverse again and rotate the crystal in the original direction to the limit set by the first arm, etc. In this way the crystal shaft is oscillated back and forth through an angle roughly equal to the angle subtended by the pins from the center of the crystal-rotation axis. The exact oscillation angle is actually this angle plus a constant

FIG. 101C. Spindle assembly of oscillating-crystal apparatus.

FIG. 101D. Motor and switch assembly of oscillating-crystal apparatus.

dependent on the throwing lag of the switch. This constant is large for ordinary toggle switches, but " micro switches " may now be obtained on which the throwing lag is vanishingly small. The oscillation angle can be changed instantly by adjusting the positions of the pins. The crystal orientation, ω, at any moment, is always given by a vernier reading against a dial attached to the crystal-rotation axis.

Photographs of an oscillating-crystal apparatus based upon this oscillation mechanism are shown in Figs. 101A and 101B. Photographs of several units of the apparatus are shown in Figs. 101C and 101D. In Fig. 102, a diagrammatic cross section is given of the mechanism for producing oscillating motion. The adjusting crystal holder screws onto the flat base, a, which rotates with shaft b. The motor-driven worm, c, drives the shaft through the gear d. The arm, e, trips the micro switch and reverses the direction of the motor. The angular opening between tripping arms, e, can be adjusted by loosening screw f and resetting e at another angular position on its drum, g. In this way the oscillation range can be controlled. The zero of the oscillation range is adjusted to its desired position by loosening milled nut h, and then rotating drum g to its desired position.   The angular position of the crystal shaft is given at all times by dial l and vernier m.

To use the apparatus as a simple rotating-crystal apparatus, screw f is loosened and the two switch-throwing arms, e, are completely removed.   By doing this, the reversing switch is not actuated, thus permitting the motor to run continuously in one direction.

When the crystal is attached to the end of the shaft at a, its orientation has already been adjusted on the optical goniometer, as described on page 183.   It still needs translation adjustment, however, to center it in the x-ray beam.   While this adjustment is being made,

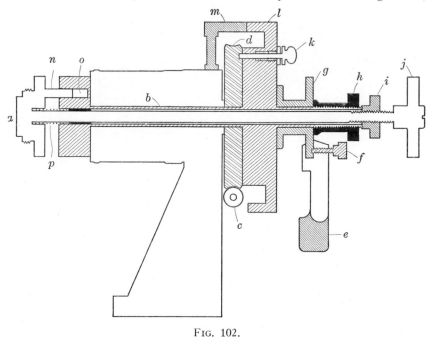

Fig. 102.

the operator looks down the pinhole system at the crystal.   The left-right translation to the beam center is then made by turning milled nut i, which permits the crystal to undergo only a horizontal translatory motion against spring p; the rotation of the shaft is prevented by pin n working in hole o.   The translation adjustments normal to the rotation axis have already been roughly made on the optical goniometer.   To refine these, the crystal-rotation shaft b is freed from pinion d by withdrawing pin k.   The crystal may then be rotated, free of the drive mechanism, by turning knob j.   If the crystal appears to move from the center of the field, as one looks down the pinhole system† and rotates the shaft, the two translation screws of the adjusting crystal holder are adjusted until the crystal remains motionless on rotation. When this adjustment has been completed, pin k is again slipped into its hole, and the crystal again has its original orientation.

The camera for this apparatus has already been described in Chapter 9, Figs. 91 and 92.

CONVERGENT RADIATION

Another scheme for producing the equivalent of crystal oscillations is to flood a stationary crystal with x-radiation converging within a limited aperture toward it from a broad source. This is the geometrical equivalent of oscillating the x-ray beam instead of the crystal. The convergent radiation is taken from an x-ray tube having a broad focal spot. With this method a plane in a position to reflect does so continuously. In the ordinary oscillation method, the plane produces a reflection only during the short time taken by its reciprocal lattice point to pass through the surface of the sphere of reflection. The convergent radiation method thus has the advantage of producing photographs in very short time, as compared with the oscillation method. It has the disadvantage that one does not know which direction, within the aperture of the incoming radiation, is responsible for producing a given reflection. It is therefore not possible to determine the x coordinate of any spot on the film, for the origin of x coordinates is at the point where that particular direct ray meets the film.

THE OSCILLATING-CRYSTAL PHOTOGRAPH

In order to index an oscillating-crystal photograph, the reciprocal lattice must already have been determined. This is accomplished by the ordinary rotating-crystal method as discussed in Chapter 8. The reciprocal lattice may then be graphically reconstructed and a plan made of any level. Figure 103A shows an elevation of a reciprocal lattice and the sphere of reflection. Figure 103B shows a plan of the zero level, and Fig. 103C shows a plan of the second level. In these plans, the full circle is the section of the sphere of reflection on that particular level; its radius, R_0, R_2, etc., and location on any level, can be obtained graphically from the elevation, Fig. 103A.

† To facilitate vision, a lens of about 20 diopters' power is clipped over the x-ray tube side of the pinhole system while the crystal is being adjusted. This enables the eye to focus directly on the crystal, which then appears framed in the circular aperture of the last hole in the pinhole system.

Suppose that the crystal is rotated counterclockwise through a limited oscillation angle.   All reciprocal lattice points are then rotated

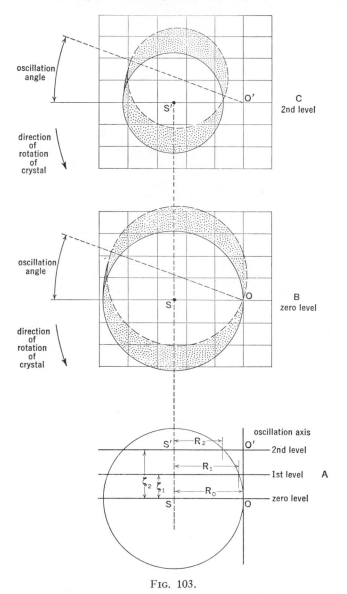

Fig. 103.

counterclockwise about the origin through this angle.   The same effect would be obtained if the crystal had remained stationary and the x-ray beam and film had been rotated together about the origin through the same angle *but in the opposite direction.* This latter situation is easier to plot than the former. The direct beam, of course, carries with it the sphere of reflection. The positions of the sections of the sphere of reflection at the end of the oscillation motion are indicated in Figs. 103*B* and *C* by dashed lines. All lattice points which have passed through the circles have been in a position to reflect. Only such lattice points have reflected, therefore, as are *included in the crescent-shaped areas between the circles, plus the shaded area farthest from the rotation axis plus, for n-layers, an area near the rotation axis too small to illustrate (outside the end positions of the circles).*

Note, in Figs. 103*B* (and *C*), that the distribution of lattice points within the upper and lower crescents is not the same. This means that the positions of the reflections on the left and right side of a film are different. Oscillation films displaying this dissimilarity in left and right sides are shown in Figs. 109*A*, *B*, and *C*. Under special conditions the left and right may be the same, for example, if the oscillation range is symmetrically disposed on each side of a symmetry element in the crystal. The matter of symmetry will be discussed later.

INDEXING AN OSCILLATION PHOTOGRAPH

An oscillation photograph is indexed in the same way as a rotating-crystal photograph, except for restrictions imposed by the limitation in number of planes which can reflect. In order to index a photograph, one must know the range of oscillation and the location of this range with respect to the orientation of the reciprocal lattice. A good way to arrange this part of the experimental work is to select some important crystal face present on the experimental crystal, and place this initially parallel with the incoming x-ray beam. The lattice points representing orders of reflection from this plane, then, all have φ coordinates of 90°, taking the direct beam coordinate as zero. (The Υ values of the reflections of these orders — their angles from the direct beam as subtended from the center of the circle — have a series of increasing values, starting at zero for the direct beam, i. e., zero-order reflection.) The crystal is then allowed to rotate counterclockwise and return, through a small oscillation angle, usually 10–15°.

All data are now available for the preparation of a series of scale drawings of Figs. 103*A*, *B*, and *C*, for the actual case, assuming that the identity periods have already been determined. A projection of the several levels of the reciprocal lattice on the zero level is first laid out on tracing cloth. No attention is paid to the possibility of lattice centering, if already discovered, for this will simply make itself apparent in missing reflections, i. e., missing lattice points. This reciprocal lattice net, without missing points due to the various extinctions, may be termed a *blank reciprocal lattice*. For the scale of the reciprocal lattice, it is convenient to use unity = 10 cm.; i. e., one unit of the cylindrical coordinates, ξ, ζ, or the reciprocal lattice coordinates, a^*, b^*, c^*, is plotted as 10 cm. The diameter of the sphere of reflection, which is 2, is thus represented as 20 cm., and it can sweep out a tore of reflection of a maximum diameter of 40 cm., so no more of the reciprocal lattice plan need be drawn than can be included in a circle 40 cm. in diameter.

From an elevation like Fig. 103A, drawn to scale for the particular crystal at hand, the radii and centers of the circles representing the sphere of reflection at various reciprocal lattice levels may be found. These circles in their initial and final positions are then all plotted

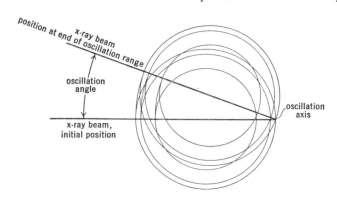

FIG. 104.

together on a second single sheet, Fig. 104, and placed under the tracing-cloth plot of the reciprocal lattice. The origin lattice point is permanently fixed to the rotation-axis position of the lower sheet by a pin. The beam direction (lower sheet) is then free to rotate to any initial oscillation position. The shaded zones, Fig. 103, immediately indicate the lattice points which could have reflected during the recording of that particular photograph.

To index the photograph, it is placed over a Bernal chart against an illuminated background, and the ξ values of each spot are listed for each layer and segregated for right and left reflection. Then a pair of dividers is opened to each of these values (using the scale unity = 10 cm.); one point is placed at the origin of the reciprocal lattice plot, and the other is swept around to find the reciprocal lattice point of identical ξ coordinate. This much is the same as the indexing of rotating-crystal photographs.   In this instance, however, only those points come up for possible consideration which lie in the appropriate level and appropriate right or left reflection of the shaded area of the lower sheet.   In this way, the choice of possible reciprocal lattice points is highly restricted.

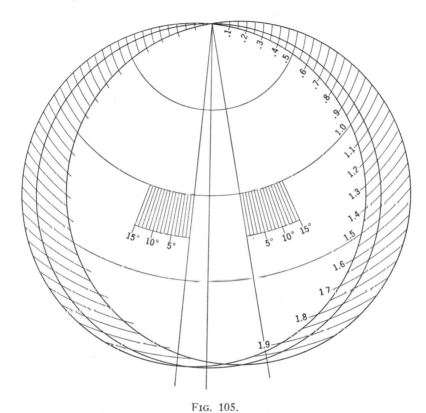

FIG. 105.

Figure 105 shows a template used by Dr. I. Fankuchen to facilitate the location of points on the zero level of the reciprocal lattice.   Zones of permitted reflections are outlined for both 5° and 10° oscillations, and within these zones a scale of ξ is engraved.   The template is made of Celluloid.

An excellent systematic way of filing the information that a reflection has been detected is to write its indices next to the reciprocal lattice point as soon as the reciprocal lattice point which gave rise to the reflection has been identified.   A convenient plan of quickly distinguishing the permissible reflecting areas of the several levels from one another, as in Fig. 104, is to shade them with different-colored crayons.

It is desirable to have as representative a set of reflections as possible for deciding the space group of the crystal being investigated. Accordingly, one single oscillation photograph is not enough. A series should be prepared, the second range starting where the first ended, with, perhaps, a slight overlap to avoid missing points at the ends of the oscillation range. The sum of the oscillation ranges need not exceed 90° for an orthorhombic crystal and may be reduced for more symmetrical crystals.

INDEXING REFLECTIONS FROM CRYSTALS WITH VERY LARGE CELLS

If a crystal has a very large cell, then there is a special oscillation technique by which its reflections in the low-θ range can be easily indexed on inspection.† The theory of this method is illustrated in

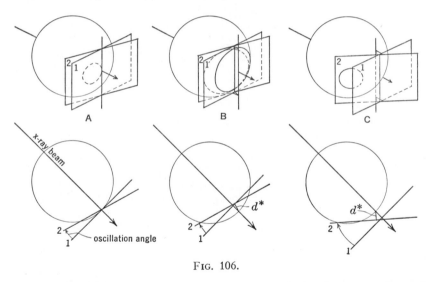

Fɪɢ. 106.

Fig. 106. In order to make use of this method, the densest reciprocal lattice net is placed initially normal to the x-ray beam. Consider, first, the particular plane of this system which contains the rotation axis, i.e., the zero plane. In the initial position 1, Fig. 106A, this plane is tangent to the reflecting sphere. If the crystal is allowed to oscillate from this position to position 2, then the plane intersects the sphere in

† This has been especially developed by Bernal and his school in their investigation of proteins, which have very large cells.

a small circle, Fig. 106A.   During the oscillation, the reciprocal lattice points of the plane contained within this circle have passed through the sphere and reflected.   Now, in the small, low-θ region near the center of the film, the reciprocal lattice plane projects from the center of the sphere to the surface of the film with little distortion.   The film will therefore contain a small circular region within which are recorded reflections arranged as they were in the reciprocal lattice plane, with but slight distortion.   This can be seen in the oscillation films shown in Fig. 107.   Note that this reflection grid can be easily indexed on inspection if something is known of crystallographic directions within it. If only one or two lattice points occur within the circle of Fig. 106A, then the photograph is not very useful.   The method is useful in proportion to the number of reciprocal lattice points which are located within the dashed circle.   This requires the reciprocal cell of the crystal to be very small.   Note that, since the edge a^* of the reciprocal cell is proportional to λ/a, the reciprocal cell edge is small if the direct cell edge is large, and also if the wavelength is small.   Speaking specifically, the reciprocal cell is about small enough to make this method come up for consideration with molybdenum radiation if the direct cell edges contained within the plane are about 10 Å, while if copper radiation is to be used, the cell edges ought to be 20 Å or longer.

It has been pointed out that, in the immediate region of the center of the film, the x-ray photograph consists of a pattern of spots which is an only slightly distorted projection of the points on the zero plane of the reciprocal lattice.   It is therefore apparent that the lengths of the reciprocal cell axes and the interaxial angle can be approximately determined from appropriate measurements of the geometry of the spots on the photograph.   Since the reciprocal cell geometry can be easily transformed into direct cell geometry, this provides an alternative method of determining the approximate direct cell constants which is very convenient.

So far, only the zero plane has been considered.   Figure 106B shows how planes nearer the x-ray source than the zero plane intersect the sphere during oscillation.   The lattice points which pass through the sphere lie within two crescents whose boundaries are the two circles of intersection of the sphere with the plane in its two end-oscillation positions.   The lattice points within such crescentic areas can be seen recorded as reflections in Fig. 107A.   Since the crescentic areas do not contain the reciprocal lattice point, 000, they are less easy to index. Note, however, that the crescentic areas contain lattice points along the two reciprocal lattice axes.   In Fig. 107, for example, the vertical axis is the a^*-axis and the horizontal line is the b^*-axis.   The indices

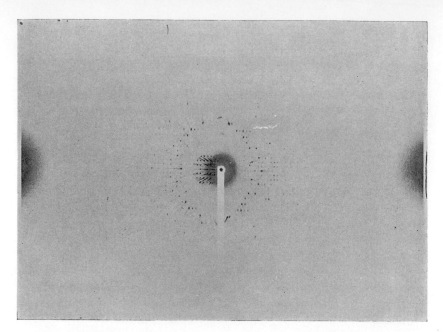

FIG. 107*A*. Oscillating-crystal photograph (recorded with cylindrical camera) of small angular range, and made with a crystal having a large cell. Note the central record of the zero reciprocal lattice plane confined within a circular area. (Compare with Fig. 106*A*.) The record of the first reciprocal lattice plane toward the x-ray source is confined to two crescentic areas. (Compare with Fig. 106*B*.)

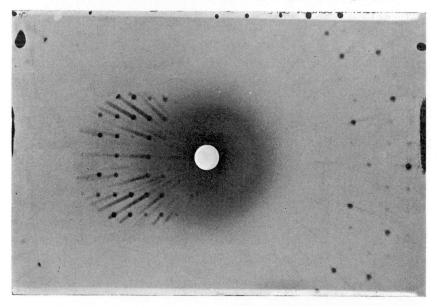

FIG. 107*B*. Same as Fig. 107*A*, but recorded on a flat film placed about 9 cm. from the crystal. (Berthierite, $FeSb_2S_4$, orthorhombic; *a*-axis oscillation, *c*-axis initially parallel with beam; Mo$K\alpha$ radiation from hot-cathode x-ray tube, filtered through zirconia. The radial streaks, which are of the same nature as those shown in Fig. 87, are due to general radiation from the sputtered tungsten of the filament, and are typical results of the "dirty" radiation from hot-cathode x-ray tubes.)

within the first crescentic areas are therefore $hk1$.   The reciprocal cell points $h01$ lie along the vertical central line, and the reciprocal lattice points $0k1$ lie along the horizontal central line.   With this information, all points lying within the crescentic areas can be indexed.

Figure $106C$ shows how a parallel reciprocal lattice plane, spaced a distance d^* from the zero plane and away from the x-ray source, gives rise to its reflections.   In order to have such a plane make contact with the sphere, it is necessary to tip the crystal beyond a definite limiting angle before it becomes tangent to the sphere and reflects.   It can easily be shown that this angle is $\omega = \cos^{-1}(1 - d^*)$.

When this method is used, it is necessary that the records of the several parallel planes do not overlap on the film.   If the reciprocal cell axes are known in advance, the oscillation angle which will give no overlap can be determined graphically.   Overlap for the zero plane and the first plane spaced a distance d^* away from it and toward the x-ray source occurs at $\omega = \sin^{-1}\sqrt{\dfrac{d^*}{2}}$, but overlaps between higher-order planes develop at smaller angles than this.   Overlap between the zero plane and the first plane away from the x-ray source occurs at larger angles.   A series of oscillations can be prepared in which the oscillation ranges are small and in which no overlaps occur.   In this way a substantial portion of the crystal's reflections can be indexed by inspection.

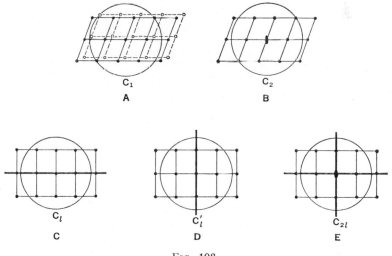

Fig. 108.

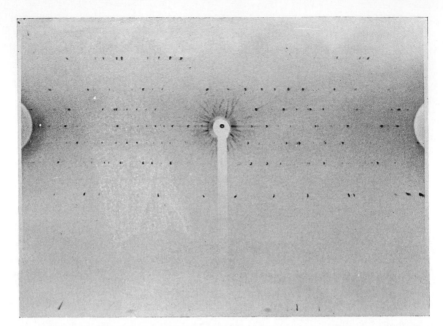

Fig. 109A. Oscillation photograph displaying no symmetry. (Realgar, AsS, monoclinic; c-axis oscillation, random oscillation range location.) All photographs of
Fig. 109 were made using CuKα radiation filtered through nickel foil.

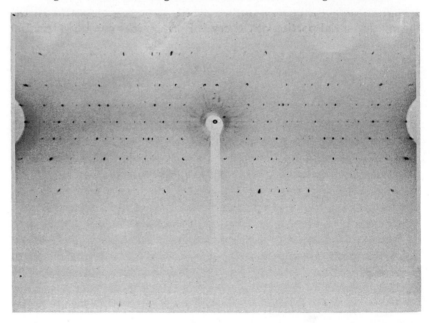

Fig. 109B. Oscillation photograph displaying a twofold rotor. (Realgar, AsS,
monoclinic; c-axis oscillation with b-axis parallel to beam in center of oscillation
range.)

202

Fig. 109C. Oscillation photograph displaying a horizontal reflection line. (Berthierite, $FeSb_2S_4$, orthorhombic; c-axis oscillation, random oscillation range location.)

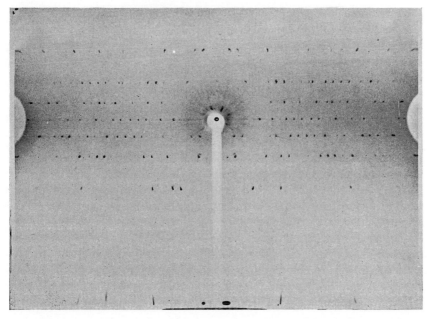

Fig. 109D. Oscillation photograph displaying a vertical reflection line. (Realgar, AsS, monoclinic; c-axis oscillation with (010) parallel to beam in center of oscillation range.)

203

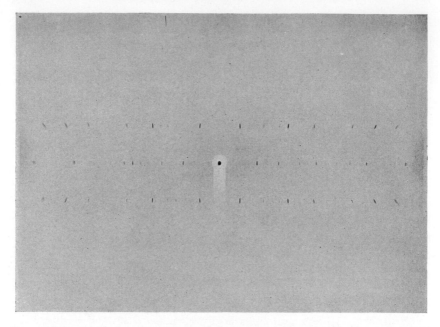

FIG. 109E. Oscillation photograph displaying the symmetry C_{2l}, i. e., a twofold rotor at the intersection of a horizontal and a vertical reflection line. (Berthierite, $FeSb_2S_4$, orthorhombic; c-axis oscillation with b-axis approximately parallel to beam in center of oscillation range.)

SYMMETRY INFORMATION DETERMINABLE FROM OSCILLATION PHOTOGRAPHS

Easily determinable symmetry. A certain amount of symmetry information can be derived from oscillation photographs. A more systematic discussion of symmetry will be given presently, but two bits of information which can be obtained from oscillation photographs very readily will be mentioned at once.

The most dependable, easily obtained, and obvious information which can be obtained from a single photograph is whether or not the centrosymmetrical crystal class of the crystal has a symmetry plane normal to the rotation axis or not. If not, then the upper and lower halves of the reciprocal lattice, Fig. 108A, strike different arrays of points, and, as a consequence, the upper and lower halves of the oscillation photograph, Fig. 109A, have different characters, i. e., either different spot locations or intensities or both. If there *is* a symmetry plane normal to the rotation axis, Fig. 108C, then the sphere of reflection strikes identical point sequences above and below the equator, and the oscillation photograph, Fig. 109C, is symmetrical with respect to the zero layer line.

The second most easily obtained bit of information requires two photographs. If the rotation axis is parallel with an n-fold axis of symmetry, then oscillation photographs taken at ω intervals of $\dfrac{360°}{n}$ will be identical, because the sphere of reflection sweeps through symmetrically equivalent lattice point fields after this interval. If such an axis is suspected, it can be confirmed or disproved by two oscillation photographs taken at an initial ω setting interval of $\dfrac{360°}{n}$.

In the above discussion, and that which follows, it should be remembered that x-ray diffraction experiments are inherently centrosymmetrical and therefore appear to introduce a center of symmetry into the symmetry combination of the crystal being investigated. The symmetry as determined by the methods mentioned in this section lead to the centrosymmetrical symmetry of the crystal, which is related to the actual symmetry as indicated in Table 1.

The possible symmetries of oscillation photographs. Oscillation photographs can have five kinds of symmetry. The reasoning leading to this conclusion is as follows: The statistical symmetry of an oscillation movement is $D_{2h} = mmm$. For example, a single plane parallel with the rotation axis sweeps out the space shown in Fig. 110A. This is the highest symmetry an oscillation can display. All other symmetries, due to the lower symmetries of the oscillated reciprocal lattice, must be subgroups of D_{2h}. Furthermore, it is not the symmetry D_{2h} which is recorded on the film, but a projection of it from the center of the sphere. This projected symmetry is the plane group, C_{2l}, Fig. 235. This is the maximum symmetry an oscillation photograph can display. Other symmetries, due to the lower symmetries of the oscillated reciprocal lattice, must be subgroups of C_{2l}. These subgroups are C_1, C_2, C_l, C'_l (a second orientation), and C_{2l}.

These several subgroup symmetries, applied to the reciprocal lattice about to be oscillated back and forth partially through the sphere of reflection, are shown in Fig. 108. Each case represents the appearance of the reciprocal lattice while it is in the mid-oscillation position, looking along the direction of the x-ray beam. As the lattices having each of these symmetries are oscillated equally to each side for the position shown, they produce the corresponding photographs in Fig. 109. Case C_1 occurs when the midpoint of the oscillation coincides with a position (line group, page 472) of no symmetry. Case C_2

occurs when the midpoint of the oscillation coincides with a twofold axis of symmetry.   Case C_l occurs when the midpoint of the oscillation coincides with a plane of symmetry perpendicular to the rotation axis.   Case C_l' occurs when the midpoint of the oscillation coincides with a plane of symmetry parallel with the rotation axis.   Case C_{2l} occurs when the midpoint of the oscillation coincides with twofold axis at the intersection of a plane of symmetry parallel with the rotation axis and a plane of symmetry perpendicular to the rotation axis.

Note that the symmetries C_3, C_{3l}, C_4, C_{4l}, C_6, and C_{6l} do not occur in oscillation photographs.   This is because, if each of these line-group symmetries, Fig. 237, of the reciprocal lattice occupy the mid-position of the oscillation, the statistical D_{2h} symmetry of the oscillation degrades these symmetries to C_1, C_l, C_2, C_{2l}, C_2, and C_{2l}, respectively.

A new method of producing symmetry-true photographs.   It is evident that the oscillation method has limited possibilities of providing symmetry information about a crystal because of the low statistical symmetry of its motion, Fig. 110A.   If a motion can be devised which has the statistical symmetry $D_{\infty h}$, Fig. 110C, then photographs which are true projections of the symmetry of a crystal axial direction can be produced.

Another way of looking at the problem is that, in the oscillation method, two radial directions in the reciprocal lattice are especially favored, Fig. 110B.   If the reciprocal lattice, instead of being rocked left and right into the sphere of reflection, is rocked into the sphere without favoritism in all radial directions from the center, Fig. 110D, then all radial lines in the reciprocal lattice will have an equal chance to reflect.   This motion is closely approximated by the scheme shown diagrammatically in Fig. 110E.   In this diagram a reciprocal lattice plane is shown free to rotate with respect to two axes.   In order to favor all lines equally and make them pass through the sphere, the normal to the reciprocal lattice plane (which is a row of the direct lattice) is caused to occupy successive generators of a cone having half-opening angle ω.   (Here ω corresponds with the oscillation range in oscillating-crystal practice.)   This can be arranged by attaching the plane-normal to the end of a radius of the base of the cone, and then causing the base to rotate about its center.   This gives the plane-normal a kind of precessing motion.

An instrument for producing this sort of motion and photographically recording the resulting x-ray diffraction is shown in Fig. 111. The crystal is mounted and oriented on an adjusting crystal holder. The entire holder is rocked with a motion like that shown diagrammati-

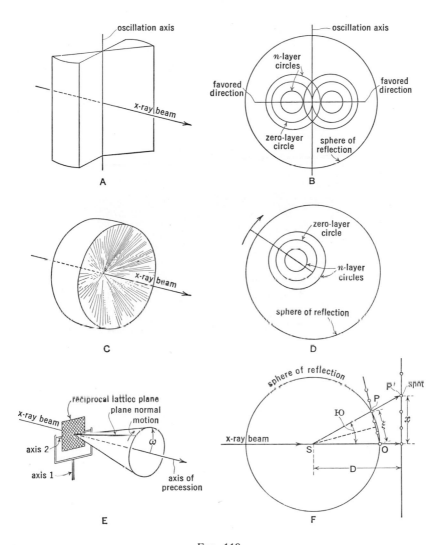

FIG. 110.

cally in Fig. 110*E*, by means of an arm attached to the inner axis. This arm bears a pin at its extremity, the pin being coincident with the plane-normal shown in Fig. 110*E*. The pin fits into a bearing which may be set at any desired angle, ω. For purposes of preliminary adjustment and centering of the crystal to the x-ray beam, the angle ω may be set at zero; the plane-normal is then parallel with the x-ray beam.

Fɪɢ. 111.   Apparatus for producing symmetry-true x-ray diffraction photographs.

Photographs taken with this instrument are shown in Fig. 112.   A photograph of this kind has very desirable characteristics.   A plane containing the x-ray beam direction and passed through a spot on the film, Fig. 110F, also contains the reciprocal lattice point responsible for the spot.   This is true for any spot on the film; consequently any central line of reciprocal lattice points is represented by a radial line of diffraction spots on the film.† This holds not only for the zero level, but also for any level (counting levels as normal to the x-ray beam when the value of ω in Fig. 110E is zero).   Consequently, this kind of photograph is angle-true at the origin.

The favorable features go farther than this, however, for such a

---

† This statement is subject to the following qualification: In Fig. 110E, points in the shaded reciprocal lattice plane which are also on axis 2 pass into and out of the sphere of reflection at the same point on the sphere.   Points not on axis 2 pass into the sphere at one point and out of the sphere at another nearby point.   This behavior consequently produces, not a single spot on the film, but a doublet.   If the angle ω is not too large, the doublet is a very close one, and if the crystal is not too small, the sizes of the reflections on the film are so large that the doublets are not resolved. Unresolved doublets are shown in Fig. 112B.   In Fig. 112A, spots at a considerable distance from the center, and not on the horizontal axis, are just perceptibly resolved into doublets.

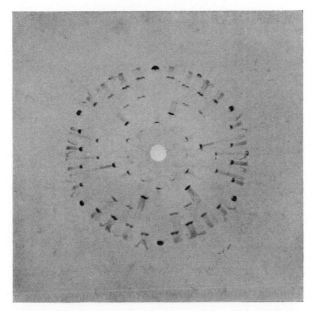

Fig. 112A. Photograph displaying plane symmetry C_6. (Willemite, Zn_2SiO_4, hexagonal; c-axis parallel to generator of cone, Fig. 110E.)

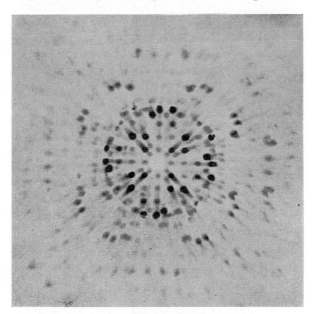

Fig. 112B. Photograph displaying plane symmetry C_4. (Meionite, $Ca_4Al_6Si_6O_{24}CO_3$, tetragonal; c-axis parallel to generator of cone, Fig. 110E.)

photograph is a picture of the reciprocal lattice projected in the direction of the x-ray beam and having only a slight distortion. The amount of distortion can be deduced with the aid of Fig. 110*F*. This diagram shows a reciprocal lattice point, *P*, whose radial distance from the origin is ξ, recording as a diffraction spot, *P′*, whose distance from the center of the film is *x*. That the distance *x* is approximately proportional to ξ is evident from the following relations:

$$x = D \tan \text{IO} \tag{1}$$

$$\sin \frac{\text{IO}}{2} = \frac{\xi}{2}. \tag{2}$$

Substituting the value of the angle, IO, given by (2) into (1) gives

$$x = D \tan \left(2 \sin^{-1} \frac{\xi}{2} \right). \tag{3}$$

For small values of IO this may be approximated

$$x \approx D \, \xi . \tag{4}$$

Since photographs of this sort are pictures of the reciprocal lattice they are very easily indexed by inspection. The only indexing difficulty arises from the possible overlaps in the recording regions for the several levels. A similar overlap has been discussed in connection with the explanation of Fig. 106. The overlap can be mitigated by limiting the angle ω. No overlap of the zero layer and the first layer away from the x-ray source occurs for angles less than $\omega = \cos^{-1}(1 - d^*)$, where d^* is spacing of the reciprocal lattice planes. No overlap of the zero layer and the first layer toward the x-ray source occurs for angles less than $\omega = \cos^{-1}(\frac{1}{2} + \frac{1}{2}\sqrt{1 - d^*})$. The overlap can be avoided entirely by using a layer line screen which prevents the recording of more than one layer at a time. The screen is a metal plate containing an annular slit which permits only one cone of rays (from the center of the sphere to one circle, Fig. 110*D*) to reach the film. Since the circle travels with the crystal, the layer line screen must also. It is conveniently attached to the follower arm of the instrument shown in Fig. 111.

This apparatus may be made to record the reciprocal lattice without any distortion whatever by applying the principle of de Jong and Bouman (Chapter 17). This is accomplished by mounting the flat film holder on a two-axis mount similar to that on which the crystal is suspended, and always keeping the film-normal and the normal to the reciprocal lattice plane parallel. The principle of this more compli-

cated arrangement will become clear after reading Chapter 17. If
n-layers are to be recorded without distortion, then the center of the
film rotation must be appropriately changed.

The advantages of this new method of recording x-ray diffraction
may be summarized as follows:

1. The symmetry of the photograph is the symmetry of the axis
normal to the reciprocal lattice plane, Fig. 110E. The photograph
is therefore symmetry-true.

2. The photograph is an angle-true picture of the reciprocal
lattice along this axis, with but little distortion.

3. The diffraction spots can be indexed by inspection.

4. The motion of the crystal is continuous and uniform. No
difficulty is experienced therefore due to lost motion such as occurs
at the end of the oscillating motion in the oscillating-crystal method.

5. By adding another motion to keep the film parallel with the
reciprocal lattice level, the radial distortion inherent in the method
and also the doubling of spots somewhat distant from the origin
may be completely removed, and the recorded range of reflections
may be increased.

THE DETERMINATION OF IDENTITY PERIODS IN TWINS

Identity periods are ordinarily determined by making complete
rotation photographs. There are occasions, however, when a complete
rotation cannot be made, and then the identity period may be determined from an oscillation photograph. Such occasions arise, for
example, when the only crystal material available is in the form of
obvious twins. The identity periods of the separate individuals of the
twin can then be determined by mounting the twin so that the direct
x-ray beam penetrates only a single individual during the course of a
limited rotation, i. e., an oscillation. This is illustrated by the mineral
sternbergite, $AgFe_2S_3$. In this mineral the crystals invariably occur
as twins, and owing to the plastic nature of the crystals, it is impossible
to separate the individuals without badly bending them, which
renders them unfit for x-ray diffraction work. The twin plane is a
prism, which makes the c-axis identity periods of the two twinned
individuals coincident. This common identity period can thus be
determined by a rotation about the common axis. The a- and b-axis
identity periods may be obtained by oscillating the crystal as shown in
Figs. 113A and B, respectively. In each case, it is arranged that
diffraction can occur from only one individual of the twin. The
oscillation range in such instances is made as large as possible without

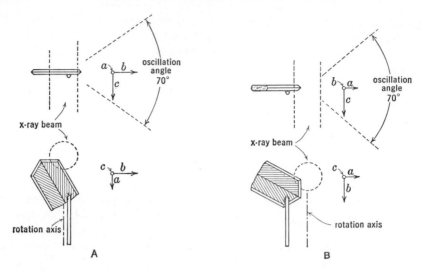

FIG. 113.

permitting the direct beam to reach the second individual.   The large range insures a large number of reciprocal lattice points contributing to the diffraction record, thus increasing the certainty that the layer-line spacing represents the true identity period.

LITERATURE

The oscillating-crystal method

J. D. BERNAL.   On the interpretation of x-ray, single crystal, rotation photographs. *Proc. Roy. Soc. London* (*A*), **113** (1926), especially 150–157.

E. SCHIEBOLD.   Die Drehkristallmethode. *Fortschr. Mineral.*, etc., **11** (1927), 111–280.

JOHN W. GRUNER.   The oscillation method of x-ray analysis of crystals. *Am. Mineral.*, **13** (1928), 123–141.

JOHN W. GRUNER.   The use of the oscillation method in determining the structure of analcite. *Am. Mineral.*, **13** (1928), 174–194.

Oscillation mechanism

RALPH W. G. WYCKOFF.   *The structure of crystals.* 2nd edition.   (Chemical Catalog Company, New York, 1931.)   Pages 108–110.

S. H. YU.   Theory and design of the cam of an oscillating crystal x-ray spectro-graph.   *Z. Krist.* (*A*), **96** (1937), 1–6.

The convergent beam method

O. KRATKY.   Über die Untersuchung von mikroskopischen Kristallen mit Rönt-genstrahlen.

I. *Z. Krist.* (*A*), **73** (1930), 567–571.

II. *Z. Krist.* (*A*), **76** (1930), 261–276.

O. KRATKY. Zwei neue Methoden zur eindeutigen Indizierung von Konvergenz-aufnahmen. Z. Krist. (A), **76** (1931), 517–524.

O. KRATKY. Konvergenzaufnahmen als Verfaren zur Herstellung von Schicht-liniendiagrammen in besonders kurzer Zeit. Z. Krist. (A), **82** (1932), 152–154.

O. KRATKY. Ein neues Goniometerkopf für die röntgenographische Einkristall-untersuchung. Z. Krist. (A), **95** (1936), 457–459.

CHAPTER 11

AN INTRODUCTION TO MOVING-FILM METHODS

Of the several problems which confront the analyst of crystal patterns, the most fundamental is that of recording the reflections from each crystal plane in such a way that they may be unequivocally identified. It has already been shown that each reflection is associated with a point of the reciprocal lattice of the crystal. Since a reciprocal lattice point is identified by three coordinates, it follows that to identify unequivocally a diffraction beam associated with such a point requires the specification of three coordinates. Now, the plane surface of a photographic film is two-dimensional, and hence can record but two coordinates. It therefore follows that one cannot, in general, unequivocally index the entire diffraction record of a crystal if it is recorded on a single film. It is for this reason that the rotating- and oscillating-crystal methods are unsatisfactory ways of recording crystal-diffraction results. The rotating-crystal method is clearly an attempt at recording the three coordinates of a reciprocal lattice point with two film coordinates, and the oscillating-crystal method is an attempt to record on a two-dimensional film the three coordinates of a reciprocal lattice point which is contained within certain limits of three-dimensional space. These attempts correspond, respectively, with attempting to solve for three unknowns with the aid of two equations, and with two equations and a pair of inequalities. The solution in both cases is indeterminate. In general, therefore, such indexing methods must fail, although they may be made to yield a certain amount of information for certain special symmetry cases.

Now, the solid reciprocal lattice can be resolved into individual layers of points, each level of which is two-dimensional. If the diffraction of each layer is singled out and recorded by itself, each of its points has a position specified by only two coordinates (the two lattice translations) in the level, and there arises the geometrical possibility of recording the diffraction due to this layer of points in such a way that each point may be unequivocally indexed. This recording of diffraction from only one layer at a time amounts to the predetermination of one of the three indices, *hkl*, of each reflection by preselecting it, because the layer to be recorded, and therefore one index of the triple,

214

hkl, is preselected. For example, if a crystal is rotated about its
c-axis, and one chooses to record only the zero layer on a film, then
every spot on the film will bear an index of the type *hk0*. If only the
first layer is recorded, then every spot on the film will bear an index of

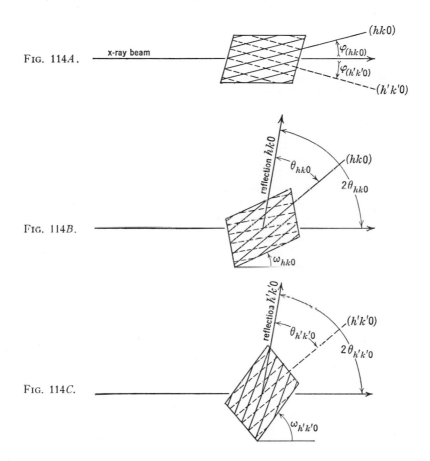

FIG. 114*A*.

FIG. 114*B*.

FIG. 114*C*.

the type *hk1*, and so on. Only *h* and *k* are unknown in these instances,
and these two variables can be spread appropriately over the two
dimensions of the film. It is only necessary to provide a physical
mechanism which will distribute them according to some known
law.

Another way of looking at the indexing problem is shown in Fig. 114,
where the reflection of a beam of x-rays by two different prism planes,
$(hk0)$ and $(h'k'0)$, is illustrated in plan view. In Fig. 114*A*, the
crystal is shown placed in an arbitrary zero orientation. Under these

conditions the plane $(hk0)$ is not, in general, in a position to reflect. Reflection can occur only after the crystal orientation has been changed by rotating the crystal through an angle $\omega_{(hk0)}$, at which setting the plane $(hk0)$ has been so placed that it makes the glancing angle, θ, with the x-ray beam which satisfies the Bragg reflection condition

$$n\lambda = 2d_{(hk0)} \sin \theta_{(hk0)}.$$

For a given experimental set-up, the characteristic glancing angle, $\theta_{(hk0)}$, depends only on the spacing $d_{(hk0)}$ of the system of planes. The total angle of rotation, $\omega_{(hk0)}$, necessary to cause reflection to occur, depends fundamentally on the slope of the planes with respect to the crystal axes, and therefore on the original angle the plane makes with the x-ray beam before rotation starts. Now, if two systems of planes, $(hk0)$ and $(h'k'0)$, should have almost identical spacings, d, then their glancing angles, θ, would be almost identical. Hence they would both send reflected beams to practically identical spots on a fixed film. There would be no possibility of finding out from the diffraction record whether the plane $(hk0)$ or $(h'k'0)$, or both, had been responsible for producing the spot on the film. This is all because the film records θ but not ω. From this point of view, a reflection has two coordinates:

θ, the glancing angle (which is a function of d only), and

ω, the angle through the crystal must be turned to place the plane in reflecting position (which is a function of the original slope, φ, of the plane with respect to its crystallographic axes, and also a function of its glancing angle, θ).

Now, if, during the rotation of the crystal, the film can be moved synchronously in any direction other than that of the plane of the angle θ, then the angular position, ω, of the crystal at the moment of reflection can be measured by the amount of film movement up to the diffraction spot. A number of mechanisms have been used for this purpose. Several of them are shown in Figs. 115–118. Figure 115 illustrates the general principle involved. As the crystal rotates, the recording film is raised parallel with the crystal-rotation axis by a coupling (a string, in the illustration, wound around a drum on the rotation axis) in such a way that the film translation is proportional to the angular rotation of the crystal. The amount of rotation, $\omega_{(hk0)}$, of the crystal which has occurred when $(hk0)$ is in a position to reflect is measured directly by the linear displacement of the spot from the film's edge, or other zero mark. The size of the glancing angle, θ, is measured by the linear displacement of the spot away from

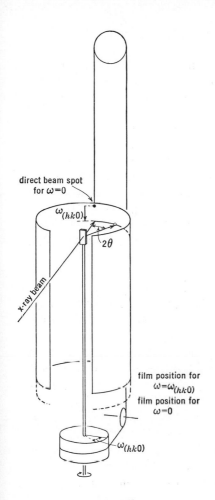

direct beam spot
for ω=0

ω$_{(hk0)}$

2θ

x-ray beam

film position for
ω=ω$_{(hk0)}$

film position for
ω=0

ω$_{(hk0)}$

FIG. 115.

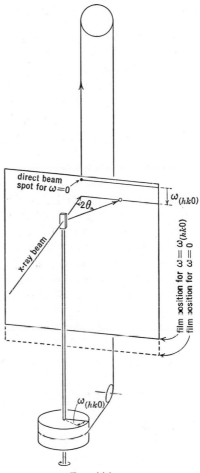

direct beam
spot for ω=0

2θ

ω$_{(hk0)}$

x-ray beam

film position for ω = ω$_{(hk0)}$
film position for ω = 0

ω$_{(hk0)}$

FIG. 116

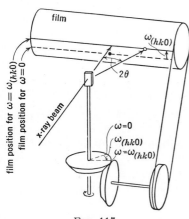

film

ω$_{(hk0)}$

2θ

film position for ω = ω$_{(hk0)}$
film position for ω=0

x-ray beam

ω=0
ω$_{(hk0)}$
ω=ω$_{(hk0)}$

FIG. 117

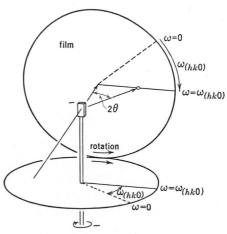

film

ω=0

ω$_{(hk0)}$

ω=ω$_{(hk0)}$

2θ

rotation

ω$_{(hk0)}$

ω=ω$_{(hk0)}$

ω=0

FIG. 118

217

the center line of the film.   The two coordinates ω and θ enable one to deduce what plane $(hk0)$ was responsible for the reflection, and thus provide the means of indexing the reflection.   The several methods of arriving at the index are treated in subsequent chapters.

Figures 115–118 indicate several methods of moving a film so as to separate ω from θ.   They differ in providing diffraction data of various degrees of completeness and in permitting interpretations of various degrees of convenience.   Figure 115 is the Weissenberg scheme of film movement.   Figure 116 shows a variation of it for recording on flat film.   This has been used by Robertson for a special instrument. Figure 117 shows a mechanism for producing the same result on a film which is rotated about a cylinder axis.   This method has been used by B. E. Warren (unpublished).   Figure 118 shows another type of movement for flat film, which is used in the Sauter apparatus.

Each of these methods has its advantages and disadvantages.   The chief disadvantage of those in Figs. 116, 117, and 118 is that they record a limited range of θ values, which is a disadvantage inherited from the fixed flat film type of recording in the rotating-crystal method. This disadvantage does not apply to the Weissenberg scheme of recording, Fig. 115, which has the same advantages as the cylindrical film type of recording in the rotating-crystal method; it permits the recording of a complete range of θ.   (Other disadvantages of the schemes shown in Figs. 116, 117, and 118 are that, though they record the zero-layer lines without difficulty, they run into certain complications in the recording of n-layer lines.)

In order to permit only one layer line at a time to record, the diffraction cone of this layer must be isolated.   The customary mechanical scheme for accomplishing this is shown in Fig. 119.   The crystal is covered with a closed hollow metal cylinder and the desired layer line cone is singled out by permitting it alone to emerge from the cylinder through a circular slit.   The particular layer line desired is preselected by raising or lowering the cylinder an appropriate amount to permit the desired cone to pass through the slit.   If the zero setting of the layer line screen is such as to permit the flat zero-layer cone to pass through the slit, then Fig. 119 indicates that the screen setting for an n-layer line, whose cone angle is ν, is given by

$$s = r_s \tan \nu_n. \tag{1}$$

The layer line cone angle, ν, is easily computed from the height of the layer line in question according to (3), Chapter 5.

A rotating-crystal photograph must, in general, be made as a preliminary to using a moving-film method.   The rotating-crystal

photograph provides data for computing the direct lattice translation parallel with the rotation axis (see Chapter 5) and also provides
data for computing the layer line screen settings needed for the moving-film method, as indicated above.   The moving-film method, in
turn, provides data for investigating the reciprocal lattice translations

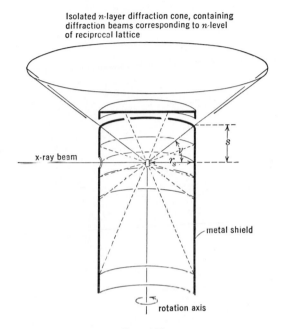

Isolated n-layer diffraction cone, containing
diffraction beams corresponding to n-level
of reciprocal lattice

x-ray beam

metal shield

rotation axis

FIG. 119.

normal to the rotation axis.   These two kinds of translations completely establish the reciprocal lattice as well as direct lattice, and thus
form the basis of a complete crystal-pattern analysis.   It will be the
purpose of the subsequent chapters to discuss the theories of the
several moving-film methods and to detail the means of interpretation
of their diffraction records.

LITERATURE

Weissenberg method
   K. WEISSENBERG.   Ein neues Röntgengoniometer.   *Z. Physik*, **23** (1924), 229–238.

Flat-film Weissenberg method
   J. MONTEATH ROBERTSON.   A two crystal moving film spectrometer for compara-
      tive intensity measurements in x-ray crystal analysis.   *Phil. Mag.*, (7) **18** (1934),
      729–745.

Dawson method

    W. E. DAWSON.   A simple method of determining the orientation and structure of crystals with x-rays.   *Phil. Mag.*, (7) **6** (1928), 756–768.

Sauter method

    ERWIN SAUTER.   Eine einfache Universalkamera für Röntgen-Kristallstruktur-analysen.   *Z. Krist.* (*A*), **85** (1933), 156–159.

CHAPTER 12

A GENERAL INTRODUCTION TO THE WEISSENBERG METHOD

WEISSENBERG MECHANISMS

In order to permit the cylindrical camera to move parallel with the crystal-rotation axis, it is convenient to mount it on a horizontal track or its equivalent. This, of course, requires the crystal-rotation axis to take a horizontal position also (Figs. 120 and 121). Note that this is a different arrangement from that sometimes employed in the rotating-crystal method, in which the rotation axis may be placed in a vertical position.

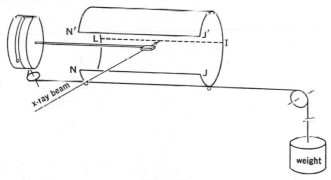

Fig. 120.

There are two main ways of coupling the camera translation to the crystal rotation. One is to connect the camera to a string which is wound around a drum on the rotation axis (Fig. 120). As the crystal rotates, the drum winds up the string and draws the camera toward it by an amount proportional to the amount of crystal rotation. An arrangement must be provided so that the crystal-rotation motion comes to a stop and reverses. When this occurs, the drum plays out the string, and the camera, pulled away from the drum by a weight, undergoes translation in the reverse direction until the crystal-rotation motion is again stopped and reversed. The reversing is usually accomplished by reversing the direction of the electric motor which drives the mechanism. When the camera comes to the end of its

221

desired motion, a reversing switch is thrown by a finger either on the crystal-rotation axis or on the camera itself.

The second method of coupling the film translation to the crystal rotation is shown in Fig. 121. In this case, the camera is driven uniformly back and forth on its track by means of a screw which is geared to the crystal-rotation axis. This is an exceedingly neat, compact, and accurate method. The first method is cumbersome because of the weight, and is inaccurate on account of the tendency of the string to stretch, especially with changes of weather.

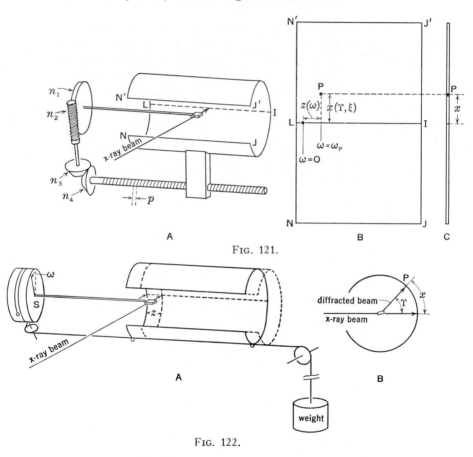

FIG. 121.

FIG. 122.

INSTRUMENTAL CONSTANTS

Rotation photographs of the same crystal taken on different instruments are the same except for scale, which is dependent solely on the diameter of the camera. Weissenberg photographs, on the other hand, may differ in scale in two different dimensions.   One of these scales (along x), Fig. 121B, depends on the diameter of the camera; the other (along z) depends on the coupling of film translation to crystal rotation.   A given Weissenberg instrument, therefore, has two instrumental constants, C_1 and C_2, corresponding to these two different scales.

Figure 122B illustrates the meaning of the first instrumental constant, C_1.   From the figure it is apparent that the distance, x, from a diffraction spot, P, to the center line of the film is proportional to the reflection angle, Υ (upsilon).   Thus,

$$\Upsilon = C_1 x. \tag{1}$$

The constant C_1 can be easily evaluated for the reflection angle 360°, in which case the following proportion may be written:

$$\frac{\Upsilon}{x} = \frac{360°}{\text{Camera circumference}} = \frac{360°}{2\pi r_F}. \tag{2}$$

The right member of (2) is evidently C_1, according to (1); hence

$$\Upsilon = \left[\frac{360°}{2\pi r_F}\right] x. \tag{3}$$

Since a millimeter scale is most convenient to use for measuring x, it is convenient to have the instrumental constant C_1 equal to some simple number for the purposes of rapid mental conversion from x to Υ according to (1).   This constant is a function of film radius, r_F, only. The camera diameters for some desirable simple conversion constants are:

Instrumental constant, $C_1 = \left[\dfrac{360°}{2\pi r_F}\right]$	Corresponding camera diameter, $2r_F$
$\frac{1}{2}°$/mm.	229.184 mm.
1	114.592
2	57.296

Since the length of the x-ray exposure varies as the square of the x-ray path, it is desirable to have the camera diameter as small as possible. At the same time it must be large enough to house within it the adjusting crystal holder and the layer line screen.   A camera diameter which has these two properties as well as giving the instrumental constant C_1 a convenient, simple value is 57.296 mm. This is thus the optimum Weissenberg camera diameter, and the corresponding instrumental constant C_1 is 2.

Figure 122A illustrates the meaning of the coupling constant, C_2. As the crystal rotates through an angle ω, the Weissenberg mechanism translates the film through a linear distance, z. The ratio

$$\frac{\omega}{z} = C_2 \text{ degrees per millimeter} \tag{4}$$

is a constant for any Weissenberg instrument, and depends on the characteristics of the coupling mechanism. In string-coupled instruments, Fig. 122, it evidently depends on the effective drum diameter and can be determined by considering the rotation per full drum of string. In this case, evidently, the following proportion holds:

$$\frac{\omega}{z} = \frac{360°}{2\pi r_D} = C_2. \tag{5}$$

Here r_D is the effective drum radius and equals the actual drum radius plus the string radius.

In screw-driven instruments, Fig. 121, the coupling constant depends on two factors, namely, the angular reduction of the rotation in the train of gears (shown as a worm and wheel plus a pair of bevel gears, in Fig. 121A), as well as the pitch of the translation screw. Specifically, the constant may be determined according to the following scheme:

$$C_2 = \frac{\omega}{z} = \frac{\text{Degrees}}{\text{Millimeter}}$$

$$= \frac{\text{Degrees}}{\text{Turn}} \cdot (\text{Turn reduction in gearing}) \cdot \frac{\text{Turns of screw}}{\text{Screw translation in mm.}} \cdot (6)$$

Using n_1, n_2, n_3, and n_4 to indicate the number of teeth in the several gears in sequence from the crystal-rotation shaft to the screw (Fig. 121A) and using p to designate the pitch (advance, $\frac{\text{mm.}}{\text{turn}}$) of the screw, then (6) may be expressed,

$$C_2 = \frac{360}{1} \cdot \left(\frac{n_2}{n_1} \cdot \frac{n_4}{n_3}\right) \cdot \frac{1}{p}. \tag{7}$$

It will appear shortly that it is desirable to have C_2 come out equal to 2. This can be accomplished by making the separate variables of (7) have the following actual values:

$$2 = \frac{360}{1} \cdot \left(\frac{1}{180} \cdot \frac{n_4}{n_4} \right) \frac{1}{1} \cdot$$

In other words, convenient, practical values for the gearing-and-screw combination consist of a worm with 1 thread playing on a gear with 180 teeth, and a pair of bevel gears having identical numbers of teeth, combined with a screw having a 1-mm. pitch.

Relation (4) is conveniently thrown into the form

$$\omega = C_2 z. \tag{8}$$

Referring back, now, to Fig. 121B, suppose that at a given zero of rotation, i.e., at $\omega = 0$, some film record is made, as, for example, an exposure of the film to the direct x-ray beam. Then the z distance from this zero to the diffraction spot, P, is a measure of the rotation, ω, through which the crystal goes until the diffraction spot is produced. This rotation can therefore be determined by measuring this z coordinate of the spot and converting it into ω with the aid of (8).

Since a millimeter scale is most convenient to use in measuring C_2, it is convenient to have C_2 some simple number for the purpose of rapid mental conversion of z to ω. Now it is desirable to have both the Υ and ω coordinates of a Weissenberg photograph measured by means of the same scale. This is desirable because (a) it is convenient to use the same millimeter scale on both coordinates, and have it mean the same angular value in both directions, thus avoiding confusion, and because (b) then all Weissenberg photographs present the same appearance except for magnification, and they can thus be examined and discussed in publications with a minimum of confusion. Such Weissenberg photographs, which have $C_1 = C_2$, may be said to have undistorted scale, whereas others, for which $C_1 \neq C_2$, may be said to have distorted scale.

It has already been shown that the optimum value for C_1 is 2. Hence the correct value for C_2 to give undistorted scale photographs is also 2.

THE MEASUREMENT OF x AND z COORDINATES

To facilitate the interpretation of a Weissenberg film, the direction of the z coordinate axis is marked out on it as follows: After the diffraction exposure is complete, but before the film is removed for development, the camera is uncoupled from its translation drive and manually pushed once along its track while the direct x-ray beam is permitted to strike the film.   In order to do this, of course, the direct beam screen must be displaced out of the line of the direct beam.   This procedure produces a black streak across the film center which gives the z-axis, i. e., the direction of the z coordinates and the zero for x.   This black streak can be seen on most of the Weissenberg films reproduced in this book.

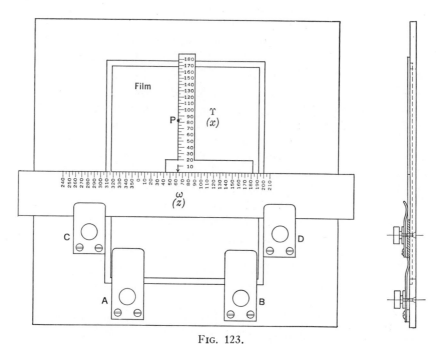

FIG. 123.

The actual measurement of the x and z coordinates of the diffraction spots on the film can be conveniently carried out with the aid of the simple apparatus shown in Fig. 123, which is placed on the front of a viewing stand (Fig. 98).   The film is slipped down behind clips A and B and then the horizontal scale is slipped over the film but behind clips C and D.   Clips C and D are temporarily tightened and the horizontal scale then marks out the direction of the z-axis.   The film is now adjusted so that its center line, as marked out by the method discussed in the last paragraph, is along the fiducial edge of the horizontal scale. Clips A and B are then tightened.   A second scale, at right angles to the first one, is arranged to move along the top of the first scale.   To determine the coordinates of any diffraction spot, P, the movable scale is moved until its fiducial edge touches P. The reading of P on this scale gives x directly, and the pointer at its base indicates the z coordinate of P on the fixed scale. Under ordinary circumstances, the z coordinate may be referred to an arbitrary zero. If it is desired to assign a specific z coordinate to any particular spot, P, clamps C and D are loosened and the z scale moved horizontally in its guides until the arrow at the foot of the x scale points to the desired value of z on the horizontal scale.

If special scales are constructed, it is convenient to have them graduated directly in degrees ω and Υ instead of in millimeters.

THE RELATION BETWEEN WEISSENBERG AND ROTATION PHOTOGRAPHS

Figure 121B shows the occurrence of a diffraction spot and its coordinates on a Weissenberg photograph; Fig. 121C shows the same layer line in a rotation photograph. Each spot on the Weissenberg photograph has two film coordinates, x and z, whereas on the rotation photograph it has only one coordinate, x. The layer line on a rotation photograph is thus the collapsed equivalent of a Weissenberg photograph in which the coordinate z has been eliminated. Another way of expressing this is that the layer line on a rotation photograph is a projection of the corresponding Weissenberg photograph in which the projection is performed in the z direction so that z vanishes. This relation is further brought out in Fig. 124, where an actual Weissenberg and corresponding layer line from a rotation photograph are shown. Looked at from another aspect, a Weissenberg photograph is the resolved equivalent of a layer line on a rotation photograph, where the resolved element is the coordinate z of each diffraction spot.

THE COMPACTION AND EXTENSION OF SPOTS ON n-LAYER PHOTOGRAPHS

A feature noticeable on many n-layer Weissenberg photographs, particularly those of high ς value, is the drawing-out of spots on one half of the film and the compaction of spots on the other half. This feature can be seen in Fig. 140B, for example. This effect has its origin in the divergence of the beam. Figure 125 illustrates the principles involved. It shows a diagrammatic view looking down on top of a Weissenberg instrument. For clearness, the crystal is shown as a cylinder. At the moment the Bragg reflection condition is satis-

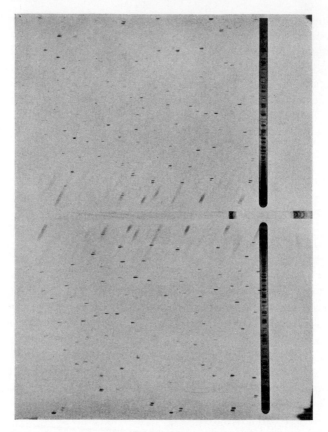

FIG. 124. The relation between Weissenberg and rotating-crystal photographs. The main part of the photograph is the Weissenberg record; the darker vertical stripe contains the rotating-crystal record of the same layer and recorded on the same film. The relation between the two is shown in Fig. 121B and C: a horizontal line may be drawn connecting any Weissenberg spot with the corresponding spot on the rotating-crystal record. Note the intense background in the rotation photograph; on the Weissenberg record this background is spread out over the entire film width and consequently appears much attenuated. (Pectolite, $Ca_2NaSi_3O_8(OH)$, monoclinic; b-axis rotation; $CuK\alpha$ radiation from hot-cathode x-ray tube, filtered through nickel foil.)

fied for the center of the crystal, plane A, with its normal at a, reflects to position (1) on the film. The crystal rotates in the direction indicated, and plane A passes out of the position instantaneously satisfying the Bragg condition because the glancing angle, 2θ, becomes too large. Owing to the different angle at which the x-ray beam strikes planes like B, farther out along the crystal, the angle 2θ is smaller here and the

Bragg condition may still be satisfied. In this way the position at which the Bragg condition is satisfied travels outward toward the end of the crystal, and the resulting reflection migrates from points (1) to (2). As the crystal continues to rotate, however, planes A and B eventually come to face downward and plane B reaches its Bragg condition first, and the position at which the Bragg condition is satisfied then travels inward along the crystal. For either the upper or lower

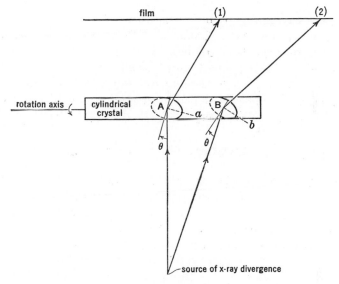

FIG. 125.

reflection, the moving film is traveling in a direction opposite to that of the migration of the reflection. On this side of the film the spot is drawn out into a streak. On the other side, the film is traveling in the same direction as the migration of the reflection, and the resulting spot record is compacted.

The distortion of the spot is greatest for spots near the center of the film and for layers of high ζ coordinate.

LITERATURE

K. WEISSENBERG. Ein neues Röntgengoniometer. *Z. Physik*, **23** (1924), 229–238.

J. BÖHM. Das Weissenbergsche Röntgengoniometer. *Z. Physik*, **39** (1926), 557–561.

W. SCHNEIDER. Über die graphische Auswertung von Aufnahmen mit dem Weissenbergschen Röntgengoniometer. *Z. Krist.*, **69** (1928), 41–48.

CHAPTER 13

THE NORMAL-BEAM WEISSENBERG METHOD

There are two main methods employing the general moving-film scheme shown in Fig. 115. The first of these, called the *normal-beam* Weissenberg method, is characterized by having the axis of crystal rotation normal to the direct x-ray beam. This method, which is treated in this chapter, is the natural outgrowth of the rotating-crystal method, and has developed from applying the idea of a moving film directly to the rotating-crystal method using a cylindrical film.

DEVELOPMENT OF A REFLECTION

Given a Weissenberg photograph, the x and z coordinates of every diffraction spot can be measured as described in the last chapter. From each x, z coordinate pair, the Υ and ω angular coordinates of the reflection can be determined from (1) and (8), Chapter 12. (If the instrumental constants are both 2, as recommended above, then the Υ and ω coordinates are just double the corresponding x and z millimeter scale readings.) From the Υ and ω values of each spot, the reciprocal lattice of the level represented by the Weissenberg photograph can be reconstructed as discussed below.

Figure 126 illustrates the theoretical basis for the reconstruction of the reciprocal lattice. The figure shows the plan and elevation of the x-ray beam, its sphere of reflection envelope, and one reciprocal lattice point, P. In the upper part of Fig. 126, the rotation axis is seen end-on. The point P is on the nth layer at a radial distance, ξ, from the rotation axis. At the beginning of the diffraction experiment, before the crystal begins rotation, point P occupies the initial position P_1, and has two-circle goniometer coordinates ρ, φ. The point P does not give rise to a reflection until the crystal rotation has brought it around to P_2, when it touches the sphere of reflection. The contact occurs on the circumference of the n-level reflecting circle. The crystal rotation necessary to bring this contact about is ω_P. At contact, P_2 develops a reflection whose direction is Υ_P. Note that a point, P, in general, passes through the sphere twice, once giving a reflection of direction Υ_P, which produces a spot on the upper half

230

of the film, and again giving a reflection of direction $360° - \Upsilon_P$, which produces a spot on the lower half of the film.

To the interpreter of Weissenberg films, the above situation presents itself in reverse order. Υ_P and ω_P can be measured for each

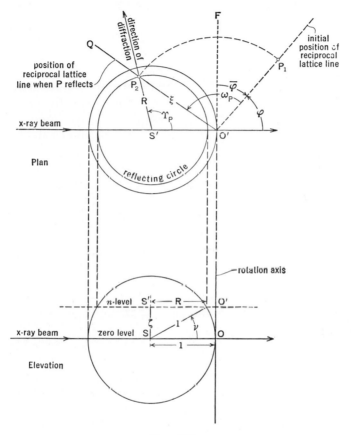

Fɪɢ. 126.

spot on the Weissenberg film. The question arising in interpreting the film is, what was the original position, P_1, of the reciprocal lattice point, P? This can be determined in several ways.

GRAPHICAL RECONSTRUCTION OF THE RECIPROCAL LATTICE: SCHNEIDER'S METHOD

The location of the original position of P before rotation can be solved graphically by a method due to Schneider. For this purpose, the bottom projection of Fig. 126 should first be constructed. The construction starts by drawing a horizontal line for the primary x-ray beam direction. At a convenient point, the profile of the sphere of reflection is constructed centering on this line, and having a radius unity (conveniently taken as 1 decimeter, i.e., 10 cm.). If the Weissenberg photograph of the nth layer is being interpreted, then the position of the nth level must be found on this drawing. This may be established by laying off the layer line cone angle, ν, where ν is computed from the y height of the layer line according to (4), Chapter 5. A still easier way of locating the layer line in the projection is to determine its ζ coordinate on the rotation photograph directly from the Bernal chart, Fig. 79. The level is then laid off as a line parallel with the x-ray beam and ζ decimeters above it. The intercept of the n-level on the sphere of reflection establishes the diameter of the reflecting circle.

Part of the upper half of Fig. 126 may now be constructed by projecting from the lower drawing. In this way the x-ray beam, the equator of the sphere of reflection, the reflecting circle, and the rotation axis (origin of the reciprocal lattice) may be established. Now, from the Υ coordinate of a spot on the Weissenberg film, the reflection direction is established. This angle is laid off at S', the center of the reflecting circle. The intersection of this reflection line with the reflecting circle locates P_2, the position of the lattice point at the moment of reflection. This point has been rotated $\omega°$ from its initial position. Hence, to find its original position, swing P_2 about the rotation axis, O, through $-\omega°$ to P_1. Note that, for every spot on the upper half of the film, Υ is less than 180°. For spots on the lower half of the film Υ is greater than 180°, or, what amounts to the same thing, it lies between 0° and $-180°$.

When this procedure has been completed for the coordinate pair of each spot on the Weissenberg photograph, then there have been reconstructed all the original positions of all the points on that layer of the reciprocal lattice which were within reflecting range. In other words, the nth level of the reciprocal lattice has been graphically reconstructed, in the orientation that level occupied before the crystal began rotating.

In a like manner, each level of the reciprocal lattice may be reconstructed from the Weissenberg photograph of that level. Under ordinary conditions, the first and second levels for rotations about a single rotation axis give sufficient data for establishing the lattice type, while the zero layer of a rotation axis gives information concerning a possible glide plane and possible screw axes lying normal to the rotation axis. If glide planes and screw axes of other locations come up for consideration because of suspected symmetry elements lying normal to other possible rotation axes, then additional zero-layer Weissenberg photographs are desirable in order to establish the space group of the crystal.

RELATIONS BETWEEN CYLINDRICAL COORDINATES AND REFLECTION COORDINATES

Transformation theory.   In Chapter 8 it was pointed out that cylindrical coordinates offer a convenient reference system for dealing with problems involving rotation of the reciprocal lattice.   When the crystal is correctly adjusted for rotation about an identity period, then the cylinder axis is taken coinciding with the rotation axis.   A point on the reciprocal lattice is then completely specified by the coordinate triple (ζ, ξ, φ), Fig. 75B.   All points on the same level are characterized by a common ζ coordinate, and, within a level, then, each point is completely located by the two polar coordinates ξ and φ.   These correspond with the customary symbols r and θ of ordinary polar coordinate practice.

Now, in the Weissenberg film, each reflection point is completely located by the two reflection coordinates Υ and ω.   It should be possible to find the relation between the (ξ, φ) system and the (Υ, ω) system of locating a point, and when this relation is found, it will be possible:

(a) to measure the reflection coordinates, Υ, ω, of each reflection spot on the film and, by means of appropriate transformations, transform these readings into cylindrical coordinates of the reciprocal lattice; or

(b) to map out the Weissenberg film in cylindrical coordinates, from which the polar coordinates ξ, φ of the points on that level can be read directly from the Weissenberg film.   From the polar coordinates obtained in either way, the level of the reciprocal lattice represented by the Weissenberg film can be plotted, point by point, on polar coordinate paper.

An easy approach to the relation between cylindrical coordinates and reflection coordinates is afforded in Fig. 127, which represents a plan of the zero layer of the reciprocal lattice, together with its Weissenberg representation.   In Fig. 127A, there is shown a line, OQ, which may be thought of as a continuous series of (potential) reflecting points, P.   The question now presents itself, how do the reflection directions, Υ, for the point P depend on ω, the angle through which the line has been rotated from its zero position, OF?   In the

Weissenberg film, which may be regarded as *reflection space*, this is equivalent to the question, how does the reflection position, P, migrate across the Weissenberg film, Fig. 121B, as the abscissa, ω, varies? This question is easily answered in Fig. 127A: Bisect angle Υ and note that its half, OST, has its two legs perpendicular to the two legs of ω. Hence,

$$OST = \omega. \tag{1}$$

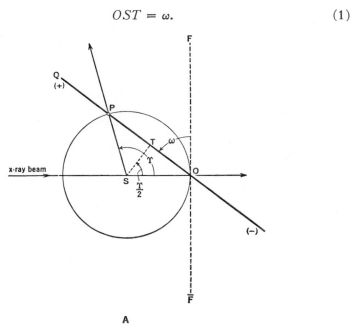

A

Fig. 127A.

In other words, no matter what the amount of rotation, ω, the direction of the reflection, Υ, is always given by the relation

$$\frac{\Upsilon}{2} = \omega$$

or

$$\Upsilon = 2\omega. \tag{2}$$

Since Υ and ω are Weissenberg film coordinates, (2) may be plotted directly on the film, Fig. 127B. The meaning of the plot is that, if a line of reciprocal lattice points passing through the origin (i. e., a *central line*) is rotated, it produces a series of reflections, $O\Xi$, on the Weissenberg film. This is a straight line having a constant slope of $\Upsilon/\omega = 2$ by virtue of (2). This line passes off the positive side of the film at $\omega = 90°$, and at the same time reappears on the negative side. This occurs because reflection angles up to 180° appear on the upper half of the film and reflection angles between 180° and 360° appear on the lower half of the film. The line $O\Xi$ passes through zero again at $\omega = 180°$ and becomes positive. In Fig. 127A, this corresponds with the line OQ occupying the position $O\overline{F}$. Further rotation causes the points on the *negative end of OQ* to reflect to the upper half of the film. The negative half of OQ then proceeds to

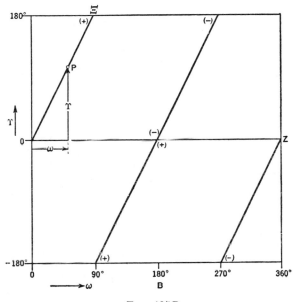

F ig. 127B.

give rise to a cycle of reflections identical with that of the positive half of the line, except that the point of beginning is 180° instead of 0°.

Consider, now, the more complicated case of an n-layer of the reciprocal lattice, Fig. 128. It is first desirable to evaluate the radius, R, of the reflecting circle on this level. From the elevation in Fig. 128A, this is evidently

$$R = \cos \nu, \tag{3}$$

or, alternatively,

$$R = \sqrt{1 - \zeta^2}. \tag{4}$$

Looking, now, at the plan, Fig. 128A, it will be observed that, as the line $O'Q$ rotates, it touches the reflecting circle, in general, in two points, P' and P'', and thus gives rise to two reflections at angles Υ' and Υ'', respectively. If the bisector, $S'T'$, of the angle $P'S'P''$ is drawn, it is again evident that the angles $O'S'T'$ and ω have their legs at right angles, and hence

$$O'S'T' = \omega. \tag{5}$$

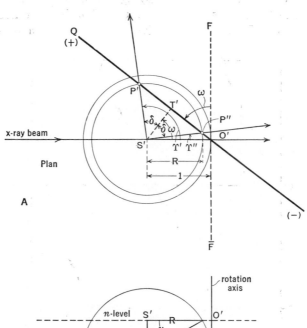

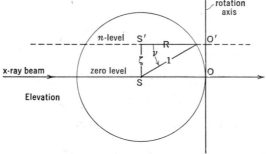

Fig. 128A.

The two reflection angles, Υ' and Υ'', may then be evaluated as

$$
\begin{aligned}
\Upsilon' &= O'S'T' + P'S'T', \\
\Upsilon'' &= O'S'T' - P''S'T'.
\end{aligned}
\tag{6}
$$

Calling

$$P'S'T' = P''S'T' = \delta, \tag{7}$$

and substituting the value of $O'S'T'$ given by (5), then (6) becomes

$$\Upsilon' = \omega + \delta, \\ \Upsilon'' = \omega - \delta. \Big\} \tag{8}$$

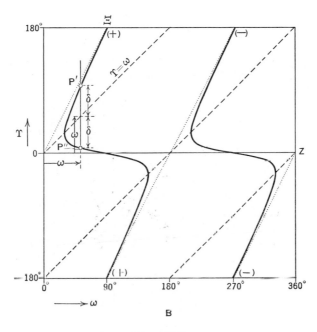

Fig. 128B.

The angle δ may be evaluated from Fig. 128A:

$$\cos \delta = \frac{S'T'}{R} \tag{9}$$

and

$$\cos \omega = \frac{S'T'}{1}, \tag{10}$$

from which

$$\cos \delta = \frac{\cos \omega}{R}. \tag{11}$$

Substituting the value of δ indicated by (11) in (8) gives

$$\Upsilon', \Upsilon'' = \omega \pm \cos^{-1}\left(\frac{\cos \omega}{R}\right). \tag{12}$$

This can be further developed by substituting, for R, the values given by (3) or (4):

$$\Upsilon', \Upsilon'' = \omega \pm \cos^{-1}\left(\frac{\cos \omega}{\cos \nu}\right) \tag{13}$$

or

$$\Upsilon', \Upsilon'' = \omega \pm \cos^{-1}\left(\frac{\cos \omega}{\sqrt{1 - \zeta^2}}\right). \tag{14}$$

The characteristics of the function expressed variously by (12), (13), or (14) may be studied graphically in Fig. 128B, which represents the

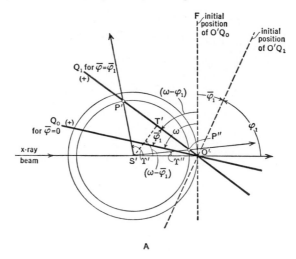

A

FIG. 129A.

Weissenberg film corresponding with the situation given in Fig. 128A. From the form of (12), it is evident that two reflection angles, Υ' and Υ'', are found symmetrically above and below a line $\Upsilon = \omega$ by an amount $\cos^{-1}\left(\frac{\cos \omega}{R}\right)$. Note that this last term of (12), (13), and (14) depends on R, ν, or, in the last analysis, on ζ, the height of the layer. For low values of ω, $\cos \omega$ is large and may exceed R. For this region, $\frac{\cos \omega}{R} > 1$ and the angle $\cos^{-1}\left(\frac{\cos \omega}{R}\right)$ is $\cos^{-1} > 1$, so this term is meaningless. There is therefore no solution for (12) under these circumstances, and accordingly there is no reflection until ω is large enough to make $\frac{\cos \omega}{R} = 1$. The physical meaning of this is evident from

Fig. 128*A*. As the line *OQ* rotates from its initial position, *OF*, it does not give rise to a reflection until it touches the reflecting circle. This occurs, as indicated above, when

$$\frac{\cos \omega}{R} = \frac{\cos \omega}{\cos \nu} = \frac{\cos \omega}{\sqrt{1 - \zeta^2}} \text{ becomes 1.} \qquad (15)$$

When contact occurs, a single degenerate reflection occurs for

$$\Upsilon' = \Upsilon'' = \omega.$$

As the rotation continues, the point of initial contact, *P*, resolves itself

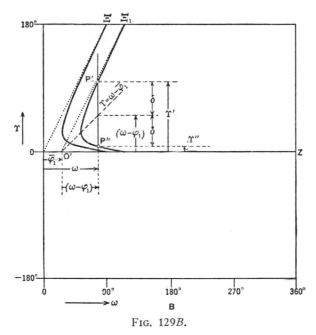

FIG. 129*B*.

into two points, *P'* and *P''*, which spread away from each other. This causes the reflection direction to resolve itself into two reflection directions Υ' and Υ'', symmetrically disposed above and below the value of

$$\Upsilon = \omega, \qquad (16)$$

which represents a straight line of slope $\dfrac{\Upsilon}{\omega} = 1$ on the film, Fig. 128*B*.

The Weissenberg *n*-layer curve,

$$\Upsilon = \omega \pm \cos^{-1}\left(\frac{\cos \omega}{R}\right),$$

always stays below the zero-layer locus, *OΞ*, up to the value $\omega = 90°$, at which point it attains the position of the zero-layer locus (dotted in Fig. 128B). Beyond $\omega = 90°$, $\Upsilon' > 180°$ and $\Upsilon'' < 0°$; i.e., both come into the region between 180° and 360°, and thus both Υ' and Υ'' reflect to the bottom of the film. The two angles approach each other and again become equal for $\cos \left(\dfrac{\cos \omega}{R} \right) = -1$. Rotation beyond this point permits no further contact of line and reflecting circle until the negative end of OQ makes contact with the upper half of the reflecting circle. Between these two contacts, no reflections develop and (12) has no meaning because $\cos \left(\dfrac{\cos \omega}{R} \right) > 1$. Further rotation causes the negative end of OQ to produce a cycle of reflection directions identical with the cycle for the positive end of OQ except that it is displaced by 180° ω.

Note that, for the zero layer, $R = 1$, and (12) becomes

$$\left.\begin{aligned} \Upsilon' &= \omega + \omega = 2\omega, \\ \Upsilon'' &= 0. \end{aligned}\right\} \tag{17}$$

This brings out the point that, even in the zero layer, there is a second reflection, Υ'', corresponding with the direction of the direct x-ray beam. This is due to the fact (Fig. 127A) that any central line contains the rotation axis which, in the zero layer, passes through the reflecting circle regardless of the value of ω.

This discussion applies to the special case of a reciprocal lattice line originally located along $\bar{\varphi} = 0$. The more general case of a lattice line of location $\bar{\varphi} = \bar{\varphi}_1$ is illustrated in Fig. 129. Note that the geometrical situation of line $O'Q_1$, in Fig. 129A is exactly the same as that for line $O'Q$ in Fig. 128A, except that, in Fig. 129A, the angle $(\omega - \bar{\varphi}_1)$ is substituted for the angle ω in Fig. 128A. In the Weissenberg field, Fig. 129B, the new line $O'Q_1$ therefore records a reflection locus $O'\Xi_1$, of exactly the same shape as $O\Xi$ in Fig. 128B, except that it has been displaced along the ω axis by $(\omega - \bar{\varphi}_1)°$. This new line is therefore represented by substituting $(\omega - \bar{\varphi}_1)$ for ω in the expressions derived in the foregoing discussion; i. e.;

$$\Upsilon = (\omega - \bar{\varphi}_1) \pm \delta, \tag{18}$$

$$\Upsilon = (\omega - \bar{\varphi}_1) \pm \cos^{-1} \left(\frac{\cos (\omega - \bar{\varphi}_1)}{R} \right), \tag{19}$$

$$\Upsilon = (\omega - \bar{\varphi}_1) \pm \cos^{-1} \left(\frac{\cos (\omega - \bar{\varphi}_1)}{\cos \nu} \right), \tag{20}$$

$$\Upsilon = (\omega - \bar{\varphi}_1) \pm \cos^{-1} \left(\frac{\cos (\omega - \bar{\varphi}_1)}{\sqrt{1 - \zeta^2}} \right). \tag{21}$$

It remains to find the cylindrical coordinate equivalent of the reflection angle, Υ. The relation between the several quantities involved is illustrated in Fig. 130. By the law of cosines,

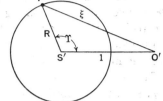

$$\xi^2 = 1^2 + R^2 - 2 \cdot 1 \cdot R \cos \Upsilon$$

$$= 1 + R^2 - 2R \cos \Upsilon, \qquad (22)$$

from which,

$$\cos \Upsilon = \frac{1 + R^2 - \xi^2}{2R}. \qquad (23)$$

Fig. 130.

This may be further developed by substituting the value of R for the nth layer, given in (3) or (4), and reducing:

$$\Upsilon = \cos^{-1} \left(\frac{1 + \cos^2 \nu - \xi^2}{2 \cos \nu} \right) \qquad (24)$$

or

$$\Upsilon = \cos^{-1} \left(\frac{2 - \zeta^2 - \xi^2}{2\sqrt{1 - \zeta^2}} \right). \qquad (24')$$

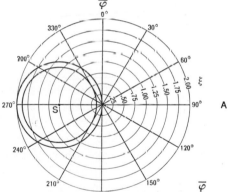

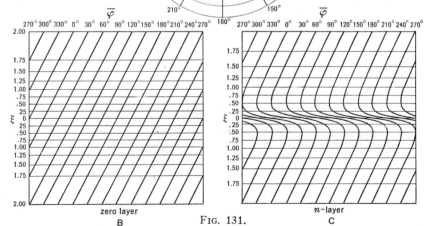

zero layer

B

n-layer

C

FIG. 131.

The mapping of polar coordinates on a Weissenberg film.   For a given layer line, a Weissenberg film can be marked out with a series of curves such as shown in Figs. 128*B* and 129*B*, along each of which $\bar{\varphi}$ = a constant, i. e., along each of which reflections lie which are due to reciprocal lattice points having the same polar coordinate, φ.   The film can also be marked out with a series of lines parallel with the center line, along each of which ξ = a constant.   Figures 131*B* and 131*C* show both zero and *n*-layer Weissenberg films marked out in ($\bar{\varphi}$, ξ) polar coordinates corresponding with the marking-out of reciprocal lattice space, Fig. 131*A*, into polar coordinates.

It was first suggested by Wooster and Wooster that Weissenberg films could be interpreted by preparing a set of charts like Figs. 131*B* and 131*C* for a series of values of the layer line coordinate, ζ, differing by intervals of 0.1.   In the construction of such charts, each $\bar{\varphi}$ = a constant curve may be constructed from the equation

$$\Upsilon = \omega \pm \cos^{-1}\left(\frac{\cos\omega}{\sqrt{1-\zeta^2}}\right).$$

Solutions of this for the plotting of such curves are given in Table 12, and a family of curves is reproduced for an undistorted Weissenberg film of camera diameter 57.3 mm. in Fig. 132.   This figure may be used for any other Weissenberg camera of undistorted scale but different diameter D, by enlarging the figure by the factor $\dfrac{D}{57.3}$.

The marking off of Wooster charts for ξ = a constant may be done either by setting ξ in (24) equal to .1, .2, .3, .4, $\cdots$, and solving for Υ, or, more easily, by using the ξ markings for the desired ζ level on the Bernal chart for cylindrical films.   The Bernal chart divisions have been plotted from the same computation.

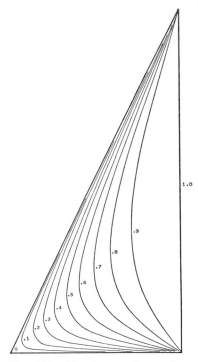

FIG. 132.   The shapes of central reciprocal lattice lines for various level coordinate values, ζ, for normal-beam Weissenberg photographs.   (Camera diameter, 57.3 mm.)

TABLE 12

DATA FOR CONSTRUCTION OF CENTRAL LATTICE LINE, NORMAL-BEAM SETTING

$$\text{Solution of } \Upsilon = \omega \pm \cos^{-1}\left(\frac{\cos \omega}{\sqrt{1 - \varsigma^2}}\right)$$

To transform Υ to x for cylindrical film, multiply by $\dfrac{1}{C_1} = \dfrac{2\pi r_F}{360°}$.

ω		$\varsigma = 0$	$\varsigma = .1$	$\varsigma = .2$	$\varsigma = .3$	$\varsigma = .4$	$\varsigma = .5$	$\varsigma = .6$	$\varsigma = .7$	$\varsigma = .8$	$\varsigma = .9$	$\varsigma = 1.0$
ω for nose of curve →		0.00°	5.74°	11.54°	17.46°	23.58°	30.00°	36.87°	44.43°	53.13°	64.16°	....
0°		0.00°										
		0.00										
5°		10.00										
		0.00										
10°		20.00	18.22°									
		0.00	1.78									
15°		30.00	28.88	24.65°								
		0.00	1.12	5.35								
20°		40.00	39.19	36.45	29.91°							
		0.00	0.81	3.55	10.09							
25°		50.00	49.37	47.33	43.18	33.56°						
		0.00	0.63	2.67	6.82	16.44						
30°		60.00	59.50	57.89	54.79	49.11	30.00°					
		0.00	0.50	2.11	5.21	10.89	30.00					
35°		70.00	69.59	68.28	65.83	61.65	53.94					
		0.00	0.41	1.72	4.17	8.35	16.06					
40°		80.00	79.65	78.57	76.58	73.30	67.80	56.75°				
		0.00	0.35	1.43	3.42	6.70	12.20	23.25				
45°		90.00	89.71	88.81	87.16	84.51	80.26	72.89	53.05°			
		0.00	0.29	1.19	2.84	5.49	9.74	17.11	36.95			
50°		100.00	99.76	99.00	97.64	95.47	92.08	86.54	75.83			
		0.00	0.24	1.00	2.36	4.53	7.92	13.46	24.17			
55°		110.00	109.80	109.17	108.04	106.26	103.52	99.19	91.57	72.07°		
		0.00	0.20	0.83	1.96	3.74	6.48	10.81	18.43	37.93		
60°		120.00	119.83	119.32	118.39	116.94	114.74	111.32	105.56	93.56		
		0.00	0.17	0.68	1.61	3.06	5.26	8.68	14.44	26.44		
65°		130.00	129.87	129.45	128.70	127.54	125.79	123.11	118.72	110.22	79.18°	
		0.00	0.13	0.55	1.30	2.46	4.21	6.89	11.28	19.78	50.82	
70°		140.00	139.90	139.57	138.99	138.09	136.74	134.69	131.38	125.25	108.31	
		0.00	0.10	0.43	1.01	1.91	3.26	5.31	8.62	14.75	31.69	
75°		150.00	149.92	149.68	149.26	148.60	147.61	146.12	143.75	139.45	128.58	
		0.00	0.08	0.32	0.74	1.40	2.39	3.88	6.25	10.55	21.43	
80°		160.00	159.95	159.79	159.51	159.08	158.43	157.46	155.93	153.18	146.52	
		0.00	0.05	0.21	0.49	0.92	1.57	2.54	4.07	6.82	13.48	
85°		170.00	169.97	169.90	169.76	169.54	169.22	168.75	167.99	166.65	163.47	
		0.00	0.03	0.10	0.24	0.46	0.78	1.25	2.01	3.35	6.53	
90°		180.00	180.00	180.00	180.00	180.00	180.00	180.00	180.00	180.00	180.00	180.00°
		0.00	0.00	0.00	0.00	0.00	0.00	0.00	0.00	0.00	0.00	0.00

The interpretation of Weissenberg films by the Wooster method consists of placing a Wooster chart printed on a transparent film, and of appropriate ζ value for the level to be interpreted, over a Weissenberg film.   The $\bar{\varphi}$, ξ coordinates of each spot are then read, and the corresponding reciprocal lattice point plotted on polar coordinate paper. One thus reads $\bar{\varphi}$, ξ for each spot on Fig. 131C, and plots it on Fig. 131A.

The Wooster method has three mechanical disadvantages: (1) it requires much preliminary labor in the preparation of a set of carefully constructed charts; (2) the charts represent discrete ζ values; charts for intermediate values are not available, and hence only an approximate and inaccurate reconstruction of the reciprocal lattice can be carried out; (3) it is comparatively easy to " lose one's place " among spots which have been plotted and those which have not yet been plotted.

Improved method of determining $\bar{\varphi}$, ξ coordinates from films.   The following method for determining $\bar{\varphi}$, ξ coordinates from a Weissenberg film makes use of the geometrical principles involved in the Wooster method but lacks the disadvantages of that method.

The film to be interpreted is clamped, as in the determination of the (x, z) coordinates, Fig. 123, to a frame on an illuminated viewing stand, and the same ω scale is adjusted to the center line of the film, also as in Fig. 123.   In place of the perpendicular Υ scale, however, a celluloid triangle, Fig. 133, is substituted.   This has a left edge having a slope of $\dfrac{\Upsilon}{\omega} = \dfrac{2}{1}\cdot$ This edge makes an angle of 63°26′ with the horizontal for Weissenberg films of undistorted scale.   Before use, the triangle is laid on Fig. 132, and the $\bar{\varphi}$ = constant curve for the correct ζ value of the layer line is traced on with soft pencil, interpolating between the curves of Fig. 132 as necessary.   The right (vertical) edge of the triangle is next laid along the ξ scale of the Bernal chart at the appropriate ζ value of the layer line, and the ξ intervals marked off.   These intervals are extended horizontally until they intersect the $\bar{\varphi}$ = constant curve.

To find the $\bar{\varphi}$, ξ coordinates of a spot, the triangle is moved over the horizontal scale until the $\bar{\varphi}$ = constant curve runs directly through the center of the spot.   The vertical scale on the triangle then gives ξ directly, and the stationary horizontal scale reading, indicated by the mark at the left lower point of the triangle, gives $\bar{\varphi}$ directly.   That the last statement is true may be easily verified from Fig. 129B.

Garrido's method of reconstructing the reciprocal lattice.   In the methods of reconstructing the reciprocal lattice just described, the polar coordinates were plotted on the Weissenberg film.   It is also possible to map the Weissenberg reflection coordinates (ω, Υ) in polar coordinate space. This is the basis of Garrido's method. In this method, one measures the ordinary Weissenberg film coordinates, ω and Υ, as in Fig. 123. Then these readings are transposed to charts, Fig. 134, in reciprocal lattice space, marked off with ω = constant, and Υ = constant, curves. This particular chart is valid only for the zero layer; for any other layer, another chart must be especially prepared.

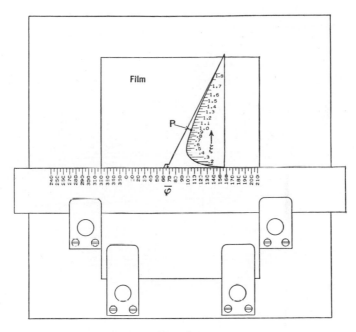

Fig. 133.

This method has the same disadvantages as the Wooster method, plus the additional disadvantage that there is no natural distribution of spots on the Weissenberg film along scale readings Υ = constant. On the other hand, all spots on central reciprocal lattice rows have $\bar{\varphi}$ = constant, and hence, in the modified Wooster method described in the last section, the $\bar{\varphi}$ reading of a whole row of spots is the same, and the row may be rapidly reconstructed. All reciprocal lattice points on zero levels of any crystal are grouped together in several central lattice rows; the same is true of the n-layers of all crystals rotated about any axis of symmetry. Thus, the modified Wooster method is the natural one for reading and plotting coordinates, while the Garrido method is a very unnatural one.

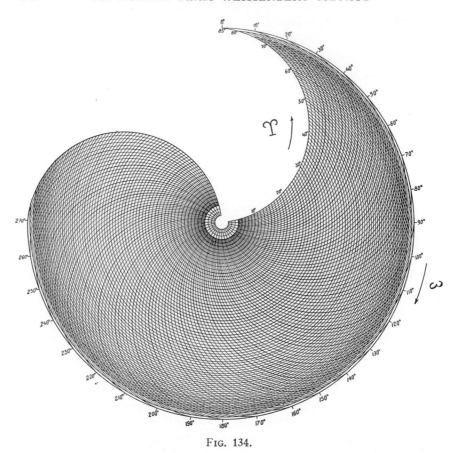

FIG. 134.

THE FORMS OF RECIPROCAL LATTICE ROWS IN NORMAL-BEAM WEISSENBERG PHOTOGRAPHS

The points of a reciprocal lattice are localized in lattice rows.   If the lattice row happens to be a *central* one, the points of the lattice produce reflections which lie along curves of the sort shown in Fig. 128. The locus of reflections for any lattice row whatever may be derived with the aid of Fig. 135.

The figure shows a heavy line, representing the reciprocal lattice row, at a distance, d, from the rotation axis.   The lattice row, in general, intersects the reflecting circle of the level in two points, P' and P'', and consequently two reflections having azimuth angles Υ' and Υ'' are developed.   Assuming, for simplicity, that the lattice row has an initial direction $\varnothing = 0$, then

$$\angle O'S'T'' = \omega,$$

and the two reflection angles are

$$\left.\begin{array}{l} \Upsilon' = \omega + \delta, \\ \Upsilon'' = \omega - \delta. \end{array}\right\} \tag{25}$$

The angle δ may be evaluated from

$$\cos \delta = \frac{S'T''}{S'P'} = \frac{D_2}{R}. \tag{26}$$

Here,

$$D_2 = D_1 + d. \tag{27}$$

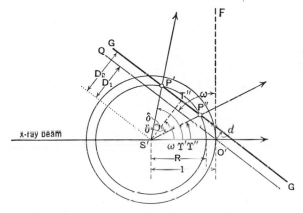

FIG. 135.

The part D_1 may be evaluated from

$$\cos \angle O'S'T'' = \cos \omega - \frac{D_1}{1},$$

from which

$$D_1 = \cos \omega; \tag{28}$$

and

$$\begin{aligned} D_2 &= D_1 + d \\ &= \cos \omega + d. \end{aligned} \tag{29}$$

Substituting, in (26), the value of D_2 found in (29) gives

$$\cos \delta = \frac{D_2}{R},$$

$$\cos \delta = \frac{\cos \omega + d}{R}. \tag{30}$$

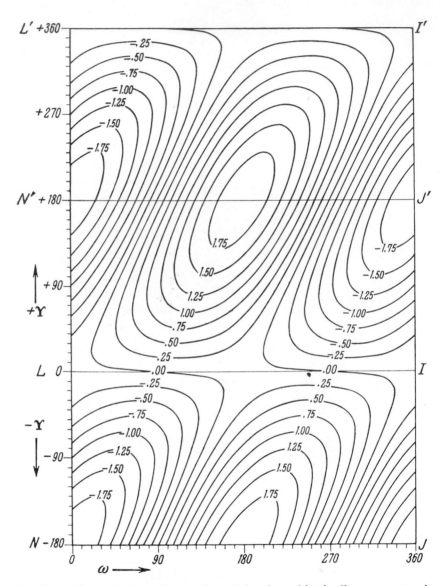

F<small>IG</small>. 136.   The appearance of a set of parallel reciprocal lattice lines on a normal-beam Weissenberg photograph ($\nu = 20°$, $\zeta = .34$).   The labeling of the curves gives the distance, d, of the reciprocal lattice line from the origin of the level.   The actual film corresponds with the square $NN'\ J'J$.

The reflection angles of (25) can now be completely evaluated as

$$\Upsilon', \Upsilon'' = \omega \pm \cos^{-1}\left(\frac{\cos\omega + d}{R}\right), \qquad (31)$$

and also further developed, with the aid of (3) and (4), giving

$$\Upsilon' = \omega \pm \cos^{-1}\left(\frac{\cos\omega + d}{\cos\nu}\right), \qquad (32)$$

and

$$\Upsilon'' = \omega \pm \cos^{-1}\left(\frac{\cos\omega + d}{\sqrt{1 - \zeta^2}}\right). \qquad (33)$$

The appearance of this kind of curve on the Weissenberg film is shown in Fig. 136. This illustration not only shows the ordinary Weissenberg film area, $NLN'J'IJ$ (letters correspond with those on the film cylinder, Fig. 121B), in which Υ varies from $-180°$ through zero to $+180°$, but also shows the area $LN'L'I'J'I$, in which Υ varies from zero through $+180°$ to $+360°$. Since the usual Weissenberg camera requires the film to end at line $N'J'$, to permit entrance of the direct beam into the film cylinder (see Fig. 121A), values of Υ from $+180°$ to $+360°$ are not ordinarily realized in continuity with the $+180°$ region of the film, but only in continuity with the $0°$ region.

Study of Fig. 135 will show that, for values of d less than $AO' = 1 - R$, the value of Υ for intersection P'' diminishes as ω increases, passes through zero, and becomes negative. For values of d greater than $AO' = 1 - R$, Υ'' diminishes to a minimum and then increases. Lattice rows of this d spacing thus form a series of closed oval curves, while those of smaller spacing form a series of open curves, Fig. 136.

One of the important properties of the curves shown in Fig. 136 can be seen by writing (31) as follows:

$$\Upsilon', \Upsilon'' = \omega \pm \cos^{-1}\left(\frac{\cos\omega}{R} + \frac{d}{R}\right). \qquad (34)$$

The entire second term of the right member is the correction angle δ. This gives the height of the curves above and below the $45°$ line $\Upsilon = \omega$. From (34), this correction, for a given ω value, varies with R and hence is different from layer to layer. Thus, in general, the shapes of these ovals are different from layer to layer. The significance of this will appear by contrast in the next chapter.

An actual n-layer Weissenberg photograph taken by the normal-beam method is shown in Fig. 137.

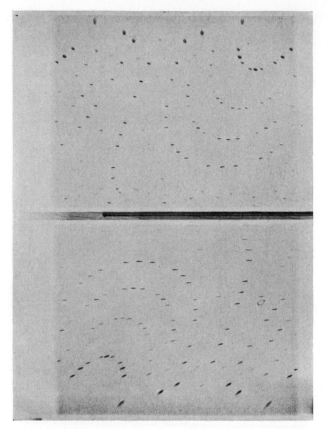

FIG. 137. n-Level Weissenberg photograph taken by the normal-beam method. Compare the spot alignment with the curve shapes shown in the lower left part of Fig. 136. (Valentinite, Sb₂O₃, orthorhombic; c-axis rotation, second level, $\zeta = 0.568$, $\nu = 34.6°$; Cu$K\alpha$ radiation from hot-cathode x-ray tube, filtered through nickel foil.)

LITERATURE

W. SCHNEIDER. Über die graphische Auswertung von Aufnahmen mit dem Weissenbergschen Röntgengoniometer. *Z. Krist. (A)*, **69** (1928), 41–48.

A. GERSTÄCKER, H. MÖLLER, and A. REIS. Über den Kristallbau des Pentaerythrit-Tetraacetates und -Tetranitates. *Z. Krist. (A)*, **66** (1928), especially 372–392.

GEORGE TUNELL. Determination of the space-lattice of a triclinic mineral by means of the Weissenberg x-ray goniometer. *Am. Mineral.*, **18** (1933), 181–186.

TOM. F. W. BARTH and GEORGE TUNELL. The space-lattice and optical orientation of chalcanthite (CuSO₄·5H₂O): an illustration of the use of the Weissenberg x-ray goniometer in the triclinic system. *Am. Mineral.*, **18** (1933), 187–194.

W. A. WOOSTER and NORA WOOSTER. A graphical method of interpreting Weissenberg photographs. *Z. Krist. (A)*, **84** (1933), 327–331.

M. J. Buerger.   The Weissenberg reciprocal lattice projection and the technique of interpreting Weissenberg photographs.   *Z. Krist.* (*A*), **88** (1934), especially 376–380.

J. Garrido.   Nota sobre la interpretación de los diagramas Weissenberg.   *Anales de la sociedad española de física y química*, **34** (1936), 399–401.

CHAPTER 14

THE EQUI-INCLINATION WEISSENBERG METHOD

INTRODUCTION

When Weissenberg photographs are taken with the x-ray beam inclined so that $\mu = -\nu$, their interpretation becomes exceedingly simple. Under these circumstances, the direct x-ray beam and the diffracted beam are equally inclined to the layers of the reciprocal lattice, and consequently this method is known as the *equi-inclination method*. This method, in common with any which utilizes an inclined x-ray beam, requires a special instrument of the general design outlined immediately below. The instrument, together with its adjustments for use with the equi-inclination method, are described in detail at the end of the chapter.

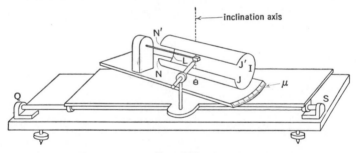

Fig. 138.

Design of Weissenberg instrument for use with inclined beams. The simple Weissenberg instrument is not suitable for taking photographs with an x-ray beam inclined to the reciprocal lattice layers. It is also not convenient to take such photographs with the simple instrument modified so that the pinhole system can be inclined from its normal position. This is because, with such an instrument, it is necessary to set the pinhole system at the required angle, μ, which is very easily done, and then readjust the entire apparatus so that the focal spot of the x-ray tube target is collinear with the pinhole system. This is a patience-trying task and it must be done anew for each photograph.

The entire difficulty of beam adjustment can be avoided by making the pinhole system integral with the base of the instrument and adjusting it to the x-ray beam once for all. An inclined beam is then achieved by building the crystal and film-carrying assembly on a separate base which may be rotated on the lower base about an axis passing through the crystal. A diagrammatic sketch of this design is given in Fig. 138. The entire pinhole system base can be translated out of the path of the x-ray beam by sliding it on the bar QS and a support in the rear (not shown). This allows one to move the line of the pinhole system enough to one side so that it clears the x-ray tube. The crystal can then be centered visually by sighting directly down the pinhole system.

General approach to equi-inclination photographs. A general viewpoint for approaching the material in this chapter is as follows: A Weissenberg photograph is simply a two-dimensional projection of a two-dimensional reciprocal lattice level. A point on the reciprocal lattice level may be referred to various kinds of two-dimensional coordinate systems, and any one selected could then be projected onto the Weissenberg photograph; or conversely, a point could be located in the Weissenberg photograph by any one of a number of possible coordinate systems, and then this coordinate system could be projected onto the reciprocal lattice level. Some of these coordinate systems are unnatural, some are more or less natural to either the photograph or to the reciprocal lattice; and the problem of interpreting a Weissenberg photograph resolves itself into choosing a natural coordinate system for plotting or identifying reciprocal lattice points.

The transformations from film to reciprocal lattice for some coordinate systems have already been discussed for the normal-beam Weissenberg photograph. The naturalness or unnaturalness of these reference systems also holds in the equi-inclination method. Briefly, it may be recapitulated that Schneider's and Garrido's methods depend on measuring the film coordinates, ω, Υ, on the film. These are natural film coordinates, but unfortunately they are thoroughly unnatural reciprocal lattice coordinates, and hence the methods are ill-adapted to interpreting a film based upon a reciprocal lattice. On the other hand, a reciprocal lattice point on a given level may be located with increasing naturalness by reference to:

(a) a polar coordinate system, or

(b) a coordinate system composed of the lattice rows themselves.

These two coordinate schemes form the basis of the interpretative methods recommended in this chapter. In transforming, point by point, from film to reciprocal lattice, that reference system which has the greatest grouping of points along coordinates of constant value gives the quickest mechanical transformation from film to lattice. From this point of view, (b) is thoroughly natural. The partial naturalness of (a) depends on the number of lattice points which occur on a line $\bar{\varphi}$ = constant, i. e., on a central line. More than one point occurs on a $\bar{\varphi}$ = constant line only if the line is rational. This is true for:

(a) all central lines on the zero layer of any crystal,

(b) certain central lines on the n-levels which, in certain crystal systems, are rational by reason of symmetry.

Thus, whereas lattice rows form a universally natural coordinate system, polar coordinates are only sometimes natural, and, even at best, points on central lattice lines are not packed as densely as they are along simple lattice rows.

The characteristics of equi-inclination Weissenberg photographs may be derived by reducing the results of the general inclination discussion, given subsequently in Chapter 15, to the special case $\mu = -\nu$. This special case, however, is of more practical importance than the general one; indeed, Weissenberg photographs taken by the equi-inclination method are so easily interpreted that this method has been utilized almost exclusively since its desirable properties have been established. For the benefit of those wishing to use this method, and to become familiar with the derivation of its characteristics by the easiest possible route, these characteristics are derived directly in this chapter.

TRANSFORMATION FROM FILM POSITIONS TO POLAR COORDINATES

The reflecting circle. The lower part of Fig. 139 is a diagram in the plane containing the rotation axis and the direct x-ray beam. It corresponds with a view from below the instrument, looking up, in Fig. 138. The circle is the profile of the sphere of reflection. *Note that the origin is at O, the intersection of the x-ray beam and the exit side of the sphere of reflection.* The zero level, of course, is normal to the rotation axis at the origin. The nth level is also normal to the rotation axis at a distance, ζ, above the zero level. The nth level cuts the sphere of reflection in the reflecting circle for that level. The radius, R, of this circle is evidently given by

$$\cos \nu = R, \tag{1}$$

and ν is controlled by the relation

$$\sin \nu = \frac{\zeta}{2}. \tag{2}$$

The value of R may thus be found as a function of the layer coordinate by combining (1) and (2):

$$R = \cos\left(\sin^{-1}\frac{\zeta}{2}\right),\qquad(3)$$

from which

$$R = \sqrt{1 - \left(\frac{\zeta}{2}\right)^2}.\qquad(4)$$

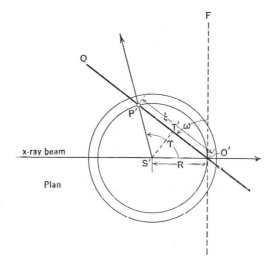

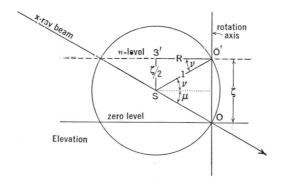

FIG. 139.

Reflection from a central lattice line. The upper part of Fig. 139 is a plan view and corresponds with a view looking toward the right end of the film cylinder of Fig. 138. The heavy line, $O'Q$, is a reciprocal lattice line passing through the rotation axis, and is therefore a

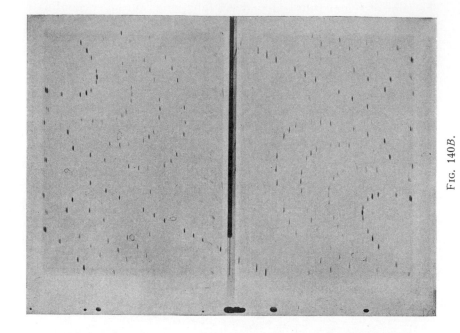

Fig. 140B.

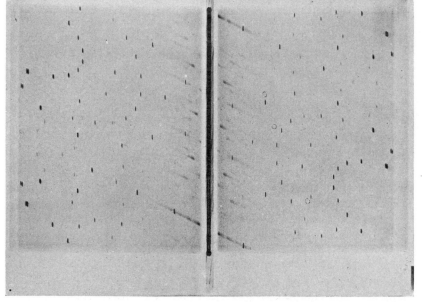

Fig. 140A.

FIG. 140. Weissenberg photographs taken by the equi-inclination method. Note the diagonal straight lines of spots starting near the left and right ends of the center lines. (All photographs taken with CuKα radiation from hot-cathode x-ray tube, filtered through nickel foil.)

FIG. 140A. Valentinite, Sb₂O₃, orthorhombic; c-axis rotation, zero level, $\zeta = 0$, $\mu = 0$.

FIG. 140B. Valentinite, c-axis rotation, 2nd level, $\zeta = .568$, $\mu = 16.4°$.

FIG. 140C. Low-tridymite, SiO₂, orthorhombic; a-axis rotation, 4th level, $\zeta = .621$, $\mu = 18.1°$.

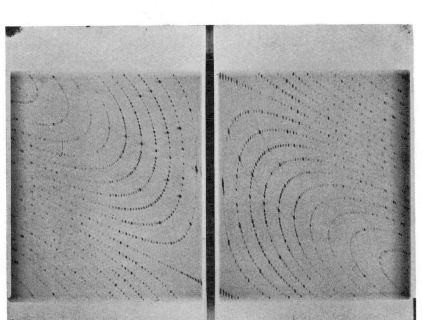

FIG. 140C.

central lattice line.   Before the crystal starts rotating, the initial position of this lattice line is assumed to be $O'F$, and rotation takes place
counterclockwise.   In general, the line $O'Q$ cuts the reflecting circle
at P', giving rise to a potential reflection at angle Υ.   To evaluate Υ,
construct its bisector $S'T'$.   Evidently,

$$\frac{\Upsilon}{2} = \omega.$$

A central lattice line thus produces a series of reflections on equi-
inclination photographs at angles

$$\Upsilon = 2\omega, \tag{5}$$

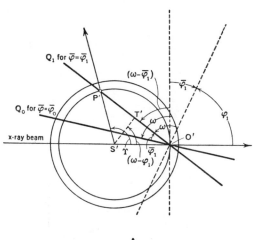

A

FIG. 141A.

regardless of what layer the photograph represents.   This is exactly
the same relation as existed specifically for the zero layer of normal-
beam Weissenberg photographs, (2), Chapter 13, and illustrated in
Fig. 127.   Thus a central lattice line, regardless of level, records on an
equi-inclination Weissenberg photograph as line of slope 2.   Photographs illustrating this are shown in Fig. 140.

It has been assumed in the above discussion that the line $O'Q$ has a
$\bar{\varphi}$ coordinate of zero.   Suppose that, as in Fig. 141A, it has a general
$\bar{\varphi}$ coordinate, $\bar{\varphi}_1$.   After rotation through angle ω, the line then attains
position $O'Q_1$.   It is evident from Fig. 141A that relation (5) then
becomes, more generally,

$$\Upsilon = 2(\omega - \bar{\varphi}_1). \tag{6}$$

The relation between (5) and (6) is illustrated in Fig. 141B. Line $O'Q_0$, Fig. 141A, of zero coordinate, records along the sloping line $\bar{\varphi} = 0$, Fig. 141B. Line $O'Q_1$, of $\bar{\varphi}$ coordinate $\bar{\varphi}_1$, records along sloping line, $\bar{\varphi} = \bar{\varphi}_1$, i. e., $\bar{\varphi}_1$ degrees along the ω scale, OZ.

If any two quantities in (6) are known, the remaining one may be solved for. Thus, if the film coordinates, Υ and ω, of a reflection are measured, the cylindrical coordinate of the point P' of the reciprocal lattice responsible for the reflection may be found by rearranging (6) thus:

$$\bar{\varphi}_1 = \omega - \frac{\Upsilon}{2}. \tag{7}$$

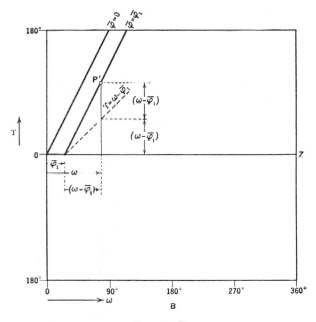

FIG. 141B.

The second cylindrical coordinate, ξ, of lattice point P', may also be easily obtained. In Fig. 139,

$$\sin (\angle O'S'T') = \frac{O'T'}{O'S'},$$

or

$$\sin \frac{\Upsilon}{2} = \frac{\xi/2}{R}, \tag{8}$$

TABLE 13

DATA FOR CONSTRUCTION OF ξ SCALE FOR EQUI-INCLINATION SETTING

Solutions of $\Upsilon = 2 \sin^{-1}\left(\dfrac{\xi}{2 \cos \mu}\right)$

To transform Υ to x for cylindrical film, multiply by $\dfrac{1}{C_1} = \dfrac{2\pi r_F}{360°}$.

ξ	Υ							μ for
	$\mu = 0°$	$\mu = 5°$	$\mu = 10°$	$\mu = 15°$	$\mu = 20°$	$\mu = 25°$	$\mu = 30°$	$\Upsilon = 180°$
0.05	2.87°	2.87°	2.90°	2.97°	3.05°	3.17°	3.30°	
0.10	5.73	5.77	5.83	5.93	6.10	6.33	6.60	
0.15	8.60	8.63	8.73	8.90	9.17	9.50	9.93	
0.20	11.48	11.53	11.60	11.87	12.20	12.67	13.27	
0.25	14.36	14.43	14.57	14.87	15.30	15.83	16.60	
0.30	17.25	17.33	17.53	17.87	18.37	19.05	19.97	
0.35	20.16	20.23	20.47	20.87	21.47	22.27	23.33	
0.40	23.07	23.17	23.43	23.90	24.57	25.50	26.70	
0.45	26.01	26.10	26.43	26.93	27.70	28.73	30.10	
0.50	28.96	29.05	29.40	30.00	30.87	32.03	33.57	
0.55	31.92	32.07	32.43	33.10	34.03	35.33	37.03	
0.60	34.92	35.03	35.47	36.20	37.23	38.67	40.57	
0.65	37.93	38.07	38.53	39.33	40.57	42.03	44.10	
0.70	40.95	41.13	41.63	42.60	43.77	45.43	47.67	
0.75	44.04	44.23	44.77	45.70	47.03	48.90	51.33	
0.80	47.16	47.33	47.93	48.93	50.40	52.40	55.03	
0.85	50.30	50.50	51.13	52.20	53.80	55.93	58.77	
0.90	53.49	53.70	54.37	55.53	57.23	59.53	62.60	
0.95	56.72	56.97	57.67	58.93	60.73	63.20	66.53	
1.00	60.00	60.27	61.03	62.33	64.30	66.97	70.53	
1.05	63.34	63.60	64.43	65.83	67.93	70.80	74.63	
1.10	66.73	67.03	67.90	69.53	71.67	74.73	78.87	
1.15	70.20	70.50	71.37	73.05	75.47	78.77	83.20	
1.20	73.74	74.07	75.05	76.80	79.37	82.90	87.70	
1.25	77.36	77.73	78.77	80.63	83.37	87.20	92.40	
1.30	81.08	81.47	82.60	84.60	87.53	91.63	97.30	
1.35	84.91	85.33	86.53	88.67	91.83	96.30	102.40	
1.40	88.85	89.30	90.60	92.90	96.30	101.13	107.87	
1.45	92.94	93.40	94.83	97.30	100.97	106.27	113.67	
1.50	97.18	97.70	99.20	101.87	105.90	111.70	120.00	
1.55	101.61	102.17	103.90	106.70	111.10	117.53	127.00	
1.60	106.26	106.87	108.63	111.83	116.70	123.93	134.93	
1.65	111.18	111.83	113.80	117.33	122.77	131.10	144.57	
1.70	116.42	117.17	118.80	123.30	129.53	139.40	157.93	
1.75	122.09	122.87	125.47	129.90	137.23	149.80	—	28.95°
1.80	128.32	129.23	132.10	137.40	146.60	166.43	—	25.83
1.85	135.34	136.40	139.90	146.53	159.73	—	—	22.33
1.90	143.61	144.97	149.47	159.17	—	—	—	18.22
1.95	154.32	156.33	163.80	—	—	—	—	12.83
2.00	180.00	—	—	—	—	—	—	0.00

from which

$$\xi = 2R \sin \frac{\Upsilon}{2}. \tag{9}$$

The value of R for any level has already been obtained in (1) and (4). Substituting these values in turn in (8) gives two new useful forms of ξ:

$$\xi = 2 \cos \nu \sin \frac{\Upsilon}{2} \tag{10}$$

and

$$\xi = 2 \sqrt{1 - \left(\frac{\zeta}{2}\right)^2} \sin \frac{\Upsilon}{2}. \tag{11}$$

In order to take an equi-inclination Weissenberg photograph, the instrument must be set for the correct inclination angle, $\mu = -\nu$. This is easily obtained as a function of ζ for the layer, according to (2), and is usually read directly from a chart, Fig. 155. Since μ is thus known for each photograph to be taken, form (10) of the above relation is the most immediately useful. An accurate plot of the scale of ξ on the film, as it varies with μ, may be prepared once for all by solving (10) for Υ and assuming decimal scale values for ξ. Such solutions are recorded with varying inclination angle, μ, in Table 13. A plot of this ξ scale as it varies with μ is given in Fig. 142, for the particular camera diameter 57.3 mm.

The mapping of polar coordinates on a Weissenberg film. The rectangular coordinates of a Weissenberg film, taken parallel with the film edges, enable one to measure directly the reflection coordinates, ω and Υ. It has just been shown that the polar reciprocal lattice coordinates may be expressed as functions of these, and that for the equi-inclination method they are very simple functions indeed, namely,

$$\bar{\varphi} = \omega - \frac{\Upsilon}{2}, \tag{7}$$

and

$$\xi = 2 \cos \nu \sin \frac{\Upsilon}{2}. \tag{10}$$

It is evident that a Weissenberg film could be mapped out with lines along which $\bar{\varphi}$ was constant, and with other lines along which ξ was constant. The locations and characteristics of these lines are easily found by setting $\bar{\varphi}$ constant in (7) and finding the resulting locus, and then setting ξ constant in (10) and finding the resulting locus. These

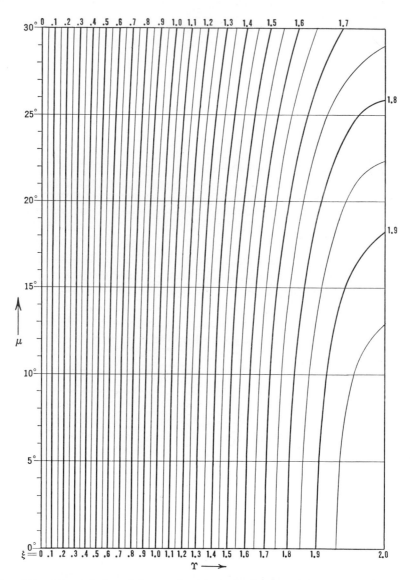

FIG. 142.   The variation of the ξ scale with the inclination angle, μ, for equi-inclination Weissenberg photographs (camera diameter, 57.3 mm.)

loci have the following forms, where the terms in parenthesis are constants for the photograph:

$$\bar{\varphi} \text{ constant:} \quad \Upsilon = 2\omega - (2\bar{\varphi}), \tag{11}$$

$$\xi \text{ constant:} \quad \Upsilon = 2 \sin^{-1}\left(\frac{\xi}{2 \cos \nu}\right). \tag{12}$$

Note that the $\bar{\varphi} = $ constant loci are straight lines of slope 2, spaced equally along the increasing ω intervals for increasing values of the constant $\bar{\varphi}$. Thus the radial polar coordinate $\bar{\varphi} = $ constant lines of Fig. 143A become, on the equi-inclination Weissenberg film, Fig. 143B, straight lines of slope 2.

The $\xi = $ constant loci are functions of Υ and independent of ω. They are therefore horizontal lines parallel with the ω-axis. Thus the equally spaced circular polar coordinates $\xi = $ constant of Fig. 143A, become on the equi-inclination Weissenberg film, Fig. 143B, horizontal straight lines. For equal ξ intervals, these lines occur with increasing spaces because the Υ level depends on the angle whose sine is $\dfrac{\xi}{2 \cos \nu}$ in (12), and this angle increases more and more rapidly with increasing ξ.

From the above discussion, it is obvious that the appropriate polar coordinate system for a Weissenberg film is one with the axes as indicated by the arrows at the left-hand center of Fig. 143B. The axis for $\bar{\varphi}$ calls for no comment. The axis for ξ is naturally taken along the line $\bar{\varphi} = 0$, i. e., along $O\Xi$, not along the vertical edge of the film. The divisions on the ξ scale along $O\Xi$ (abbreviated ξ_{Ξ}) are longer than those on the ξ scale along OX (abbreviated ξ_X) by the ratio

$$\frac{\xi_{\Xi}}{\xi_X} = \frac{1}{\sin (O\Xi X)} = \frac{1}{\sin (\Xi OZ)}. \tag{13}$$

The angle (ΞOZ), here designated Я (pronounced " ya "), is such that, according to (5) the slope is always

$$\tan Я = \frac{\Upsilon}{\omega} = 2, \tag{14}$$

referred to the natural ω, Υ coordinate system of the Weissenberg photograph. Measured in absolute units, x and z, this slope depends upon the instrumental constants defined by (1) and (4), Chapter 12:

$$\tan Я = \frac{x}{z} = \frac{\Upsilon/C_1}{\omega/C_2} = \frac{C_2}{C_1} \cdot 2. \tag{15}$$

Combining (13) with (15), therefore, the absolute scale distances along $O\Xi$ become

$$\xi_\Xi = \frac{\xi_X}{\sin \tan^{-1}\left(\dfrac{C_2}{C_1} \cdot 2\right)} . \tag{16}$$

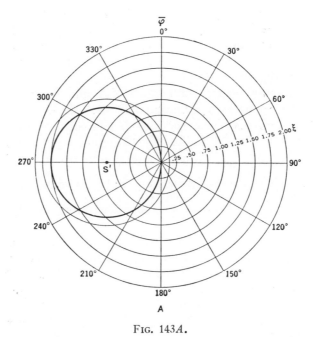

FIG. 143A.

For Weissenberg instruments with undistorted scale, $C_1 = C_2$, and (16) reduces to

$$\xi_\Xi = \frac{\xi_X}{\sin \tan^{-1} 2}$$

$$= \frac{\xi_X}{\sin 63° 26'} . \tag{17}$$

The variation of ξ divisions along $O\Xi$, with inclination angle μ, is accurately given in Fig. 144 for a camera of undistorted scale and standard diameter, 57.3 mm.

Reconstruction of the reciprocal lattice.   The easiest way to determine the cylindrical coordinates of the reflections on a Weissenberg photograph is with the aid of the coordinate measuring device shown in Fig. 145. The film to be interpreted is clamped to a frame on an illuminated background in the same manner as if the reflection coordinates, ω, ϒ, were to be measured as in Fig. 123. The ω scale, which now acts as the $\bar{\varphi}$ scale, is adjusted to the center line of the film, also as in Fig. 123, and shifted left or right so that it gives the desired

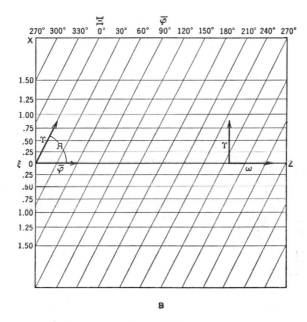

FIG. 143B.

reciprocal lattice row a zero reading (this row is ordinarily a row of pinacoid or dome reflections). In place of the perpendicular ϒ scale, however, there is substituted a celluloid triangle whose left edge has a slope $ϒ/ω = 2/1$. This edge makes an angle of 63° 26′ with the horizontal for Weissenberg films of undistorted scale. Before using, the edge of this triangle is laid along the chart of Fig. 144 and the ξ_{Ξ} scale marked out on it with soft pencil. These marks should be continued so that they are horizontal when the triangle is in position, Fig. 145; i. e., they should not be continued perpendicular to the edge of the triangle.

To find the φ, ξ coordinates of any spot, it is only necessary to move the triangle over the horizontal scale until the graduated edge of the triangle cuts the center of the x-ray diffraction spot in question.

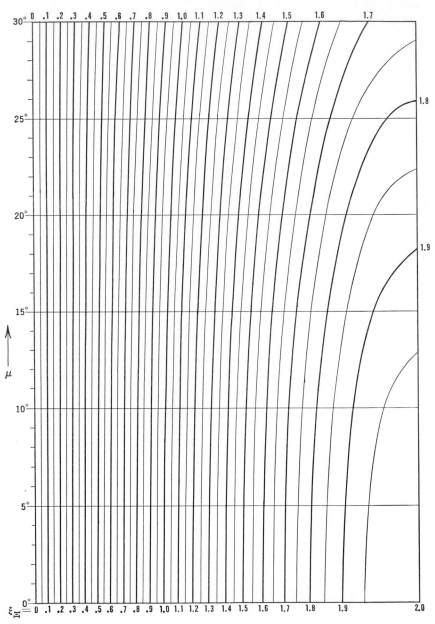

FIG. 144.   The variation of the ξ scale along $O\Xi$, with the inclination angle, μ, for equi-inclination Weissenberg photographs (camera diameter, 57.3 mm.).

The graduation along this edge gives the ξ coordinate of the spot directly, while the graduation indicated on the horizontal scale by the point of the triangle gives the φ coordinate of the spot directly. These two coordinates, φ, ξ, may then be plotted directly on polar coordinate paper or the equivalent, and the reciprocal lattice point giving rise to the reflection is then located in its initial position before the lattice began to rotate. When this process has been carried out for each of the points on the upper half of the film, the reciprocal lattice within range of this half of the film has been reconstructed.

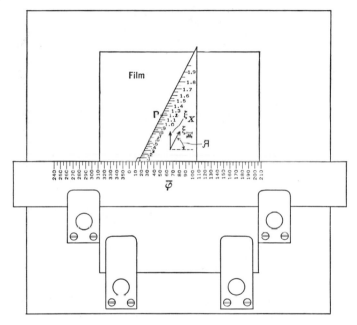

Fig. 145.

It is desirable to have, as equipment, two triangles; one of these is reserved for the zero layer, the other for n-layers. The zero-layer triangle always has the same scale, and this may be accurately and permanently engraved or scratched on its surface at vertical intervals taken from Table 13. This accurate zero-layer scale permits a very accurate reconstruction of the zero layer of the reciprocal lattice, and from this, rather accurate cell dimensions may be derived. The n-layer triangle is left blank for the marking process mentioned above. The main function of this triangle is to determine ξ coordinates with sufficient accuracy to permit the n-layer point to be located on the grid already accurately established for the zero layer.

THE WEISSENBERG RECIPROCAL LATTICE PROJECTION

Introduction. It is desirable, from this point on, to adopt the viewpoint that a Weissenberg photograph is a projection of a level of the reciprocal lattice. The actual reciprocal lattice has straight lattice

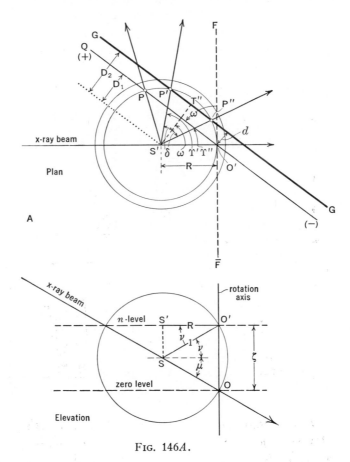

FIG. 146A.

rows which appear, by transformation onto what may be termed the Weissenberg projection, as curved lines. If the forms of these curves are known, it is just as easy to draw lattice lines, and therefore assign indices, directly on the Weissenberg projection as it is to draw lattice lines on the reciprocal lattice level itself and assign indices. In order to be able to use this technique, it is necessary to be able to recognize a reciprocal lattice row on a Weissenberg photograph.

The form of a reciprocal lattice row on equi-inclination Weissenberg photographs. If the reciprocal lattice row happens to be a central one, it projects on an equi-inclination Weissenberg photograph, regardless of layer, as a straight line of slope 2, as discussed in the earlier sections

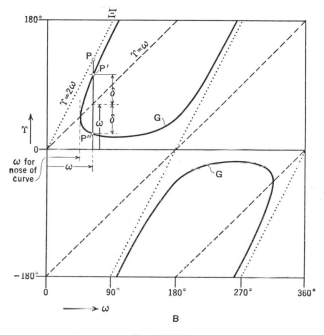

FIG. 146B.

of this chapter. If the reciprocal lattice row is not a central one, then the derivation of its Weissenberg projection is somewhat more complicated:

Figure 146A is similar to Fig. 139. The projection along the rotation axis, however, shows not only a central line, $O'Q$, but also a parallel lattice row, GG, which is a non-central one, and at a distance d from the rotation axis. Assume, for simplicity, that the original orientation of these rows is parallel with direction $O'F$. Provided that the non-central lattice row intersects the circle, it does so, in general, at two points, P' and P'', and two potential reflections at angles Υ' and Υ'' are continually developed. If $S'T''$ is drawn bisecting $P'S'P''$, then

$$O'S'T'' = \omega, \tag{18}$$

because these angles have their legs at right angles.   The two reflection angles may be evaluated as

$$\left.\begin{array}{l} \Upsilon' = O'S'T'' + P'S'T'', \\ \Upsilon'' = O'S'T'' - P''S'T''. \end{array}\right\} \quad (19)$$

Calling

$$P'S'T'' = P''S'T'' = \delta, \quad (20)$$

and substituting the value of $O'S'T''$ given by (18), then (19) becomes

$$\left.\begin{array}{l} \Upsilon' = \omega + \delta, \\ \Upsilon'' = \omega - \delta. \end{array}\right\} \quad (21)$$

The angle δ may be derived thus:

$$\cos \delta = \frac{S'T''}{S'P'} = \frac{D_2}{R}, \quad (22)$$

and

$$D_2 = D_1 + d. \quad (23)$$

The value of D_1 is given by

$$\cos (\angle O'S'T'') = \frac{D_1}{S'O'} = \frac{D_1}{R}, \quad (24)$$

from which (23) becomes

$$D_2 = R \cos \omega + d, \quad (25)$$

and (22) becomes

$$\cos \delta = \frac{R \cos \omega + d}{R} = \cos \omega + \frac{d}{R}. \quad (26)$$

Finally, (21) may be evaluated as

$$\left.\begin{array}{l} \Upsilon' = \omega + \cos^{-1}\left(\cos \omega + \dfrac{d}{R}\right), \\[2mm] \Upsilon'' = \omega - \cos^{-1}\left(\cos \omega + \dfrac{d}{R}\right). \end{array}\right\} \quad (27)$$

This may be further developed by substituting the value of R given by (1):

$$\Upsilon', \Upsilon'' = \omega \pm \cos^{-1}\left(\cos \omega + \frac{d}{\cos \nu}\right). \quad (28)$$

The two reflections are thus distributed at distances equally above and below the line $\Upsilon = \omega$ (Fig. 146B).

Note that, during the rotation, the central lattice line, $O'Q$, is continually in contact with the circle and gives rise to a continuous series of potential reflections, $S'P$, throughout the rotation. The parallel, non-central lattice line, however, does not come in contact with the reflecting circle during the very first part of the rotation and hence cannot give rise to a series of potential reflections at first. The corresponding situation, in (28), is that, for small values of ω, $\cos \omega$ is nearly unity, and when $\dfrac{d}{\cos \nu}$ is added to it, the term $\left(\cos \omega + \dfrac{d}{\cos \nu}\right)$ exceeds unity. Since this is the value of the cosine of the added angle, δ, this angle is meaningless and there can be no reflection. Reflection starts when

$$\delta = \cos^{-1}\left(\cos \omega + \frac{d}{\cos \nu}\right)$$

just equals unity, and at this rotation angle contact has just occurred between the non-central lattice line of Fig. 146A and the circle of reflection. At this rotation angle, P' and P'' coalesce, and there is a single degenerate reflection. As the rotation angle, ω, increases, the points P' and P'' migrate around the circle in opposite directions. P' never catches up with the central lattice line contact, P, and P'' never reaches O'. Consequently, in Fig. 146B, the non-central lattice curve is always contained entirely between the corresponding central lattice line and the center line of the film. Continued rotation eventually brings the end of the vector d to the circumference of the reflecting circle. At this rotation angle, P'' stops migrating clockwise and reverses its direction; therefore Υ'' attains a minimum and thereafter starts to increase. This occurs shortly after the vector d points straight up, i. e., shortly after $\omega = 90°$. Meanwhile, P' continues to migrate clockwise until the end of the vector d almost attains a downward direction. Just before this occurs (i. e., just before $\omega = 270°$), the end of the vector d makes contact with the circumference of the circle; P' simultaneously attains this intersection, and subsequent rotation causes it to reverse its direction and migrate clockwise; therefore Υ' attains a maximum and starts to decrease. P' and P'' are now approaching each other. They coalesce again in a single point when

$$\delta = \cos^{-1}\left(\cos \omega + \frac{d}{R}\right)$$

has increased to unity.   After this it becomes meaningless, the lattice
line no longer makes contact with the reflecting circle, and no reflection
develops.   From this discussion it is evident that the form of a reciprocal lattice line is, in general, a closed, oval curve on the ω, Υ field.   For
central lattice lines, the oval form becomes degenerate.   The two
reflections then become:

$$
\left.
\begin{aligned}
\Upsilon' &= \omega + \cos^{-1}\left(\cos\omega + \frac{0}{R}\right) = 2\,\omega, \\
\Upsilon'' &= \omega - \cos^{-1}\left(\cos\omega + \frac{0}{R}\right) = \quad 0.
\end{aligned}
\right\}
\tag{29}
$$

The first reflection thus traces out a straight line of slope 2, while the
second traces out a straight line of slope 0 along the center line of the
film.   The oval curve of a central lattice line thus degenerates into a
parallelogram.

It will be observed that in the discussion of the curves it is desirable
to deal with a continuous ω, Υ field.   This is labeled $LN'L'I'J'I$ in
Fig. 147 and corresponding figures.   Unfortunately, the necessity for
introducing the direct beam at $\Upsilon = 180°$, Fig. 138, makes it a practical
necessity to split a Weissenberg film at this Υ value.   For this reason,
the actual Weissenberg film has a continuous Υ range from $-180°$
through 0° to $+180°$, corresponding with the area $NLN'J'IJ$.

A plot of a set of reciprocal lattice curves for different values of d
is given in Fig. 147.   Both the continuous Υ range of 0 to 360° and the
actual Weissenberg range of $-180°$ to $+180°$ are shown.

Invariance of reciprocal lattice row curve shape with layer line.
Attention is now directed to one of the most important practical
properties of the curve,

$$
\Upsilon = \omega \pm \cos^{-1}\left(\cos\omega + \frac{d}{R}\right).
\tag{30}
$$

This curve describes the shape of a reciprocal lattice row as it appears
on an *equi-inclination Weissenberg photograph of any layer*.   The only
part of (30) in which anything characteristic of the level enters is in
the size of the radius, R, of the reflecting circle on that level.   This
enters (30) only in the single term $\frac{d}{R}$.   Therefore, the shapes of all
lattice row curves of identical $\frac{d}{R}$ value are identical, regardless of layer
height.   This means that, if a set of parallel reciprocal lattice rows of
spacing $d = d_0$ on the zero layer is plotted on a Weissenberg photo-

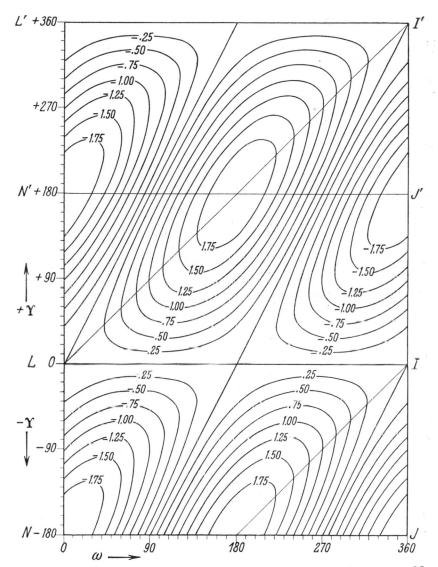

FIG. 147.   The appearance of a set of parallel reciprocal lattice lines, spaced $\dfrac{.25}{\cos \mu}$ apart, on an equi-inclination Weissenberg photograph.   The labeling of the curves gives the distance, $d/\cos \mu$, of the reciprocal lattice line from the origin of the level. The actual film corresponds with the square $NN'J'J$.

graph, they would look exactly the same as a set having a spacing $d = Rd_0$ on an n-layer, because in both, the value of the term $\frac{d}{R}$ equals d. The practical significance of this is that a Weissenberg projection of a series of parallel lattice lines can be carefully prepared, and the *shapes* of these lattice line curves will hold for any layer. Only the meaning of the interval between curves will change from level to level, and the measure of this spacing varies by the simple factor R (or its equivalent, $\cos \nu$) for that level. Data for the construction of such a series of curves are given in Table 14. The actual curves accurately drawn for a Weissenberg instrument of undistorted scale and standard camera diameter 57.3 mm. are given in Fig. 148. This figure can be used for any other camera diameter for an instrument of undistorted scale, by simply enlarging it by the ratio $\frac{\text{Diameter of camera in millimeters}}{57.3}$. For cameras of distorted scale, the data of Table 14 must be plotted anew.

The d labeling for the lines on Fig. 148 is unimportant. The function of this figure is to give the shapes of reciprocal lattice line curves on an equi-inclination Weissenberg photograph so that these reciprocal lattice lines can be easily identified. In order to use Fig. 148 for this purpose most effectively, a positive print of Fig. 148 on photographic film should be prepared and laid on or below the Weissenberg film, Fig. 149. Any reflections on the Weissenberg photograph which line up along a curve of Fig. 148 are then identified as occurring on the same reciprocal lattice row, and other rows will be found parallel with it, Fig. 149, identified by spots lining up along other curves with the same setting of the chart. Note that a plane lattice has an infinite number of sets of parallel lattice rows of different $\bar{\varphi}$ directions. Corresponding with this, the template of Fig. 148 may be moved parallel with itself along the ω or $\bar{\varphi}$ axis, and different sets of spots will be found to line up along the reciprocal lattice row curves at different $\bar{\varphi}$ settings. The density of spots along the lattice row curves increases with the simplicity of the indices of the lattice row, and this density is greatest when the straight line edges of Fig. 148 are set on the film along the shortest axes of the plane lattice of the Weissenberg film. These axes are identical with the solid reciprocal lattice axes except for the possible complications of extinctions on the zero layer.

If it is necessary to determine the spacing of the lines in Fig. 148 for any purpose, it is merely necessary to lay the scale, $\xi_{\bar{z}}$ (taken from Fig. 144 for the appropriate layer angle, ν), along the middle of Fig. 148.

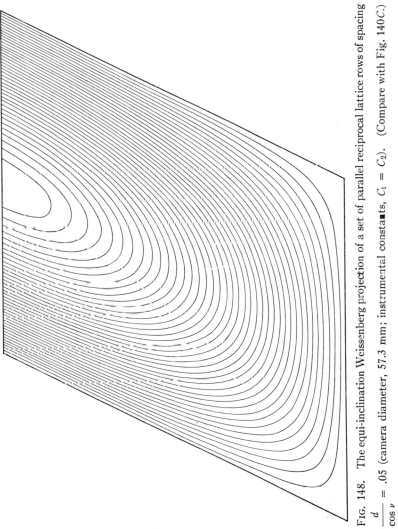

Fig. 148.   The equi-inclination Weissenberg projection of a set of parallel reciprocal lattice rows of spacing $\frac{d}{\cos \nu} = .05$ (camera diameter, 57.3 mm; instrumental constants, $C_1 = C_2$).   (Compare with Fig. 140C.)

TABLE 14

DATA FOR CONSTRUCTION OF RECIPROCAL LATTICE LINES, LOCATED AT DISTANCES, d, FROM ORIGIN, EQUI-INCLINATION SETTING

Solution of $\Upsilon = \omega \pm \cos^{-1}(\omega + d)$

ω	Υ									
	$d=$ 0.00	$d=$ 0.05	$d=$ 0.10	$d=$ 0.15	$d=$ 0.20	$d=$ 0.25	$d=$ 0.30	$d=$ 0.35	$d=$ 0.40	$d=$ 0.45
0°	0.00									
	0.00									
10	20.00									
	0.00									
20	40.00	28.23								
	0.00	11.77								
30	60.00	53.65	44.98							
	0.00	6.35	15.02							
40	80.00	75.32	70.00	63.65	54.98					
	0.00	4.68	10.00	16.35	25.02					
50	100.00	96.15	92.03	87.55	82.57	76.77	69.48	56.88		
	0.00	3.85	7.97	12.45	17.43	23.23	30.52	43.12		
60	120.00	116.63	113.13	109.48	105.58	101.40	96.87	91.78	85.83	78.20
	0.00	3.37	6.87	10.52	14.42	18.60	23.13	28.22	34.17	41.80
70	140.00	136.92	133.77	130.52	127.18	123.70	120.05	116.22	112.10	107.62
	0.00	3.08	6.23	9.48	12.82	16.30	19.95	23.78	27.90	32.38
80	160.00	157.08	154.12	151.12	148.07	144.95	141.73	138.43	135.00	131.42
	0.00	2.92	5.88	8.88	11.93	15.05	18.27	21.57	25.00	28.58
90	180.00	177.13	174.27	171.37	168.48	165.52	162.55	159.52	156.42	153.25
	0.00	2.87	5.73	8.63	11.52	14.48	17.45	20.48	23.58	26.75
100	—	—	—	—	—	—	—	179.83	176.92	173.95
	0.00	2.90	5.78	8.65	11.52	14.38	17.27	20.17	23.08	26.05
110	—	—	—	—	—	—	—	—	—	—
	0.00	3.02	6.00	8.93	11.83	14.72	17.60	20.45	23.32	26.20
120	—	—	—	—	—	—	—	—	—	—
	0.00	3.25	6.42	9.52	12.55	15.52	18.48	21.37	24.27	27.13
130	—	—	—	—	—	—	—	—	—	—
	0.00	3.65	7.13	10.48	13.72	16.87	19.85	22.97	25.95	28.88
140	—	—	—	—	—	—	—	—	—	—
	0.00	4.27	8.25	11.97	15.52	18.93	22.23	25.42	28.52	31.58
150	—	—	—	—	—	—	—	—	—	—
	0.00	5.32	10.00	14.27	18.23	21.97	25.53	28.93	32.22	35.42
160	—	—	—	—	—	—	—	—	—	—
	0.00	7.17	12.88	17.83	22.28	26.38	30.23	33.87	37.33	40.63
180	—	—	—	—	—	—	—	—	—	—
	0.00	18.20	25.85	31.78	36.87	41.42	45.57	49.48	53.13	56.63
200	—	—	—	—	—	—	—	—	—	—
	40.00	47.17	52.88	57.83	62.28	66.38	70.23	73.87	77.33	80.68
220	—	—	—	—	—	—	—	—	—	—
	80.00	84.27	88.25	91.97	95.52	98.93	102.23	105.42	108.52	111.58
240	—	—	—	—	—	—	—	—	—	—
	120.00	123.25	126.42	129.52	132.55	135.52	138.48	141.37	144.27	147.13
260	—	—	—	—	—	—	—			
	160.00	162.90	165.78	168.65	171.52	174.38	177.27			
270	180.00									
ω for $\Upsilon = 180°$	90.00	91.43	92.87	94.30	95.73	97.18	98.63	100.08	101.52	103.00
	270.00	268.57	267.13	265.70	264.27	262.82	261.37	259.92	258.48	257.00
ω for nose of curve	0.00	18.20	25.85	31.78	36.87	41.42	45.57	49.45	53.13	56.63

(This solution is for zero layer line. For n-layers, instead of d, read $\dfrac{d}{\cos \mu}$ ·)

To transform Υ to x for cylindrical film, multiply by $\dfrac{1}{C_1} = \dfrac{2\pi r_F}{360°}$.

Υ

d = 0.50	d = 0.55	d = 0.60	d = 0.65	d = 0.70	d = 0.75	d = 0.80	d = 0.85	d = 0.90	d = 0.95	d = 1.00
60.00										
60.00										
102.65	96.87	89.60	77.25							
37.35	43.13	50.40	62.75							
127.65	123.65	119.33	114.55	109.12	102.53	93.20				
32.35	36.35	40.67	45.45	50.88	57.47	66.80				
150.00	146.63	143.13	139.48	135.57	131.42	126.87	121.78	115.85	108.20	90.00
30.00	33.37	36.87	40.52	44.43	48.58	53.13	58.22	64.15	71.80	90.00
170.95	167.88	164.77	161.55	158.23	154.80	151.22	147.43	143.42	139.07	134.27
29.05	32.12	35.23	38.45	41.77	45.20	48.78	52.57	56.58	60.93	65.73
—	—	—	—	179.02	175.92	172.75	169.47	166.08	162.55	158.85
29.08	32.00	34.45	37.93	40.98	44.08	47.25	50.53	53.92	57.45	61.15
—	—	—	—	—	—	—	—	—	—	180.00
30.00	32.87	35.73	38.63	41.52	44.48	47.45	50.48	53.58	56.75	60.00
31.78	34.68	37.55	40.42	43.28	46.15	49.05	51.97	54.90	57.88	60.93
34.57	37.52	40.45	43.33	46.21	49.08	51.95	54.82	57.70	60.60	63.52
38.53	41.58	44.57	47.52	50.45	53.33	56.22	59.08	61.97	64.82	67.70
43.92	47.07	50.13	53.17	56.13	59.07	61.97	64.85	67.73	70.58	73.45
60.00	63.25	66.42	69.52	72.55	75.52	78.48	81.37	84.27	87.13	90.00
83.92	87.07	90.15	93.17	96.13	99.07	101.97	104.85	107.73	110.58	113.45
114.57	117.52	120.45	123.33	126.22	129.08	131.95	134.82	137.70	140.60	143.52
150.00	152.87	155.73	158.63	161.52	164.48	167.45	170.48	173.58	176.75	180.00
104.48	105.97	107.45	108.97	110.48	112.02	113.58	115.15	116.75	118.37	120.00
255.52	254.03	252.55	251.03	249.52	247.98	246.42	244.85	243.25	241.63	240.00
60.00	63.25	66.42	69.52	72.55	75.52	78.47	81.37	84.27	87.13	90.00

TABLE 14 — *Continued*

DATA FOR CONSTRUCTION OF RECIPROCAL LATTICE LINES, LOCATED AT DISTANCES, *d*, FROM ORIGIN, EQUI-INCLINATION SETTING

Solution of $\Upsilon = \omega \pm \cos^{-1} (\omega + d)$

ω	ϒ									
	d = 1.05	d = 1.10	d = 1.15	d = 1.20	d = 1.25	d = 1.30	d = 1.35	d = 1.40	d = 1.45	d = 1.50
0°										
10										
20										
30										
40										
50										
60										
70										
80										
90										
100	128.78 71.22	122.12 77.88	112.47 87.53							
110	154.93 65.07	150.72 69.28	146.10 73.90	140.92 79.08	134.77 85.23	126.67 93.33				
120	176.63 63.37	173.13 66.87	169.48 70.52	165.58 74.42	161.40 78.60	156.87 83.13	151.78 88.22	145.83 94.17	138.20 101.80	120.00 120.00
130	— 64.03	— 67.20	— 70.48	— 73.87	— 77.38	178.92 81.08	175.00 85.00	170.78 89.22	166.17 93.83	161.00 99.00
140	— 66.50	— 69.52	— 72.58	— 75.72	— 78.95	— 82.28	— 85.73	— 89.35	— 93.17	— 97.23
150	— 70.60	— 73.53	— 76.50	— 79.52	— 82.58	— 85.72	— 88.95	— 92.27	— 95.73	— 99.35
160	— 76.33	— 79.20	— 82.13	— 85.08	— 88.08	— 91.12	— 94.22	— 97.40	— 100.68	— 104.08
180	— 92.87	— 95.73	— 98.63	— 101.52	— 104.48	— 107.45	— 110.48	— 113.58	— 116.75	— 120.00
200	— 116.33	— 119.22	— 122.13	— 125.08	— 128.08	— 131.12	— 134.22	— 137.40	— 140.68	— 144.08
220	— 146.50	— 149.52	— 152.58	— 155.72	— 158.95	— 162.28	— 165.73	— 169.35	— 173.17	— 177.23
240										
260										
270										
ω for ϒ = 180°	121.67 238.33	123.37 236.63	125.10 234.90	126.87 233.13	128.68 231.32	130.55 229.45	132.45 227.55	134.43 225.57	136.47 223.53	138.58 221.42
ω for nose of curve	92.87	95.73	98.63	101.52	104.48	107.45	110.48	113.58	116.75	120.00

(This solution is for zero layer line. For n-layers, instead of d, read $\dfrac{d}{\cos \mu}$.)

To transform Υ to x for cylindrical film, multiply by $\dfrac{1}{C_1} = \dfrac{2\pi r_F}{360°}$.

					Υ				
$d =$ 1.55	$d =$ 1.60	$d =$ 1.65	$d =$ 1.70	$d =$ 1.75	$d =$ 1.80	$d =$ 1.85	$d =$ 1.90	$d =$ 1.95	$d =$ 2.00
154.88	146.82								
105.12	113.18								
178.37	173.48	167.87	160.93	150.25					
101.63	106.52	112.13	119.07	129.75					
—	—	—	—	177.88	170.93	160.27			
103.15	107.22	111.63	116.50	122.12	129.07	139.73			
—	—	—	—	—	—	—	176.20		
107.62	111.32	115.27	119.48	124.13	129.35	135.55	143.80		
—	—	—	—	—	—	—	—	—	180.00
123.37	126.87	130.52	134.43	138.58	143.13	148.22	154.15	161.80	180.00
—	—	—	—	—	—	—			
147.62	151.33	155.27	159.50	164.12	169.35	175.55			
140.80	143.13	145.58	148.22	151.05	154.15	157.67	161.80	167.17	180.00
219.20	216.87	214.42	211.78	208.95	205.85	202.33	198.20	192.83	180.00
123.37	126.87	130.55	134.43	138.58	143.13	148.22	154.15	161.80	180.00

The distance, d, of any line from the center line of the film is then read along the slope of Fig. 148.

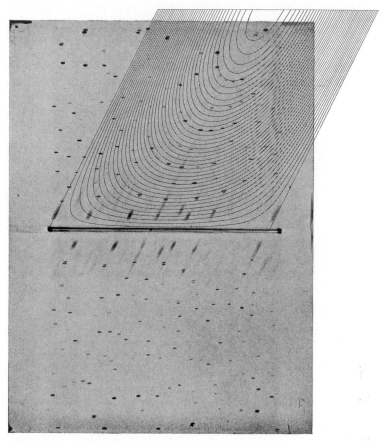

FIG. 149.   The use of the lattice row template in interpreting equi-inclination Weissenberg photographs.   Note the alignment of the spots of the photograph along the curves of the template.   Each such " festoon " of spots represents a possible reciprocal lattice row.   (Pectolite, same data as Fig. 124.)

Indexing with the aid of lattice row template.   The purpose of reconstructing a reciprocal lattice from a set of films is to enable one to visualize the reciprocal lattice, in order:

(a) to choose appropriate reciprocal lattice axes,
(b) to index the reflections on the basis of these axes, and
(c) to measure the lengths of the axes.

With equi-inclination photographs, it is unnecessary to reconstruct the reciprocal lattice, for all this may be accomplished directly on the film itself, with the aid of the several equi-inclination charts, specifically,

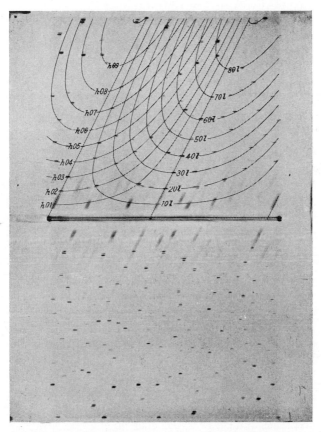

FIG. 150.   Axial reciprocal lattice rows as drawn in on the Weissenberg photograph with the aid of the template.   The rows labeled 10*l*, 20*l*, 30*l*, etc., have been drawn in with the template setting shown in Fig. 149.   The rows labeled *h*01, *h*02, *h*03, etc., have been drawn in with a second setting.

Fig. 148 for indexing, and Fig. 144 for measuring the lengths of the reciprocal lattice axes.

Although the selection of reciprocal cell axes may be somewhat complicated and deserves further discussion, especially for the oblique systems, one can ordinarily select them from symmetry on the Weissenberg photograph.   Under any circumstances, in order to index an equi-inclination Weissenberg photograph of some layer referable to axes which are central lattice rows, one first selects the straight lattice lines of the two pinacoid reflections (dome reflections on n-layers) to be used as reciprocal lattice axes. (These will be the straight lines most densely bestrewn with reflections for a primitive lattice, but may be the second densest rows for non-primitive lattices.) The photograph (or dot copy made on glass or cellophane) is then placed on top of the template so that the first of the two pinacoid (or dome) lines chosen as axes corresponds with the sloping left and right margins of the template. When the template is centered, the film spots will be found to line up along or between curves, Fig. 149. The spots of each such line, representing lattice points arranged in a lattice row parallel to the first coordinate lattice axis, may be connected by means of a line drawn with a soft pencil line directly on the film. The template is then shifted so that its left and right margins line up with the second line of pinacoid (or dome) reflections. Lines are again drawn connecting reflections from lattice points parallel with the second coordinate lattice axis. After these two sets of lines are drawn, each spot is at the intersection of two lines, Fig. 150. (For the n-layer lines of monoclinic crystals rotated about other than the b-axis, and for triclinic crystals, this procedure must be slightly modified to take account of the non-existence of central lattice lines. In these cases, one (for a monoclinic crystal not rotated about b-axis) or two (for a triclinic crystal) reciprocal lattice axes are non-central, and hence appear on the Weissenberg photograph as curves. In such instances a curve (not the straight-line margins) on the template should be found to fit the curve on the photograph; otherwise the procedure is the same.

When this procedure has been followed on each of the Weissenberg photographs representing the several layers, then the Weissenberg equivalent of the reciprocal lattice net has been laid out, with the possible exceptions of missing rows due to extinctions. Extinctions due to glide planes and screw axes occur on the zero levels, and others due to lattice centerings occur on the odd n-levels (except in rhombohedral crystals). It is best, therefore, to index first some n-layer photograph, preferably the second layer. As an example of indexing, suppose that the crystal is rotated about b, [010], and that the zero layer photograph is being analyzed. One system of lines, drawn as outlined above, may then be labeled $10l$, $20l$, $30l$, $40l$, $50l \cdots$, precautions being taken against omitting weak or even missing lines by observing the equal d-interval spacing. The other system is labeled $h01$, $h02$, $h03$, $h04$, $h05 \cdots$. (Which system shall be labeled $10l$, etc., and which shall be labeled $h01$, etc., can be controlled by a knowledge of the original orientation of the crystal when the camera translation is zero.) Each spot is now at the intersection of two of the above lines, and has an index $h0l$, where h and l are determined by the labels of the lines intersecting at the spot, Fig. 150.

In indexing the zero layer line, it is not known, in general, whether h or l values are missing because of extinctions. The lines may, however, be sketched in as above, and then, with the aid of Figs. 144 and 145, the ξ value of the d spacing of each system of lines may be compared with the corresponding ξ value of the second layer, already indexed. Missing lattice lines can then be sketched in, or allowed for, and the photograph indexed. The correct ξ value of the spacing can also be determined from other means, say the ξ value of an identity period found on the rotation photograph for a rotation about a different axis.

Advantages of taking Weissenberg photographs by the equi-inclination method. It should be apparent from the discussion of equi-inclination Weissenberg photographs that they are comparatively simple to interpret. Fundamentally, this simplicity is due entirely to the fact that the equi-inclination brings the crystal-rotation axis, Fig. 146, exactly on the circumference of the reflecting circle of the n-layer. For any other inclination of the x-ray beam, this is true only for the zero layer. The equi-inclination geometry of any layer is thus identical with that of the zero layer of any method, except for the absolute size of the radius, R, of the reflecting circle. The plan in Fig. 146A might equally well be an illustration for the zero layer of the equi-inclination method except for the appearance of the profile of the sphere of reflection, of radius 1. The plan could be completely transformed into the zero-layer case by enlarging it by the factor $\frac{1}{R}$, and removing the outer circle. This enlargement would also enlarge the only other linear element, d, by the factor $\frac{1}{R}$. The quantity d would thus enlarge into $\frac{d}{R}$. Since the reflection angles, Υ, depend in ultimate analysis on the two linear quantities, R and d, they will evidently be identical for all layers for equal values of $\frac{d}{R}$.

This situation is not true for any other inclination angle μ. For general inclination (Fig. 157) and normal-beam inclination (Fig. 126), the rotation axis does *not* pass through the circumference of the reflecting circle. For this reason, it is impossible for the geometry of any layer to resemble that of the zero layer, where this *is* the case. For all other methods than the equi-inclination method, therefore, central lines cannot project on n-level Weissenberg photograph as straight lines, but rather as curves of variable shape, Fig. 132; nor is there more than a general resemblance between reciprocal lattice rows on the different levels, Fig. 136.

Another way of looking at this is that the several equations of reflection angle for the different methods are:

General inclination method: $\Upsilon = \omega \pm \cos^{-1}\left(\dfrac{\cos \mu}{\cos \nu} \cos \omega + \dfrac{d}{\cos \nu}\right).$

Normal-beam method: $\Upsilon = \omega \pm \cos^{-1}\left(\dfrac{1}{\cos \nu} \cos \omega + \dfrac{d}{\cos \nu}\right).$

Equi-inclination method: $\Upsilon = \omega \pm \cos^{-1}\left(\ 1\ \ \cos \omega + \dfrac{d}{\cos \nu}\right).$

Only in the equi-inclination case does the expression for the reflection angle, Υ, simplify sufficiently to permit of any resemblance for different layers. The simplification is due in the final analysis to the intersection of the rotation axis with the circumference of the reflecting circle.

Owing to this simplification and the attendant resemblance of photographs for different layers, it is uniquely possible for the equi-inclination method:

(a) to record central lattice rows as straight lines, and thus permit easy reconstruction of the reciprocal lattice, and also, more generally,

(b) to record the lattice rows of all layers as curves of similar shape, and consequently permit indexing directly on the film.

AN APPARATUS FOR TAKING EQUI-INCLINATION WEISSENBERG PHOTOGRAPHS CONVENIENTLY

Introduction. The enormous advantages which accrue from taking Weissenberg photographs by the equi-inclination ($\mu = -\nu$) method have just been emphasized. The ordinary Weissenberg apparatus is either not equipped with a beam collimating system adjustable for inclination with respect to the axis of crystal rotation, or the adjustment is possible at the extreme cost of requiring the entire apparatus to be realigned to the x-ray beam each time the beam angle is changed, which, in general, occurs whenever a new photograph is taken. Quite apart from the difficulty of adapting the ordinary Weissenberg apparatus to equi-inclination beam technique, several designs have many inherent inconveniences, such as adjustments and releases placed in positions where they cannot be operated, or films exposed to scattered direct radiation. In designing the apparatus, described herewith, especially suitable for the equi-inclination beam technique, an effort was made to build into the instrument as many of the minor conveniences as possible.

General design. The general design appropriate for equi-inclination beam technique has already been indicated in diagrammatic fashion, Fig. 138. The instrument based on this design and set up ready for use is shown in Fig. 151. The important features of the instrument are described below:

Of fundamental importance is the triple-base construction. The function of the lowest base is to provide a track on which the rest of the instrument may be translated so as to bring the collimating system in and out of the x-ray beam. The normal operating position is shown in Fig. 151. The position when the collimating system is out of the beam is illustrated in Fig. 152, which shows the instrument stripped of most of the accessory apparatus normally removed when adjusting a crystal for a series of runs. It will be observed that this translation feature enables one to make final centering adjustments of the crystal by eye, allowing light to follow the same path followed by the x-ray beam during the run. When this adjustment is completed, the upper part of the instrument can be returned accurately to operating position. The " track " is actually a sleeve and shaft combined with a screw arrangement. Since this Weissenberg apparatus is used in connection with a vertical x-ray tube, from which the x-ray beam is ordinarily taken at a slant of 4° to 12° from the horizontal, it is necessary to provide for adjustment of the collimating system to this slope. It is highly desirable to take this function away from the leveling screws, so it is incorporated in the screw assembly. This permits adjustment of the slope of the entire upper part of the instrument so as to bring the collimating system rapidly and conveniently to the slope of the x-ray beam.

The second and third bases are iron castings with certain carefully machined sliding surfaces and bosses. The collimating system is an integral part of the middle base. The crystal-rotation and camera-translation assemblies are doweled to the third or upper base. The upper base may be rotated about an axis, which is supported by the middle base, such that the center of the crystal is left unmoved while the angle between the x-ray beam and the crystal-rotation axis may be made to take any value from 90° to 60°, i. e., μ may be varied from 0° to 30° (see page 294). The value μ is indicated by a vernier reading against a large circle in the middle base. As the upper base rotates, it is supported in all positions by sliding on the middle base. This is

FIG. 151 (above). Apparatus for taking equi-inclination Weissenberg photographs, in position ready for taking an n-level photograph.

FIG. 152 (below). Equi-inclination apparatus, with μ setting returned to zero, translated out of working position and stripped of accessories for adjustment of crystal.

arranged by having middle and upper bases fitted with carefully machined, circular, working or sliding faces.   When the upper base is clamped at any desired inclination angle, μ, therefore, the middle and upper bases make a very solid integral unit which is free from any unsteadiness; this makes a vibration-free support.

The arrangement of parts on the upper base can be well seen in the stripped instrument illustrated in Fig. 152.   The moving parts are driven by a synchronous reversing motor mounted in a position selected to reduce vibration as much as possible, i. e., almost over a working surface contact between second and third bases, and close to the sleeve-shaft translation support.   The motor is mounted on Bakelite and drives through a Bakelite coupling so as to avoid metallic contact with the instrument.   This precaution prevents a possible burnout of the motor in the event of any sparking from the x-ray tube during times of temporarily erratic operation.

The Weissenberg translation of the cylindrical camera is accomplished by mounting the camera on a carriage which slides along two steel shafts.   The carriage is driven by means of two half-nuts and screw, the screw being cut directly on the main driveshaft.

The shaft is also coupled to the crystal-rotation spindle through a pair of bevel gears and a worm.   The coupling gives a 2° rotation of the crystal for each 1-mm. translation of the camera.   An equi-inclination lattice row template has already been given for this coupling factor, Fig. 148.

Reversal of the translation motion is effected by means of a reversing switch operated by the movement of the translation carriage.   The switch itself is one of the new " Micro " switches.

Since the success of an instrument intended for routine service depends to a great extent on minor conveniences worked into the design, some of the parts of this instrument are described below in a certain amount of detail.

Pinhole system.   The system defining the x-ray beam consists of a tubular brass turning containing two lead disks, each perforated with a hole 1 mm. in diameter.   The direct beam scattered by the last disk is prevented from reaching the film by means of a long, tapered, hollow brass cone.   The dimensions of the metal parts are arranged so that all the high-θ reflections reach the camera without being screened by the pinhole system.   At the same time, the air scattering of the direct x-ray beam within the camera is cut down by extending the above-mentioned cone as far as possible toward the crystal (see Fig. 90).

The tubular turning containing the pinhole system, just described, must be removed frequently in order to remove the layer line screen, change the crystal, etc. This is quickly effected by removing a nut and pulling the tube back out of its housing, into which it may be returned so as to recover its original position exactly. The housing is supported on an upright which can be easily adjusted so that the pinhole axis includes the point of intersection of the inclination axis and crystal-rotation axis.

Adjusting crystal holder. The crystal, stuck by means of glue to a glass capillary, which in turn is fastened to a short section of brass rod by means of piceïn (Fig. 96), is oriented exactly with the aid of an adjusting holder (Fig. 97B). This consists of fours ledges, adjustable by means of milled knobs, two adjusting for two translations at right angles, the other two adjusting for two angles whose planes are normal to one another. If the instrument is correctly designed, the radii of the two arcs adjusting the two angles are equal to their distances from the crystal. In this manner, the crystal, being at the center of the circles of each of the arc sledges, remains unmoved on tilting. This adjusting holder must be of especially small dimensions to fit inside the Weissenberg layer line screen.

The base of the holder is arranged to fit interchangeably onto all devices in the laboratory using single crystals, i. e., Weissenberg apparatus, oscillating-crystal apparatus, goniometers, etc. The adjustment of the crystal is carried out on a two-circle goniometer, and when it is completed, the adjusting holder together with its attached crystal is removed and replaced on the Weissenberg instrument.

Crystal-rotation assembly. Figure 153 gives a diagrammatic cross section of the crystal-rotation spindle. The adjusting crystal holder fits in place on the left of the spindle. The base of the holder rests against the base of the spindle, to which it is held by a screw ring. The holder itself maintains a fixed azimuth with respect to the spindle with the aid of a small pin (not shown) on the spindle base, the pin fitting into a keyway on the holder base.

The entire spindle is rotated in its bearing by a worm which works against a gear on the spindle. The rotation coordinate, ω, may be read at any time by means of the setting of the graduated dial against the vernier. The graduated dial maintains a permanently fixed relation with respect to the azimuth of the adjusting crystal holder.

The dial has a number of uses. For example, it gives an accurate measure of the ω range of the rotation. It also enables one to start the ω range at any desired point with respect to the orientation of the crystal, which has been fixed during its adjustment on the two-circle goniometer. This orientation being known, the position at which each reciprocal lattice row starts to reflect can be determined from the dial reading. In making Weissenberg photographs, especially of triclinic crystals, it is an advantage for subsequent interpretation of the photographs to have this information marked directly on the film while it is being taken. This record may be made as follows:

In Fig. 154A, note that a reciprocal lattice row is normal to each prism plane in the zone of the rotation axis. The initial position of the plane before rotation is shown dashed. In this position, the plane has

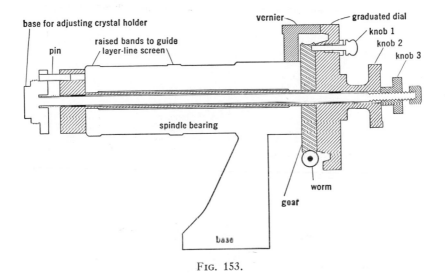

Fɪɢ. 153.

the azimuth coordinate, φ, which can be measured by the two-circle goniometer when the crystal is being adjusted to parallelism with the rotation axis. When the crystal rotation (ω) has passed through $\bar{\varphi}°$, where $\bar{\varphi} = 90° - \varphi$, then the central lattice row normal to the plane in question just starts to reflect. For the zero layer, this particular row represents the orders of reflection of the plane in question, and the indices of the points in the row will be multiples of the indices of the reflecting plane; e. g., if the index of the plane is (120), then the indices of the lattice row are 000, 120, 240, 360, 480, etc. (unless extinctions eliminate some of these). The reflection 000 represents the zero-order beam at the foot of the series, Fig. 154B. This can be selectively recorded, and therefore serves to identify the series as due to reflections from some particular plane—(120) in this case—by removing the direct beam screen when the rotation dial setting is $\bar{\varphi}_{(120)}$, and momentarily recording the direct beam. If, before the Weissenberg recording starts, three such planes are selected, and three such direct beams recorded — say for (100), (110), and (010) for an oblique crystal — then sufficient data are present on the film to correlate the reciprocal lattice level derived from the film with the planes of the crystal origi- nally observed by means of the optical goniometer.   In this manner, the surface morphology of the crystal and its reciprocal lattice may be correlated.

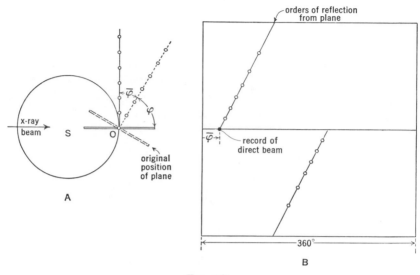

Fig. 154.

The rotation spindle assembly shown in Fig. 153 has two other important features, both concerned with the accurate adjustment of the crystal to the center of the x-ray beam.   When the crystal adjust- ing head is first transferred from the optical goniometer to the Weissen- berg spindle, it is not yet adjusted left and right with respect to the center of the x-ray beam.   This is accomplished by rotating knob 3, Fig. 153.   Rotation of this knob communicates a pure left-right trans- lation to an inner shaft, which slides within the main rotation shaft, and mutual rotation of these shafts is prevented by a pin-keyway shown at the left of the assembly.

Furthermore, before the crystal is rotated, it is desirable that its center of gravity stay in the exact center of the x-ray beam.   This adjustment should have been roughly made by means of the transla- tion screws of the adjusting crystal holder while the crystal was on the optical goniometer.   It should be checked for exactness after the crystal is transferred to the Weissenberg instrument, and possibly refined. In order to make this check, the crystal must be freely rotated.   Now, the crystal shaft is normally fixed to the rotating mechanism through the pin attached to knob 1. When this pin is withdrawn, the crystal may be rotated free of the gear by turning knob 2. The exact rotation setting of the crystal can subsequently be recovered by permitting the pin of knob 1 to engage.

The two adjustments just mentioned are made when the Weissenberg apparatus is in the out-of-beam position, Fig. 152, and while the operator is looking down the pinhole system at a well-illuminated white surface. In order to facilitate accurate centering, a lens is clipped over the x-ray-tube end of the pinhole system. This lens has a focal length equal to its distance from the crystal axis, and, with its aid, the eye can easily focus on the crystal and see it framed in the farther pinhole opening. For the instrument described, the power of this lens is about 15 diopters.

Screens. The layer line screen slips over the crystal and its adjusting holder, and is held in place by a neat friction fit on two raised bands turned on the enlarged rotation bearing housing, Fig. 153. When pushed home, the layer slit is automatically in position to permit recording the zero layer line on the film. It may be set for other layer lines with the aid of a millimeter scale on the housing. A setting of 45 mm. beyond the zero layer line is permitted by the design. Using the normal-beam ($\mu = 0$) method this is sufficient for recording even a layer line appearing on the very top of a rotation film. Only about 15 mm. of the scale are required by the equi-inclination method (Fig. 156).

The layer line slit is 2 mm. wide (but may be reduced by use of an auxiliary band for crystals whose identity period along the axis of rotation is too great to permit isolation by this slit width). The slit is arranged to permit recording at maximum angles as well as at extremely small angles, Υ.

The direct beam is caught on a piece of lead in the bottom of a slot in a brass wedge. The wings of this hollow wedge prevent the radiation scattered by the lead from reaching any part of the film through the layer line slit. The slot form of the direct ray receiver permits the lead to receive the direct beam no matter what the crystal inclination angle, μ. The entire hollow wedge slides along a slit in the layer line screen parallel with the crystal axis and may be clamped in any desired position by a small screw.

The receiver for the direct beam when taking ordinary rotation photographs is supported in the same manner as the layer line screen. The receiver proper is a small hollow brass cone with lead at the bottom of the hole.

Camera. The film, contained within a black paper envelope, is supported against the inside of a brass cylinder (Fig. 92). This method of camera construction has several advantages for the present purpose: It allows the recording of the maximum possible values of Υ. Since, for zero layer lines, $\Upsilon = 2\theta$, this means that the maximum number of orders of reflection are recorded — a great advantage in crystal-structure analysis, where the accuracy of parameter fixing usually depends on the number of orders available. This camera type also is partially self-screening against stray radiation, for the film is *inside* of a metal cylinder. The camera is slotted sufficiently to receive the pinhole assembly.

The camera cylinder may be adjusted for parallelism and concentricity with the crystal-rotation axis by loosening four screws in its supports. This allows two translation and two slight rotation movements of the cylinder with respect to the camera carriage.

The camera is further shielded from scattered stray radiation by two means: The end of the camera is capped with a brass cup which prevents stray radiation from entering the open end of the camera. The greatest source of stray radiation, namely, the open slit in the camera which permits entry of the pinhole system and which faces the x-ray tube, is shielded by means of a long brass-cased lead strip shown protruding beyond the left end of the camera in Fig. 151. This is supported on two upright fingers which are attached to the upper base and rotate with it when μ is set, thus keeping the lead strip always parallel to the camera. As the upper base swings, the pinhole tube pushes the lead strip along its support fingers. With these several screening precautions, no fogging is experienced with copper or softer radiation with any exposures, but with the harder molybdenum radiation it is sometimes advisable to put additional lead sheets between x-ray tube and instrument for the longer exposures.

The camera has an effective diameter of 57.3 mm. This is a very convenient size for reasons already pointed out on page 223. The camera takes a 5-by-$6\frac{11}{32}$-inch film, which is conveniently prepared by trimming one end of a standard 5-by-7-inch x-ray film.

In order to permit the camera to record reflections so that ω starts at the same point on each film, regardless of which layer line is being analyzed, a millimeter scale is provided at the camera base (seen in Fig. 151). This allows the camera to be moved forward by an amount equal to the distance from the zero layer line to the layer line being analyzed.

Camera carriage. The dovetail base of the camera clamps to the carriage by means of a short lever operating from the near side of the carriage.   A lever, also in the same region, releases the carriage from translation by disengaging the half-nuts from the drive screw.   The released position is maintained during the taking of pure rotation photographs and also while the position of the camera is being adjusted to the desired translation zero.

Pins on the carriage operate the trip fingers which, in turn, operate the reversing switch.   These can be set for any total ω range up to about 200°, the maximum which can be recorded without considerably increasing the size of the entire apparatus.

The carriage is designed to carry the camera as nearly as physically possible over the shaft tracks in order to reduce friction and possible vibration.   The shaft mounts and carriage shaft sleeves are carefully bored together.   When properly made, this gives an excellent translation device free from lost motion.   The method of driving by means of a screw of reasonably small pitch (1 mm.) gives the most uniform of any of the several possible Weissenberg translation drives.

Adjustments, releases, etc.   The primary alignments of several important geometrical features are permanently built into the instrument by exact machining and fitting.   Thus, the camera translation and crystal rotation, exactly machined to parallelism with the base, are adjusted to mutual parallelism during assembling by clamping, indicating, and doweling.   The μ inclination axis is then bored to intersect the crystal-rotation axis.   The pinhole system is adjusted after assembly.   The camera is adjusted to parallelism and concentricity with the crystal rotation axis by the instrument maker with the aid of a template fitting both camera and crystal rotation bearing.   Housing for the small shaft which couples crystal-rotation spindle to the driveshaft is arranged so that the gearing clearance may be adjusted without disturbing the correct positions of the gears.

All releases, operating adjustments, clamps, scales, etc., are located conveniently for use during routine operation of the apparatus.

Inclination and screen settings.   From the elevation of Fig. 146A, it can be seen that the proper inclination angle, μ for equi-inclination technique is given by

$$\mu_e = \sin^{-1}\left(\frac{\zeta}{2}\right), \tag{31}$$

where ζ is the coordinate, parallel to the rotation axis, of the layer of the reciprocal lattice under examination, and equal to the product of the layer number, n, and the reciprocal lattice level spacing, ζ_1, parallel to the rotation axis.   The maximum possible value of ζ which can be explored by the normal-beam technique is 1.   Using the equi-inclination technique, however, it is possible to explore to ζ values of 2. This corresponds with an upper limit of 90° for μ.   Mechanical diffi-culties prevent the attaining of this maximum limit without increasing the length of the pinhole system so much as to require greatly increased exposures.   The practical upper limit of μ is indicated below.

It is often convenient to set the inclination angle, μ, directly from the height, y, of the layer line under examination, as measured on a pure rotation film.   Figure 126 indicates that, under these conditions,

$$\sin \nu_\perp = \zeta, \tag{32}$$

where $\nu_\perp$ is the complement of the interior angle of the cone of diffracted rays which forms the layer line under examination, and Fig. 55 shows that

$$\nu_\perp = \tan^{-1}\left(\frac{y}{r_F}\right). \tag{33}$$

Substituting in (31) the value indicated by (32) gives

$$\mu_e = \sin^{-1}\left(\frac{\sin \nu_\perp}{2}\right). \tag{34}$$

This can be expressed in terms of film constants by substituting the value of $\nu_\perp$ given by (33),

$$\mu_e = \sin^{-1}\left(\frac{\sin \tan^{-1} (y/r_F)}{2}\right). \tag{35}$$

The maximum possible value of y is ∞.   This value is for an infinitely long cylindrical camera, and requires a μ setting of 30°.   In an ordinary Weissenberg-type camera, however, y cannot possibly exceed the length of the film cylinder, which is set by the translation correspond-ing to the maximum value of ω desired.   This is about 200°, corre-sponding to 100-mm. translation in the camera described here.   In standard rotating-crystal practice, in which the zero layer line is recorded along the center of the film, the upper limit of y is half this, or about 50 mm., which calls for an ordinary maximum value of μ in the region of 25–26°.

The appropriate layer line screen, scale setting, s, corresponding with this inclination angle, is given by Fig. 119 as

$$s = r_s \tan \nu, \tag{36}$$

where r_s is the effective radius of the layer line screen (the mean of its internal and external diameters).

Fig. 7.6. Chart for deriving the contact resistance voltage-angle if from the nine-thousand inter-couple.

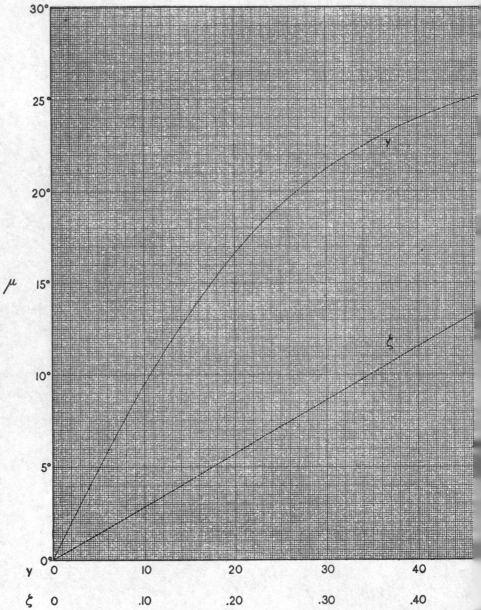

FIG. 155. Chart for deriving the equi-inclination setting angle, μ, from eit[...]
on a rotation photograph n[...]

cal lattice level coordinate, ζ, or the height, y, of the layer line as it appears
mera of diameter 57.3 mm.

It is inconvenient to make these calculations, simple as they are, for the various settings each time a photograph is to be taken. A graphical solution of (31) sufficiently accurate for setting the inclination angle is given in Fig. 155. Since the camera diameter, 57.3 mm., used in the instrument described here, is already a Weissenberg standard, a graphical solution of (35) for this camera is also given in Fig. 155.

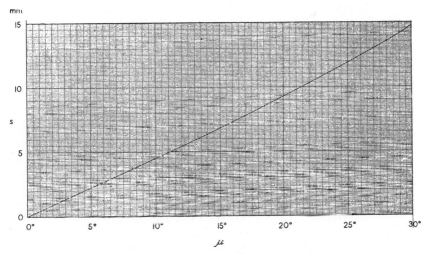

Fig. 156. Chart for deriving the layer line screen setting, s, from the equi-inclination setting, μ (layer line screen diameter, 51 mm.).

For this standard camera diameter, the optimum clearance to both camera interior and adjusting crystal holder is realized by using a layer line screen of internal and external diameters 50 and 52 mm., respectively. A graphical solution of (36) for this effective layer line screen diameter of 51 mm. is given in Fig. 156, which is sufficiently accurate for setting the screen.

LITERATURE

W. Schneider. Über die graphische Auswertung von Aufnahmen mit dem Weissenbergschen Röntgengoniometer. *Z. Krist.* (*A*), **69** (1928), 41–48.

M. J. Buerger. The Weissenberg reciprocal lattice projection and the technique of interpreting Weissenberg photographs. *Z. Krist.* (*A*), **88** (1934), 356–380.

M. J. Buerger. An apparatus for conveniently taking equi-inclination Weissenberg photographs. *Z. Krist.* (*A*), **94** (1936), 87–99.

CHAPTER 15

MOVING-FILM PHOTOGRAPHS TAKEN WITH THE X-RAY BEAM INCLINED TO THE LAYERS OF THE RECIPROCAL LATTICE

It is not customary to take Weissenberg photographs with the x-ray beam inclined at a general angle, μ, to the layers of the reciprocal lattice. Nevertheless, such photographs are of theoretical interest, and the general theory of reflection directions is of importance in the development of some of the other moving-film methods. Several special cases, in addition to $\mu = 90°$ (the normal-beam method) and $\mu = -\nu$ (the equi-inclination method), may be reduced from the relations for general inclination.

REFLECTION RELATIONS

Figure 157 shows a view normal to the plane containing the direct x-ray beam and the rotation axis. The intersection of the beam and the rotation axis determines the origin, O, of the reciprocal lattice. The zero level, OS_0, is normal to the rotation axis at the origin, *not at the " equator " of the sphere*, while the n-level is normal to the rotation axis at height ζ above the zero level.

According to Fig. 157, the radius of the zero-level reflecting circle is

$$R_0 = \cos \mu, \tag{1}$$

and the radius of the n-level reflecting circle is

$$R_1 = \cos \nu. \tag{2}$$

The height of the zero level above the horizontal equator of the sphere is

$$H_0 = \sin \mu, \tag{3}$$

and the height of the n-level above the horizontal equator of the sphere is

$$H_1 = \sin \nu. \tag{4}$$

The coordinate of the nth level is the difference between these, namely,

$$\zeta = H_1 - H_0;$$

296

therefore

$$\zeta = \sin \nu - \sin \mu. \tag{5}$$

If the direct x-ray beam angle μ is selected, then the nth layer, of coordinate ζ, will record, according to (5), at an angle

$$\nu = \sin^{-1}(\sin \mu + \zeta). \tag{6}$$

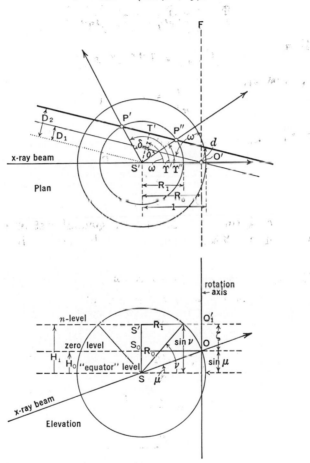

Fig. 157.

The upper part of Fig. 157 gives a view of this situation looking parallel with the rotation axis. The heavy line is a reciprocal lattice row on the nth level, and at a distance d from the origin. As the lattice rotates, the heavy line is carried with it and eventually makes contact with the reflecting circle of the nth level. As rotation continues, the line cuts the circle, in general, at two points, P' and P'', and thus gives rise to two potential reflections at angles Υ' and Υ'', respectively. If bisector $S'T'$, of angle $P'S'P''$, is drawn, it is again evident that, since $O'S'T'$ and ω have their legs at right angles,

$$O'S'T' = \omega. \tag{7}$$

The two reflection angles are, therefore,

$$\left.\begin{array}{l} \Upsilon' = O'S'T' + P'S'T', \\ \Upsilon'' = O'S'T' - P''S'T'. \end{array}\right\} \tag{8}$$

Abbreviating

$$P'S'T' = P''S'T' \text{ as } \delta, \tag{9}$$

then (8) can be written

$$\left.\begin{array}{l} \Upsilon' = \omega + \delta, \\ \Upsilon'' = \omega - \delta. \end{array}\right\} \tag{10}$$

The angle δ can be evaluated from

$$\cos \delta = \frac{S'T'}{S'P'} = \frac{D_2}{R_1}. \tag{11}$$

Here

$$D_2 = D_1 + d. \tag{12}$$

Up to this point, the derivation for the general case resembles that for the normal-beam case. From here on, however, it becomes more complicated.

D_1 is evidently

$$D_1 = R_0 \cos \omega. \tag{13}$$

Note that

$$R_0 \neq 1 \text{ (normal-beam case)}, \tag{14}$$

and

$$R_0 \neq R_1 \text{ (equi-inclination case)}. \tag{15}$$

Therefore

$$\cos \delta = \frac{D_2}{R_1} \tag{16}$$

$$= \frac{D_1 + d}{R_1} \tag{17}$$

$$= \frac{R_0 \cos \omega + d}{R_1}. \tag{18}$$

Therefore (10) becomes

$$\left.\begin{array}{l} \Upsilon_1 = \omega + \cos^{-1}\left(\dfrac{R_0 \cos \omega + d}{R_1}\right), \\[2mm] \Upsilon_2 = \omega - \cos^{-1}\left(\dfrac{R_0 \cos \omega + d}{R_1}\right). \end{array}\right\} \tag{19}$$

Substituting the values of R_0 and R_1 given by (1) and (2), (19) becomes

$$\Upsilon', \Upsilon'' = \omega \pm \cos^{-1}\left(\frac{\cos \mu \cos \omega + d}{\cos \nu}\right). \tag{20}$$

With the aid of (5), this can also be expressed in terms of the layer height coordinate, ζ, and either μ or ν:

$$\Upsilon', \Upsilon'' = \omega \pm \cos^{-1}\left(\frac{\cos \mu \cos \omega + d}{\cos [\sin^{-1} (\sin \mu + \zeta)]}\right), \tag{21}$$

$$\Upsilon', \Upsilon'' = \omega \pm \cos^{-1}\left(\frac{\cos [\sin^{-1} (\sin \nu - \zeta)] \cos \omega + d}{\cos \nu}\right). \tag{22}$$

These equations express the variation of the deflection angle, Υ, as a function of ω and other constants. For certain purposes, notably reconstructing the reciprocal lattice by the Wooster method, it is desirable to provide a ξ scale by expressing Υ in terms of ξ, and the reverse. This can be done by substituting R_0 for 1 in Fig. 130. Relation (22), Chapter 13, then becomes

$$\xi^2 = R_0^2 + R_1^2 - 2R_0R_1 \cos \Upsilon. \tag{23}$$

According to (1) and (2), R_0 and R_1 may be expressed in terms of $\cos \mu$ and $\cos \nu$, respectively. Making this substitution, (23) becomes

$$\xi^2 = \cos^2 \mu + \cos^2 \nu - 2 \cos \mu \cos \nu \cos \Upsilon, \tag{24}$$

and

$$\Upsilon = \cos^{-1}\left(\frac{\cos^2 \mu + \cos^2 \nu - \xi^2}{2 \cos \mu \cos \nu}\right). \tag{25}$$

SPECIAL CASES

In order to approach a study of the characteristics of moving-film photographs taken with an x-ray beam of general inclination, it is desirable first to discuss certain special cases. Two of these have already been studied, namely, the normal-beam case and the equi-inclination case. The third case is new.

$\mu = 0$ (**normal-beam case**), Fig. 158A. Important relations in the general case reduce to:

General lattice line: $\quad \Upsilon = \omega \pm \cos^{-1}\left(\dfrac{\cos \omega + d}{\cos \nu}\right);$ $\qquad$ (26)

Central line: $\quad \Upsilon = \omega \pm \cos^{-1}\left(\dfrac{\cos \omega}{\cos \nu}\right);$ $\qquad$ (27)

ξ scale: $\quad \Upsilon = \cos^{-1}\left(\dfrac{1 + \cos^2 \nu - \xi^2}{2 \cos \nu}\right).$ $\qquad$ (28)

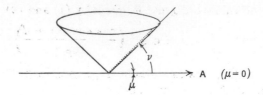

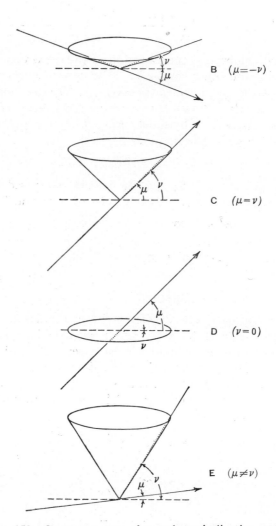

FIG. 158. Important cases of x-ray beam inclination angle, μ.

$\mu = -\nu$ (**equi-inclination case**), Fig. 158B.   Important relations in the general case reduce to:

General lattice line: $\Upsilon = \omega \pm \cos^{-1}\left(\cos \omega + \dfrac{d}{\cos \nu}\right)$;   (29)

Central line: $\Upsilon = 2\omega$;   (30)

ξ scale: $\Upsilon = 2 \sin^{-1}\left(\dfrac{\xi}{2 \cos \nu}\right)$.   (31)

$\mu = +\nu$ (**anti-equi-inclination case**), Fig. 158C.   In this case, μ has any value, and ν is equal to it in both absolute value and sign; i. e., the direct beam, after meeting the crystal, continues on as a generator of the diffraction cone in question, of cone angle $\bar{\nu}$.   According to Fig. 15, this cone can be only the zero-order diffraction cone of the rotation axis identity period, and this is equivalent to the zero layer of this identity period.   This relation also directly follows from the fact that the zero layer is located by the point of impingement of the direct beam on the sphere of reflection.   In other words, it is possible to record only the zero layer with the special case $\mu = +\nu$.   Since $\cos (+\nu) = \cos (-\nu)$, the relations for this type of recording are identical with the relations just given for the equi-inclination method, and this inclination-angle relationship carries with it all the properties of the equi-inclination method.

$\nu = 0$ (**flat diffraction-cone case**), Fig. 158D.   Important relations in the general case reduce to:

General lattice line: $\Upsilon = \omega \pm \cos^{-1} (\cos \mu \cos \omega + d)$;   (32)

Central line: $\Upsilon = \omega \pm \cos^{-1} (\cos \mu \cos \omega)$;   (33)

ξ scale: $\Upsilon = \cos^{-1}\left(\dfrac{\cos^2 \mu + 1 - \xi^2}{2 \cos \mu}\right)$.   (34)

This interesting case, which has apparently never been employed for making Weissenberg photographs, might be termed the flat-cone case. It bears an interesting relation to the normal-beam case.   Instead of moving the layer line screen so that it isolates the nth layer line, the angle of the direct beam is adjusted until the nth layer line becomes a flat cone and emerges from the layer line screen in the zero (flat cone) position.

A plot of (32) for $\mu = 30°$ ($\zeta = .5$) and d intervals of .25 is given in Fig. 159A.   This diagram illustrates the appearance of parallel reciprocal lattice lines on a Weissenberg photograph, Fig. 159B, taken by

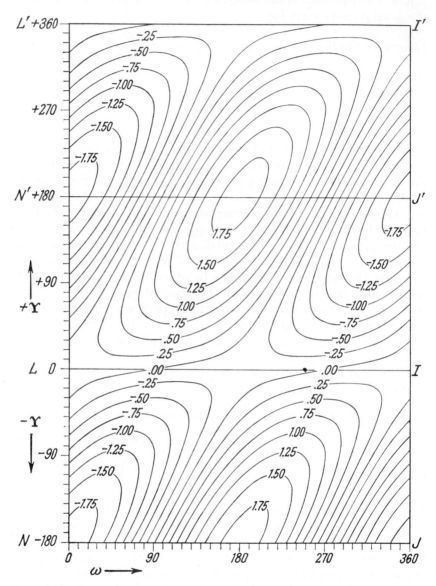

FIG. 159A.   The appearance of a set of parallel reciprocal lattice lines on a flat-cone Weissenberg photograph ($\mu = 30°$, $\zeta = .5$).   The labeling of the curves gives the distance, d, of the reciprocal lattice line from the origin of the level.   The actual film corresponds with the square $NN'J'J$.

this method. The shapes of central lattice lines on this kind of Weissenberg photograph, for various values of ζ, are shown in Fig. 160. Data for the construction of these curves are given in Table 15. The shapes of these curves may be used for templates for the interpretation of flat-

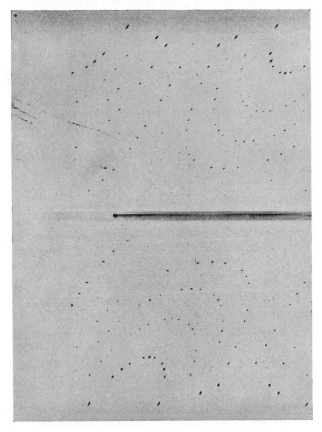

FIG. 159B. n-Level Weissenberg photograph taken by the flat-cone method. Compare the spot alignment with the curve shapes shown in the lower left part of Fig. 159A. (Valentinite, Sb_2O_3, orthorhombic, c-axis rotation, 2nd level, $\zeta = .568$, $\mu = 34.6°$; $CuK\alpha$ radiation from hot-cathode x-ray tube, filtered through nickel foil.)

cone Weissenberg photographs by the Wooster method. The ξ scale can be computed from (34). The actual scale variation with the direct beam angle, μ, is given in Fig. 161. Data for the construction of Fig. 161 are given in Table 16. In connection with this scale, the following relation is of interest: The scale relation for the flat-cone method (34) was derived from the more general relation (25). Note that in

(25) the right-hand member is symmetrical with respect to cos μ and cos ν. Thus, if $\mu = 0$ and $\nu = $ some value, m, the expression is the same as if $\nu = 0$ and $\mu = m$. In other words, the ξ scale is the same for the normal-beam case, $\nu = m$, and for the flat-cone case, $\mu = m$. The flat-cone scale as a function of ζ can therefore be obtained directly from a Bernal rotation chart.

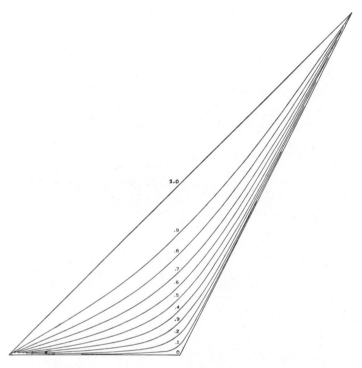

FIG. 160. The shapes of central reciprocal lattice lines for various level coordinate values, ζ, for flat-cone Weissenberg photographs (camera diameter, 57.3 mm.).

$\nu = $ **constant** (**equal-cone-case**). Another special case of inclined beam is that in which the diffraction cone angle is kept constant for a series of different layer photographs. This is a more generalized version of the flat-cone case, in which the constant was zero. An advantage to the use of this general case is that, since the diffraction cones for each layer are made the same (by appropriately varying the angle of the incident x-ray beam) the path of each and every diffracted beam, regardless of layer, is along the surface of an identical cone from the crystal to the film. All diffracted beams, therefore, have equal air paths, and strike the film at the same angle. The intensities of all diffracted beams as recorded on the film are therefore comparable with regard to air path and angle of impingement on the film.

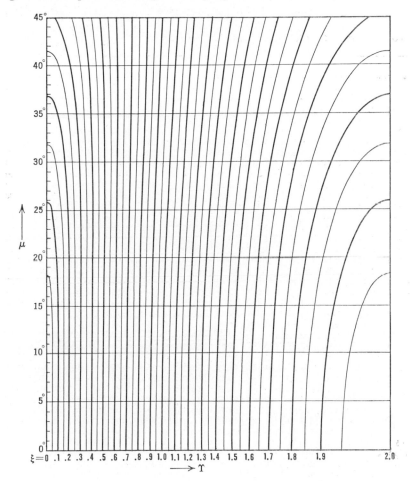

FIG. 161.   The variation of the ξ scale with the inclination angle, μ, for flat-cone Weissenberg photographs (camera diameter, 57.3 mm.).

If the cone angle is thus kept constant, the relation between the values of μ, ν, and ζ is given in (5), namely,

$$\zeta = \sin \nu - \sin \mu. \tag{5}$$

This may be written

$$\mu = \sin^{-1} (\sin \nu - \zeta)$$

$$= \sin^{-1} (\text{const.} - \zeta). \tag{35}$$

TABLE 15

DATA FOR CONSTRUCTION OF CENTRAL LATTICE LINE, FLAT-CONE SETTING

Solution of $\Upsilon = \omega \pm \cos^{-1} (\sqrt{1 - \zeta^2} \cos \omega)$

To transform Υ to x for cylindrical film, multiply by $\dfrac{1}{C_1} = \dfrac{2\pi r_F}{360°}$.

ω	Υ										
	$\zeta = 0$	$\zeta = .1$	$\zeta = .2$	$\zeta = .3$	$\zeta = .4$	$\zeta = .5$	$\zeta = .6$	$\zeta = .7$	$\zeta = .8$	$\zeta = .9$	$\zeta = 1.0$
−90°	0.00°	0.00°	0.00°	0.00°	0.00°	0.00°	0.00°	0.00°	0.00°	0.00°	0.00°
−85	0.00	0.02	0.10	0.21	0.42	0.67	1.01	1.43	2.00	2.82	5.00
−80	0.00	0.05	0.21	0.46	0.85	1.35	2.02	2.87	4.02	5.66	10.00
−75	0.00	0.07	0.31	0.71	1.28	2.05	3.05	4.35	5.06	8.52	15.00
−70	0.00	0.10	0.42	0.96	1.75	2.76	4.12	5.86	8.16	11.43	20.00
−65	0.00	0.14	0.54	1.23	2.21	3.47	5.23	7.42	10.31	14.34	25.00
−60	0.00	0.16	0.66	1.51	2.78	4.34	6.41	9.07	12.54	17.41	30.00
−55	0.00	0.20	0.80	1.83	3.29	5.21	7.69	10.82	14.86	20.52	35.00
−50	0.00	0.24	0.96	2.18	3.90	6.17	9.05	12.68	17.31	23.73	40.00
−45	0.00	0.28	1.14	2.59	4.60	7.25	10.55	14.67	19.89	27.05	45.00
−40	0.00	0.34	1.36	3.05	5.41	8.44	12.21	16.83	22.64	30.49	50.00
−35	0.00	0.40	1.62	3.62	6.35	9.82	14.06	19.20	25.56	34.08	55.00
−30	0.00	0.49	1.95	4.30	7.47	11.40	16.15	21.80	28.69	37.82	60.00
−25	0.00	0.60	2.37	5.17	8.84	13.29	18.53	24.67	32.09	41.73	65.00
−20	0.00	0.77	2.97	6.30	10.55	15.54	21.25	27.85	35.68	45.82	70.00
−15	0.00	1.03	3.83	7.87	12.72	18.23	24.40	31.38	39.58	50.10	75.00
−10	0.00	1.50	5.20	10.05	15.50	21.48	28.01	35.31	43.77	54.58	80.00
− 5	0.00	2.60	7.55	13.14	19.08	25.38	32.15	39.65	48.29	59.25	85.00
0	0.00	5.74	11.54	17.45	23.58	30.00	36.87	44.43	53.12	64.15	90.00
5	10.00	12.60	17.55	23.14	29.08	35.38	42.15	49.65	58.29	69.25	95.00
10	20.00	21.50	25.20	30.05	35.50	41.48	48.01	55.31	63.77	74.58	100.00
15	30.00	31.03	33.83	37.87	42.72	48.23	54.40	61.38	69.58	80.10	105.00
20	40.00	40.77	42.97	46.30	50.55	55.54	61.25	67.85	75.68	85.82	110.00
25	50.00	50.60	52.37	55.17	58.84	63.29	68.53	74.67	82.09	91.73	115.00
30	60.00	60.49	61.95	64.30	67.47	71.40	76.15	81.80	88.69	97.82	120.00
35	70.00	70.40	71.62	73.62	76.35	79.82	84.06	89.20	95.56	104.08	125.00
40	80.00	80.34	81.36	83.05	85.41	88.44	92.21	96.83	102.64	110.49	130.00
45	90.00	90.28	91.14	92.59	94.60	97.25	100.55	104.67	109.89	117.05	135.00
50	100.00	100.24	100.96	102.18	103.90	106.17	109.05	112.68	117.31	123.73	140.00
55	110.00	110.20	110.80	111.83	113.29	115.21	117.69	120.82	124.86	130.52	145.00
60	120.00	120.16	120.66	121.51	122.72	124.34	126.41	129.07	132.54	137.41	150.00
65	130.00	130.14	130.54	131.23	132.21	133.47	135.23	137.42	140.31	144.34	155.00
70	140.00	140.10	140.42	140.96	141.75	142.76	144.12	145.86	148.16	151.43	160.00
75	150.00	150.07	150.31	150.71	151.28	152.05	153.05	154.35	155.06	158.52	165.00
80	160.00	160.05	160.21	160.46	160.85	161.35	162.01	162.87	164.02	165.66	170.00
85	170.00	170.02	170.10	170.21	170.42	170.67	171.01	171.43	172.00	172.82	175.00
90	180.00	180.00	180.00	180.00	180.00	180.00	180.00	180.00	180.00	180.00	180.00

The solutions of this for μ as a function ζ, and for various cone angles, ν, are shown graphically in Fig. 162.

Further properties of the equal-cone case are the same as those for the general inclination case, which will now be discussed.

THE GENERAL CASE

The completely general inclination case is most easily approached by a comparison between the cases $\mu = 0$, $\mu = -\nu$, and $\nu = 0$ (normal-

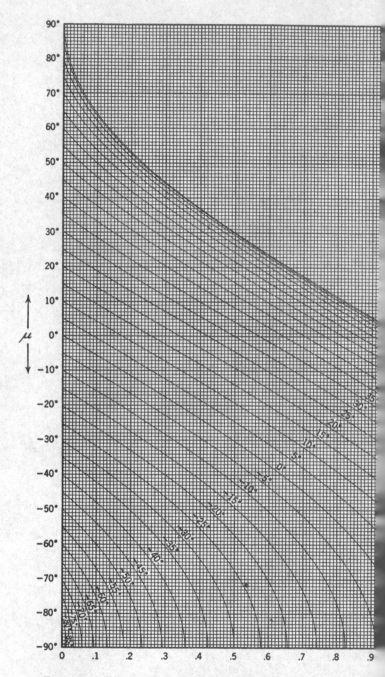

FIG. 162. Chart for deriving the inclination setting angle, μ, for a
(Each curve represents a

1.1   1.2   1.3   1.4   1.5   1.6   1.7   1.8   1.9   2.0

rocal lattice level coordinate, ζ, and a selected cone angle, ν.
lue of ν, as labeled.)

TABLE 16

DATA FOR CONSTRUCTION OF ξ SCALE FOR FLAT-CONE SETTING

$$\text{Solution of } \Upsilon = \cos^{-1}\left(\frac{1 + \cos^2 \mu - \xi^2}{2 \cos \mu}\right)$$

To transform Υ to x for cylindrical film, multiply by $\dfrac{1}{C_1} = \dfrac{2\pi r_F}{360°}$.

ξ					Υ						Value of μ for Υ = 0°	Value of μ for Υ = 180°
	μ = 0°	μ = 5°	μ = 10°	μ = 15°	μ = 20°	μ = 25°	μ = 30°	μ = 35°	μ = 40°	μ = 45°		
0.00	0.00°										00.00°	
0.05	2.87	2.85°	2.85°	2.10°							18.20	
0.10	5.73	5.73	5.77	5.47	4.72°	2.10°					25.84	
0.15	8.60	8.59	8.66	8.52	8.13	7.05	4.15°				31.79	
0.20	11.48	11.50	11.59	11.50	11.29	10.65	9.15	5.27°			36.87	
0.25	14.36	14.39	14.47	14.47	14.38	13.98	13.02	10.88	5.77°		41.41	
0.30	17.25	17.28	17.38	17.44	17.44	17.22	16.58	15.16	12.32	4.42°		
0.35	20.16	20.19	20.31	20.41	20.49	20.40	20.01	19.03	17.10	13.08		
0.40	23.07	23.12	23.25	23.39	23.54	23.57	23.36	22.72	21.36	18.64		
0.45	26.01	26.06	26.21	26.39	26.60	26.73	26.68	26.30	25.37	23.44		
0.50	28.96	29.01	29.18	29.40	29.67	29.90	30.00	29.80	29.24	27.88		
0.55	31.92	31.99	32.17	32.43	32.76	33.07	33.31	33.35	32.99	32.14		
0.60	34.92	34.98	35.19	35.48	35.86	36.27	36.63	36.86	36.80	36.28		
0.65	37.94	38.01	38.23	38.56	39.00	39.49	39.96	40.36	40.53	40.37		
0.70	40.97	41.06	41.30	41.67	42.17	42.73	43.32	43.88	44.28	44.42		
0.75	44.05	44.14	44.40	44.89	45.36	46.01	46.71	47.43	48.04	48.48		
0.80	47.16	47.25	47.54	47.98	48.59	49.45	50.20	51.02	51.83	52.55		
0.85	50.30	50.40	50.71	51.20	51.87	52.68	53.61	54.64	55.66	56.65		
0.90	53.49	53.60	53.93	54.46	55.18	56.08	57.13	58.32	59.53	60.80		
0.95	56.72	56.84	57.19	57.76	58.55	59.54	60.70	62.05	63.47	65.01		
1.00	60.00	60.13	60.51	61.12	61.98	63.05	64.34	65.85	67.48	69.30		
1.05	63.34	63.47	63.88	64.54	65.46	66.63	68.05	69.73	71.57	73.74		
1.10	66.73	66.88	67.28	68.02	69.02	70.29	71.83	73.70	75.76	78.17		
1.15	70.20	70.36	70.82	71.58	72.03	74.02	75.70	77.76	80.07	82.79		
1.20	73.74	73.90	74.40	75.21	76.43	77.85	79.69	81.94	84.50	87.57		
1.25	77.36	77.54	78.10	78.94	80.18	81.79	83.78	86.25	89.09	92.53		
1.30	81.08	81.27	81.84	82.77	84.11	85.84	88.01	90.78	93.72	97.72		
1.35	84.91	85.11	85.72	86.72	88.15	90.03	92.40	95.37	98.85	103.18		
1.40	88.85	89.07	89.72	90.80	92.35	94.39	96.97	100.23	104.10	108.98		
1.45	92.94	93.11	93.87	95.03	96.70	98.92	101.75	105.35	109.67	115.22		
1.50	97.18	97.43	98.18	99.44	101.14	103.68	106.79	110.78	115.65	122.03		
1.55	101.61	101.88	102.70	104.06	106.05	108.70	112.14	116.62	122.17	129.65		
1.60	106.26	106.55	107.44	108.94	111.11	114.05	117.89	122.98	129.43	138.55		
1.65	111.18	111.30	112.47	114.12	116.53	119.81	124.16	130.06	137.84	149.82		
1.70	116.42	116.78	117.86	119.69	122.40	126.12	131.17	138.24	148.28	169.38		
1.75	122.09	122.40	123.70	125.78	128.87	133.21	139.27	148.36	164.40			41.41°
1.80	128.32	128.77	130.20	132.57	136.22	141.50	149.35	163.71				36.87
1.85	135.34	135.87	137.53	140.44	145.00	152.04	164.94					31.79
1.90	143.61	144.29	146.39	150.23	156.77	170.35						25.84
1.95	154.32	155.30	158.53	165.40								18.20
2.00	180.00											00.00

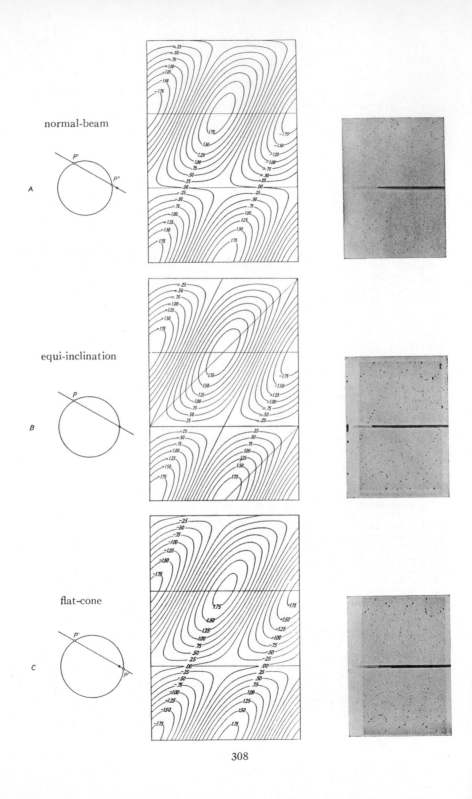

normal-beam

equi-inclination

flat-cone

A

B

C

beam, equi-inclination, and flat-cone, respectively). This comparison is shown graphically in Figs. 163A, B, and C. The transitions between the three cases are evident. The most obvious differences in curve shapes are to be found in the central lattice lines. These differences depend on the location of the rotation axis for a level with respect to the reflecting circle on that level. If the rotation axis passes through the circle of reflection, Fig. 163B, there is only one reflection, corresponding with P', which varies with the rotation angle, ω. The other, corresponding with P'', is constant and continually strikes the center of the film. If the rotation axis is not on the reflecting circle, then there are, in general, two reflections, corresponding with P' and P'', whose directions change with the rotation angle, ω. If the rotation axis is outside the circle ($\mu < \nu$), then P' and P'' start together and travel in opposite directions. If, however, the rotation axis falls within the circle ($\mu > \nu$), then P' and P'' can never coalesce but always occupy different quadrants and travel in the same direction.

This transition may also be studied analytically from the general reciprocal relation given in (20). For present purposes this may be written in the form

$$\Upsilon', \Upsilon'' = \omega \pm \cos^{-1}\left(\frac{\cos \mu}{\cos \nu} \cos \omega + \frac{d}{\cos \nu}\right). \tag{36}$$

For central lattice lines, $d = 0$, and (36) reduces to:

$$\Upsilon', \Upsilon'' = \omega \pm \cos^{-1}\left(\frac{\cos \mu}{\cos \nu} \cos \omega\right). \tag{37}$$

For equal values of the ratio $\dfrac{\cos \mu}{\cos \nu}$ this equation is the same, and hence

for equal values of the ratio $\dfrac{\cos \mu}{\cos \nu}$ the shape of central lattice line curves is the same on the film. This means that all central lattice lines record on a film as curves identical with one of either the normal-beam family, Fig. 132, or the flat-cone family, Fig. 160. The choice

between these depends on whether the ratio $\dfrac{\cos \mu}{\cos \nu}$ exceeds 1 or not. If

it does, then, as shown on page 238, (37) has no solution until ω is large

FIG. 163. Comparison of the geometrical properties of normal-beam, equi-inclination, and flat-cone photographs. At the left are shown the reflection relations of a central reciprocal lattice line. In the middle are shown the theoretical Weissenberg projections of a set of parallel reciprocal lattice rows, and on the right are shown actual Weissenberg photographs of the same layer of the same crystal taken by the several methods. Middle and right columns are reduced to identical scales.

enough to make $\cos \omega$ scale down the product $\dfrac{\cos \mu}{\cos \nu} \cos \omega$ to 1 or below.

If $\dfrac{\cos \mu}{\cos \nu}$ is less than 1, there is always a solution to (37).   The former case resembles a normal-beam case; the latter, an equal-cone case.

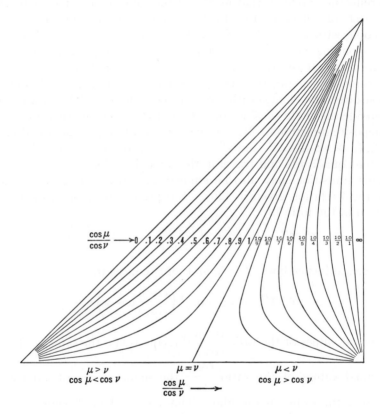

FIG. 164.   The shapes of central reciprocal lattice lines for general-inclination Weissenberg photographs, as a function of $\dfrac{\cos \mu}{\cos \nu}$ (camera diameter, 57.3 mm.).

Evidently the shape of the curve represented by (37) is a function of the single composite variable $\left(\dfrac{\cos \mu}{\cos \nu}\right)$.   The change in shape with this variable is shown in Fig. 164.

Returning now to (36), it is evident that the shape of all the curves is dependent upon the variable $\left(\dfrac{\cos \mu}{\cos \nu}\right)$, provided that no importance is attached to the d labeling of the lines.   This labeling varies with cos ν. In other words, all Weissenberg photographs, taken with any inclination whatever, resemble one another in reciprocal lattice line curve shape, provided that the comparison is made for equal values of $\left(\dfrac{\cos \mu}{\cos \nu}\right)$.

INTERPRETATION OF GENERAL-INCLINATION WEISSENBERG PHOTOGRAPHS

Weissenberg photographs taken with the beam inclined at a general angle can be interpreted graphically by Schneider's method (pages 231–232) but based upon a graphical use of Fig. 157 rather than of Fig. 126.   The rotation axis should be first drawn, Fig. 157, elevation, then the direct x-ray beam inclined $90° - \mu$ to the rotation axis. The intersection of the beam with the rotation axis locates the origin. The zero layer is located normal to the rotation axis through the origin, and the nth layer at a height ζ above the zero layer.   This construction provides dimensions for Fig. 157, plan.   The Weissenberg photograph now supplies Υ and ω for each point seen in plan in Fig. 157, plan, and these can be located exactly as in Schneider's method applied to the normal beam.

General-inclination films may also be interpreted by the Wooster method.   After μ and ν have been selected consistent with the layer coordinate, ζ, to be analyzed, according to (5), then the ratio $\left(\dfrac{\cos \mu}{\cos \nu}\right)$ enables one to select the appropriate curve for $\varphi =$ constant from Fig. 164.   The ξ scale must be computed for each new level with the aid of (25) (unless the special case of flat cone, $\nu = 0$, is being followed, when the scale is given by Fig. 161).

CHAPTER 16

THE SAUTER AND THE SCHIEBOLD METHODS

The method of recording x-ray diffraction data, commonly known as the Sauter method, was actually first suggested by Schiebold in 1931 in an obscure publication. It was rediscovered and described by Sauter in 1932. This, like the Weissenberg method, is a moving-film method, but differs in the nature of the film motion. In the Weissenberg method, the film is moved at right angles to the layer line by translation; in the Sauter method, it is moved at right angles to the layer line by rotation. The general scheme of this method is indicated in Fig. 165. Note that, in both Weissenberg and Sauter methods, the measurement of Υ is orthogonal to the measurement of ω and $\bar{\varphi}$.

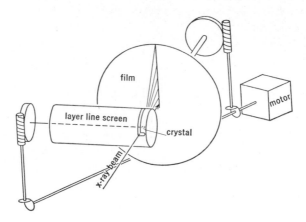

FIG. 165. Schematic representation of Sauter apparatus.

It is evident from Fig. 165 that the Sauter method is a moving-film method which corresponds with the flat-plate recording used in the simpler rotating-crystal method. It accordingly suffers from the several defects of the flat-plate recording, probably the most serious of which is that it is capable of recording only a limited range of reflections. To remedy this situation, Schiebold subsequently suggested a modification of it, Figs. 173 and 174, corresponding with the cylindrical camera recording of the simpler rotating-crystal method. This variation, known as the Schiebold method, presents mechanical difficulties but has many theoretical advantages.

Both the Sauter and Schiebold methods are comparatively little used, and the few attempts at applying them have not been seriously concerned with the recording of n-layers, without which no complete x-ray analysis can be accomplished.

The treatment of both Sauter and Schiebold methods may be much shortened by making use of relations developed for the Weissenberg methods.

THE SAUTER METHOD

Although the Sauter method has the disadvantage of limited range, it has the advantage of mechanical simplicity. A photograph of Sauter's instrument is shown in Fig. 166. The most satisfactory mechanical drive is diagrammatically indicated in Fig. 165. In this scheme, a motor simultaneously actuates two worm gears, one of

FIG. 166. Simple form of the Sauter apparatus, seen looking along the axis of crystal rotation. In this particular design, the circular film camera, seen at the right, is arranged so that the film may be inclined to the x-ray beam. The pinhole system is suspended on the inverted L-shaped piece at the upper left. This is a kind of hinge which permits the x-ray beam to be inclined to the axis of crystal rotation.

which turns a gear which rotates the crystal while the other turns a gear which rotates the film. The layer line to be analyzed is screened out in exactly the same way as in Weissenberg photographs, except that only the upper (or, alternatively, the lower) half of the slit is used. The reason for this is that the two halves of the film in the simple rotating-crystal method record duplicate but separate records (i. e., for every reciprocal lattice point, there is a reflection at xy and $\bar{x}y$ on the film). The layer line screen on the several Weissenberg methods thus passes duplicate but separate records. In the Weissenberg method, these two records are received and recorded on separate and distinct halves of the film; but in the Sauter method, both record on the very same film. They do not, however, coincide. If, therefore, both sets were permitted to reach the same film, they would constitute a double but not superposed record. The general reason why the reflections developed by the upper and lower halves of the reflecting circle do not superpose is that they occur at ω rotation settings which are related to one another by a plane of symmetry, Fig. 167A; their Sauter records are therefore mirror images of one another, Fig. 167B.

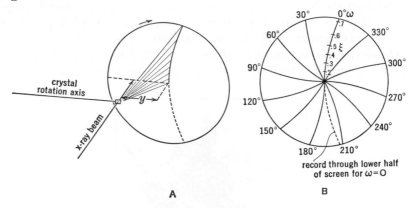

record through lower half
of screen for $\omega=0$

A   B

Fig. 167.   Relations for the normal-beam Sauter method.

Sauter photographs may be taken with the x-ray beam inclined to the axis of crystal rotation by any of the possible schemes.

Normal-beam method.   Figure 167 schematically illustrates the relations for normal-beam Sauter photographs. In order to understand the situation most easily, it should first be imagined that Fig. 167 represents the scheme for taking a simple rotating-crystal photograph, with the axis of rotation horizontal, and recording on a flat plate. The nth layer line then records along a hyperbolic curve which is symmetrical about a horizontal line. The upper part of this is shown as a full line in Fig. 167A, and the lower part dotted. For a normal-beam Sauter photograph, the center of rotation of the film is placed at the nose of the hyperbola, and the film rotates clockwise. The upper part of the layer line curve is fanned out by this rotating motion, Fig. 167B, so that reflections occurring at different ω positions of the crystal appear on the film on different curves. Note that reflections recorded through the lower half of the layer line (dotted,

Fig. 167A) are fanned out in an entirely different manner (dotted, Fig. 167B), because of the different curvature of the lower part of the layer line.

The above considerations make it a simple matter to map out a film in ω and ξ coordinates, and hence to index a film by measuring these coordinates by appropriate scales on the Sauter film. The curve of Fig. 167A is that particular hyperbola which is recorded for the ζ value of the desired layer on the Bernal flat plate chart, Fig. 80. Along each of these curves in Fig. 167B, the value of ω is constant. Thus, when the film is in the position shown in Fig. 167A, the value of ω is zero. When the crystal has rotated from this position by ω_1 degrees, then the film has also rotated by the same number of degrees, and the layer line occupies a position on the film along a curve of identical shape, but rotated ω_1 degrees counterclockwise from the zero curve. In this way the film may be thought of as marked out in curves radiating from the center, and along each of which ω is constant.

The shape of curve for each film may be traced directly from the hyperbola of appropriate ζ value on the Bernal flat plate chart, Fig. 80. On the Bernal chart the ξ scale is also correctly marked out along the particular hyperbola. Thus, for the indexing of any normal-beam Sauter film, it is only necessary to lay the film on a tracing of the correct hyperbola with its ξ divisions marked out. These divisions give the ξ values of each reflection directly, while the angle through which it is necessary to turn the film in order to bring the required reflection to the zero hyperbola gives the ω value of the reflection.

Normal-beam Sauter photographs may also be interpreted by other methods discussed under normal-beam Weissenberg photographs.

Flat-cone Sauter photographs. The disadvantage of normal-beam Sauter photographs is that the n-layer lines record as hyperbolas, all different. By setting the x-ray beam at an angle, μ, from the plane of the layer lines, such that $\nu = 0$, each hyperbola is flattened out, and thus each layer line trace on the flat film becomes a straight line as indicated in Fig. 168A. The lines of constant ω then become straight radial lines, Fig. 168B, which compares with Fig. 167B of the normal-beam method.

The lines of constant ξ are concentric circles, as indeed they are in the normal-beam case. The variation of the radius, x, of these circles, with the inclination angle μ of the x-ray beam is graphically shown in Fig. 170, and will be derived subsequently.

Flat-cone Sauter photographs may be indexed exactly as described above for normal-beam Sauter photographs, except that all hyperbolas become straight lines. In other words, the film may be treated as a polar coordinate plot having a non-uniform radial scale selected from the appropriate μ value of Fig. 170.

There are, however, more appropriate ways of interpreting flat-cone Sauter photographs.

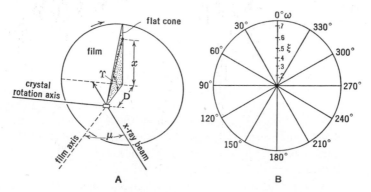

FIG. 168.   Relations for the flat-cone Sauter method.

The shape of central lattice lines on flat-cone Sauter photographs. The equation for a central lattice line, as developed for the Weissenberg method, is, according to (33), Chapter 15,

$$\Upsilon = \omega \pm \cos^{-1} (\cos \mu \cos \omega). \tag{1}$$

Figure 168A shows that the distance from the center of the film, x, at which a reflection is recorded, is

$$x = D \tan \Upsilon, \tag{2}$$

where D is the crystal-to-film distance. Substituting in (2) the value of Υ provided by (1) gives

$$x = D \tan\{\omega \pm \cos^{-1} (\cos \mu \cos \omega)\}. \tag{3}$$

In this relation, D is an instrumental constant and μ is a constant for the given layer line; thus (3) expresses the variation of the central distance, x, as a function of the rotation angle, ω, of the film. This is accordingly the polar coordinate equation of a central lattice line on a flat-cone Sauter photograph. A set of such curves for various values of the layer line coordinate, ζ [which is a function of μ according to (5) Chapter 15], is shown in Fig. 169. Data for the construction of such curves are given in Table 17. Figure 169 may be used as a template for the construction of central reciprocal lattice lines for any level height, ζ, by easy interpolation.

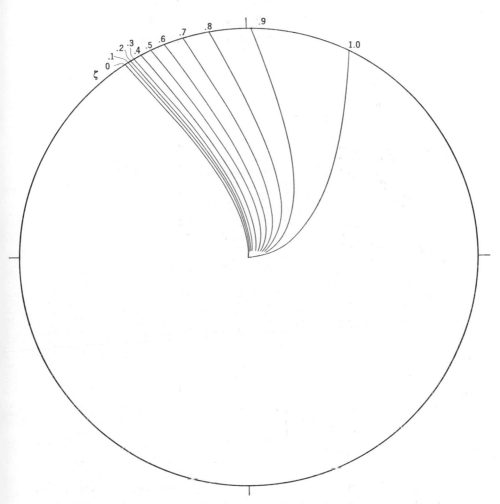

FIG. 169. The shapes of central reciprocal lattice lines for various level coordinate values, ζ, for flat-cone Sauter photographs (crystal-to-film distance, $D = 3$ cm.).

The ξ scale for flat-cone Sauter photographs. The ξ-constant lines, as already pointed out, are concentric circles. The ξ scale for the flat-cone method, according to (34), Chapter 15, is

$$\Upsilon = \cos^{-1}\left(\frac{1 + \cos^2\mu - \xi^2}{2\cos\mu}\right).$$ (4)

Substituting in (2) the value of Υ given by (4) leads to

$$x = D\,\tan\,\cos^{-1}\left(\frac{1 + \cos^2\mu - \xi^2}{2\cos\mu}\right).$$ (5)

TABLE 17

DATA FOR CONSTRUCTION OF CENTRAL LATTICE LINE, FLAT-CONE
SETTING, SAUTER CAMERA

Solution of $x = D \tan \{\omega \pm \cos^{-1} (\sqrt{1 - \zeta^2} \cos \omega)\}$

Crystal-to-film distance, D, taken as unity. For any other distance, multiply x by D.

						x					
ω	$\zeta = 0$	$\zeta = .1$	$\zeta = .2$	$\zeta = .3$	$\zeta = .4$	$\zeta = .5$	$\zeta = .6$	$\zeta = .7$	$\zeta = .8$	$\zeta = .9$	$\zeta = 1.0$
−90°	0.000	0.000	0.000	0.000	0.000	0.000	0.000	0.000	0.000	0.000	0.000
−85	0.000	0.000	0.002	0.004	0.007	0.011	0.017	0.025	0.035	0.050	0.087
−80	0.000	0.001	0.004	0.008	0.015	0.024	0.035	0.050	0.070	0.099	0.176
−75	0.000	0.001	0.005	0.012	0.022	0.036	0.053	0.076	0.088	0.150	0.268
−70	0.000	0.002	0.007	0.017	0.031	0.048	0.072	0.102	0.143	0.202	0.364
−65	0.000	0.002	0.009	0.021	0.038	0.060	0.091	0.130	0.182	0.255	0.466
−60	0.000	0.003	0.012	0.026	0.048	0.076	0.107	0.160	0.222	0.313	0.577
−55	0.000	0.003	0.014	0.032	0.057	0.091	0.135	0.191	0.265	0.374	0.700
−50	0.000	0.004	0.017	0.038	0.068	0.108	0.159	0.225	0.312	0.440	0.839
−45	0.000	0.005	0.020	0.045	0.080	0.127	0.186	0.262	0.362	0.511	1.000
−40	0.000	0.006	0.024	0.053	0.095	0.148	0.216	0.302	0.417	0.589	1.192
−35	0.000	0.007	0.028	0.063	0.111	0.173	0.250	0.348	0.478	0.676	1.428
−30	0.000	0.008	0.034	0.075	0.131	0.202	0.289	0.400	0.547	0.776	1.732
−25	0.000	0.010	0.041	0.090	0.156	0.236	0.336	0.459	0.629	0.892	2.145
−20	0.000	0.013	0.052	0.110	0.186	0.278	0.389	0.528	0.718	1.029	2.747
−15	0.000	0.018	0.067	0.138	0.226	0.329	0.454	0.610	0.827	1.196	3.732
−10	0.000	0.026	0.091	0.177	0.277	0.394	0.532	0.708	0.958	1.406	5.671
− 5	0.000	0.045	0.133	0.233	0.346	0.475	0.629	0.829	1.122	1.681	
0	0.000	0.101	0.205	0.314	0.436	0.577	0.750	0.980	1.333	2.064	
5	0.176	0.224	0.316	0.427	0.556	0.710	0.905	1.177	1.618	2.639	
10	0.364	0.394	0.471	0.579	0.713	0.884	1.111	1.445	2.029	3.626	
15	0.577	0.601	0.670	0.778	0.923	1.120	1.397	1.833	2.686	5.730	
20	0.839	0.862	0.931	1.046	1.215	1.457	1.823	2.457	3.918		
25	1.192	1.217	1.300	1.437	1.653	1.987	2.543	3.647	7.191		
30	1.732	1.767	1.877	2.078	2.411	2.971	4.056	6.940			
35	2.747	2.808	3.009	3.401	4.118	5.567					
40	5.671	5.871	6.586	8.204							

The inclination angle, μ, is constant for any layer line. Relation (5) thus gives the distance, x, from the center of the film at which a reflection of coordinate ξ will record. The variation of this scale with the beam setting, μ, is given in Fig. 170. Data for the construction of this diagram are listed in Table 18.

Interpretation of flat-cone Sauter photographs. A flat-cone Sauter photograph may be interpreted by an adaptation of the Wooster method used in reconstructing the reciprocal lattice from Weissenberg photographs. In the two preceding sections, the shape of the central reciprocal lattice line, and the ξ scale for any beam setting, μ, have been provided. A reciprocal lattice may be reconstructed, central lattice line by central lattice line, from any photograph. It is only necessary to mark out the appropriate central lattice line with the

TABLE 18

DATA FOR CONSTRUCTION OF ξ SCALE FOR FLAT-CONE SETTING, SAUTER CAMERA

$$x = D \tan \cos^{-1}\left(\frac{1 + \cos^2 \mu - \xi^2}{2 \cos \mu}\right)$$

Crystal-to-film distance, D, taken as unity. For any other distance, multiply x by D.

ξ	x										Value of μ for $x = 0$
	$\mu = 0°$	$\mu = 5°$	$\mu = 10°$	$\mu = 15°$	$\mu = 20°$	$\mu = 25°$	$\mu = 30°$	$\mu = 35°$	$\mu = 40°$	$\mu = 45°$	
0.00	0.000										0.00°
0.05	0.050	0.050	0.050	0.037							18.20
0.10	0.100	0.100	0.101	0.096	0.083	0.037					25.84
0.15	0.151	0.151	0.152	0.150	0.143	0.124	0.073				31.79
0.20	0.203	0.204	0.205	0.204	0.200	0.188	0.161	0.092			36.87
0.25	0.256	0.257	0.258	0.258	0.256	0.249	0.231	0.192	0.101		41.41
0.30	0.310	0.311	0.313	0.314	0.314	0.310	0.298	0.271	0.218	0.077	
0.35	0.367	0.367	0.370	0.372	0.374	0.372	0.364	0.345	0.308	0.232	
0.40	0.426	0.427	0.430	0.433	0.436	0.436	0.432	0.419	0.391	0.337	
0.45	0.488	0.489	0.492	0.496	0.501	0.504	0.503	0.494	0.474	0.434	
0.50	0.553	0.555	0.558	0.564	0.570	0.575	0.577	0.573	0.560	0.529	
0.55	0.623	0.625	0.629	0.635	0.644	0.651	0.657	0.658	0.649	0.628	
0.60	0.698	0.700	0.705	0.713	0.723	0.731	0.744	0.750	0.748	0.734	
0.65	0.779	0.782	0.788	0.797	0.810	0.824	0.838	0.850	0.855	0.850	
0.70	0.869	0.871	0.879	0.890	0.906	0.924	0.943	0.962	0.975	0.980	
0.75	0.967	0.970	0.979	0.996	1.013	1.036	1.061	1.089	1.112	1.130	
0.80	1.078	1.082	1.093	1.110	1.134	1.169	1.200	1.236	1.272	1.306	
0.85	1.205	1.209	1.222	1.244	1.274	1.312	1.357	1.409	1.464	1.519	
0.90	1.351	1.356	1.373	1.400	1.438	1.487	1.547	1.620	1.700	1.789	
0.95	1.524	1.531	1.551	1.586	1.635	1.700	1.782	1.885	2.003	2.145	
1.00	1.732	1.741	1.768	1.813	1.870	1.967	2.081	2.230	2.412	2.646	
1.05	1.992	2.003	2.039	2.100	2.190	2.314	2.481	2.708	3.001	3.429	
1.10	2.326	2.342	2.388	2.478	2.608	2.791	3.047	3.420	3.940	4.774	
1.15	2.778	2.802	2.875	3.003	3.201	3.492	3.923	4.610	5.712	7.905	
1.20	3.428	3.465	3.582	3.788	4.143	4.645	5.497	7.062	10.39	23.56	
1.25	4.461	4.526	4.745	5.116	5.777	6.931	9.175	15.28	62.96		
1.30	6.374	6.512	6.974	7.883	9.693	13.75	28.78				
1.35	11.23	11.69	13.36	17.15	30.96						
1.40	50.00	61.60									

aid of Fig. 169, and divide it by radial marks, with the aid of Fig. 170. This template is laid under the photograph which is rotated until the reflection in question falls on the standard central reciprocal lattice line marked out on the template beneath. The amount of rotation of the film necessary to bring this about is the cylindrical reciprocal lattice coordinate, $\bar{\varphi}$, and the other coordinate is given directly by the ξ scale.

The shape of general lattice lines in flat-cone Sauter photographs. By substituting in (2), the value of Υ given by (32) Chapter 15, namely,

$$\Upsilon = \omega \pm \cos^{-1} (\cos \mu \cos \omega + d), \tag{6}$$

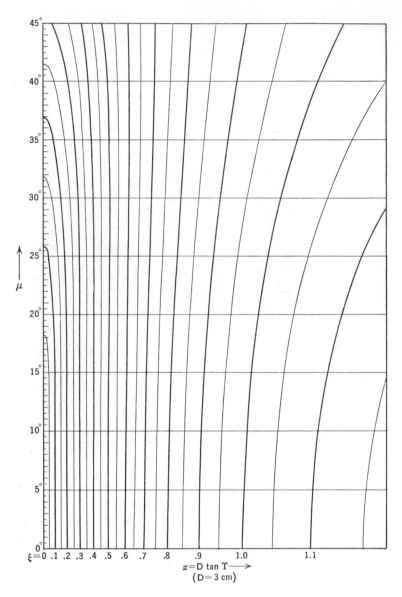

FIG. 170.   The variation of the ξ scale with the inclination angle, μ, for flat-cone
Sauter photographs (crystal-to-film distance, $D = 3$ cm.).

the shape of the general reciprocal lattice line on a flat-cone Sauter photograph may be obtained. This is

$$x = D \tan\{\omega \pm \cos^{-1} (\cos \mu \cos \omega + d)\}. \qquad (7)$$

These curves for the zero layer, and for an n-layer of ζ coordinate of 0.5, are shown in Figs. 171A and 171B, respectively. The shapes of these curves vary with level, and consequently they cannot be predicted except with the aid of an infinite number of templates of the type shown in Fig. 171. Sauter photographs are shown in Fig. 172.

Equi-inclination Sauter photographs. One might well inquire whether Sauter photographs taken by the equi-inclination method have some advantage over those taken by the normal-beam and flat-cone methods. They have one advantage, namely, that the equi-inclination method explores the reciprocal lattice clear in to the rotation axis, which is true of no other method. Photographs taken by either of the two methods just discussed do not start at the center with $\xi = 0$ but with some finite value of ξ.

Similar reciprocal lattice curve shapes for different levels do not occur in the equi-inclination Sauter method, as they do in the equi-inclination Weissenberg method. This is due to the fact that the Sauter method records reflections on a flat film, and with the equi-inclination method a layer line appears, not as a straight line, but as an hyperbola similar to that shown in Fig. 167. The shape of the hyperbola is different for each layer line.

THE SCHIEBOLD METHOD

The obvious lack of reflection range in the flat-film Sauter method called forth an attempt to devise a cylindrical-film equivalent of it. The result, the Schiebold method, is a device full of mechanical obstacles. It appears to have been used only by its inventor, Schiebold, and only to take zero-layer photographs.

The Schiebold method is best illustrated by starting with a Sauter instrument, Fig. 173A. The dimensions of this particular Sauter apparatus are such that the film disk is tangent to the back of the layer line screen. The rotation of the crystal is accompanied by a clockwise rotation of the film disk, which rubs against the layer line screen along the line of tangency. In the Schiebold modification, the film disk is wrapped forward around the layer line screen, Fig. 173B. Ideally, the diameter of the film disk is equal to the circumference of the layer line screen, so that the upper and lower flaps of the disk just touch each other and cover the complete Υ range. The film is still disk-shaped but is now wrapped in the form of a cylinder. As the disk rotates about the exit port of the layer line slit as a center, its entire area is in contact with the layer line screen cylinder, and it is consequently always bent in cylindrical form, but the axis of the cylinder is

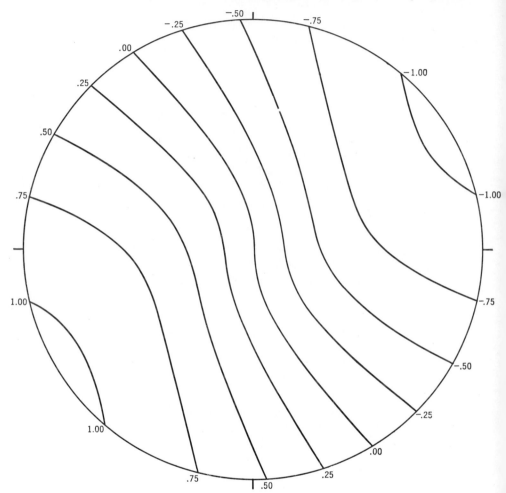

FIG. 171A. The appearance of a set of parallel reciprocal lattice lines on a zero-level Sauter photograph (crystal-to-film distance, $D = 3$ cm.). The labeling of the curves gives the distance, d, of the reciprocal lattice line from the origin of the level.

continually changing with respect to a reference direction in the surface of the disk. A diameter of the disk, when lying parallel with, and at, the layer line slit, is bent into the form of a circle. As the film is made to rotate, this circle opens up into a segment of a helix.

Continued rotation increases the pitch of the helix until it is a straight line. At this point in the rotation, the original diameter is now parallel with the crystal-rotation axis. Further rotation generalizes

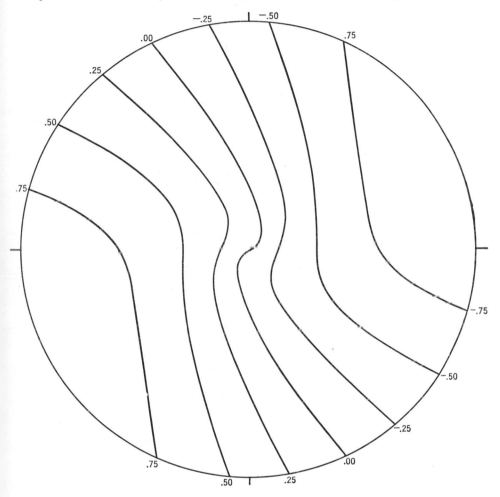

FIG. 171B. The appearance of a set of parallel reciprocal lattice lines on an n-level Sauter photograph of level coordinate $\zeta = .5$ (crystal-to-film distance, $D = 3$ cm.). The labeling of the curves gives the distance, d, of the reciprocal lattice line from the origin of the level.

the straight line into a segment of a helix having a curvature opposite in sense to that of the first helix, and continued rotation reduces its pitch until the helical segment degenerates into a circle parallel with the slit. This history occupies 180° of rotation range.

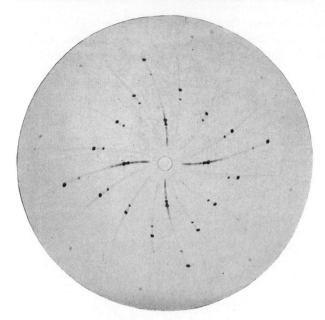

FIG. 172A.   Zero-level Sauter photograph.   (Urea, NH$_2$CONH$_2$, tetragonal, c-axis rotation, zero level.)

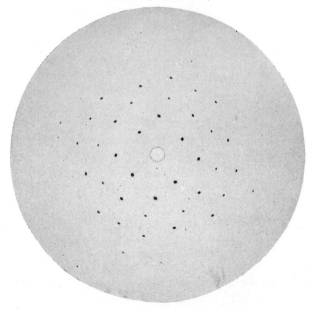

FIG. 172B. n-Level flat-cone Sauter photograph.   (Urea, c-axis rotation, 1st level; film inclined at 45° to x-ray beam, as in Fig. 166.)

From a mechanical point of view, there would be a great deal of friction plus attendant abrasion of the film, if the film were left in direct contact with the layer line screen. The best plan is to encase the film in an envelope, which is needed for light-tightness anyway, and thus eliminate a badly scratched photograph. The film is held in cylindrical shape, especially immediately on each side of the layer

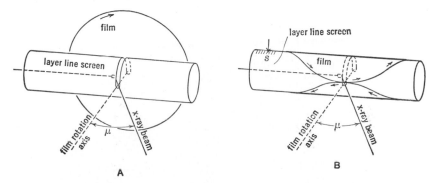

Fig. 173. Relation between Sauter method, A, and Schiebold method, B.

line slit, with the aid of a series of rollers. The most difficult mechanical feature is an appropriate mechanism for driving the film. Schiebold used a roller friction drive, but such a device is subject to slippage which would cause cumulative disregistry. A photograph of Schiebold's instrument is shown in Fig. 174.

In many respects Schiebold photographs have properties very similar to Weissenberg photographs. In the Weissenberg film, the layer line slit occupies successive vertical lines, Fig. 175B. In the Schiebold film, the layer line slit occupies successive radii, Fig. 175C. Schiebold photograph radii are therefore comparable directly with Weissenberg photograph vertical lines, and, since these lines occupied identical positions during the recording, their scales (Υ, ξ) are identical. Again it should be noted that only the positive direction is recorded in Schiebold films, while Weissenberg films record a double record, one for each half of the layer line slit. This situation will receive further discussion beyond. In Weissenberg films, the position of the layer line slit at different angular settings of the crystal is represented by a linear horizontal separation of the vertical layer line positions. In Schiebold films, it is represented by an angular separation. It is in this angular rather than linear representation of the crystal setting, ω, at the moment of reflection, that a Schiebold photograph differs from a Weissenberg photograph. It follows that any and all techniques applicable to the taking of Weissenberg photographs are also applicable to the taking of Schiebold photographs. Thus, normal-beam, equi-inclination, flat-cone, and general-inclination photographs are all available for the taking of Schiebold photographs, although hitherto n-layer photographs do not appear ever to have been recorded.

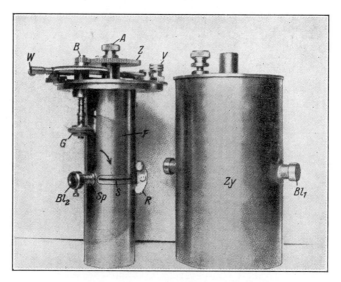

Fig. 174.   Schiebold apparatus.

Since Weissenberg and Schiebold photographs both have as a basis the theory outlined in the chapters on Weissenberg photographs, no additional fundamental theoretical discussion is necessary.   It will suffice to point out some of the properties of the several possible varieties of Schiebold photographs in comparison with their Weissenberg equivalents, Fig. 175.   In this comparison, it should be remembered that Weissenberg representation can be easily transformed to Schiebold representation by taking Weissenberg verticals, from left to right, and placing them as successive radii counterclockwise on the Schiebold diagram.   The uniform left-right (ω) Weissenberg scale is equivalent to the uniform counterclockwise Schiebold scale.

Figure 175A shows a plan view of the reflecting circle, together with a central lattice line in each of the three special positions of flat-cone, equi-inclination, and normal-beam.   The positive and negative ends of each of these lines are indicated.   These record on the Weissenberg and Schiebold film as shown in Figs. 175B and 175C, respectively.   In order to show how duplicate recording occurs through the opposite

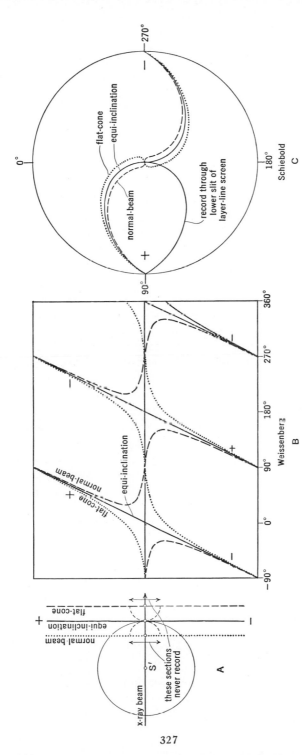

Fig. 175. Relations between Weissenberg and Schiebold records.

side of the layer line screen, the positive end of the line, only, is shown recorded through the lower half of the layer line screen in the Schiebold record. Note that in the Weissenberg record, this records on a separate half of the film, whereas in the Schiebold record it records on the same region of the film, but as the mirror image of the original record.

In Fig. 175 it should be noted that, first of all, each of the three straight lattice lines of Fig. 175A record as curves on the Schiebold photograph. In the Weissenberg photograph, one method, the equi-inclination, uniquely gives rise to a straight line in projection. In Fig. 175C, it should also be noted that the equi-inclination representation of the central lattice line enters the origin at the same angle as that in the reciprocal lattice itself, Fig. 175A. The flat-cone representation enters the origin 90° clockwise of the lattice line, and the normal-beam representation 90° counterclockwise of the lattice line. This relation, of course, holds true for every central reciprocal lattice line.

This brings up for consideration one of the most important properties of Schiebold (and Sauter photographs): they are angle-true at the origin. It should also be observed that, in the small section of the photographs in the immediate vicinity of the origin, the curves approximate straight lines. These photographs are thus radially similar to the reciprocal lattice itself for equi-inclination photographs. This is not true for the normal-beam or flat-cone cases, however, for in these instances the sections of the lines indicated in Fig. 175A can never touch the reflecting circle during rotation, and consequently there is no record of these portions. The Weissenberg and Schiebold representations of these cases start with, not $\xi = 0$, but $\xi =$ some finite value, at the origin.

It should further be noted that, in the equi-inclination case, (9), Chapter 14,

$$\xi = 2R \sin \frac{\Upsilon}{2}. \tag{8}$$

For small angles, the sine of $\Upsilon/2$ increases very approximately linearly with $\Upsilon/2$, so (8) may be approximated

$$\xi \approx 2R \cdot \frac{\Upsilon}{2} \sim \Upsilon. \tag{9}$$

That is, near the origin, equi-inclination photographs have a very approximately linear ξ scale. This, taken with the discussion of the last paragraph, shows that Schiebold equi-inclination photographs give

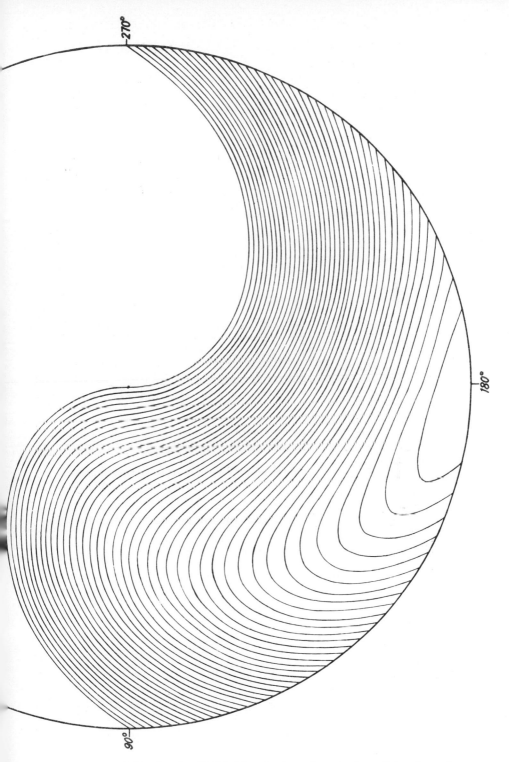

FIG. 176. The equi-inclination Schiebold projection of a set of parallel reciprocal lattice rows of spacing $\dfrac{d}{\cos \nu} = .05$ (camera diameter 57.3 mm.).

a nearly true picture of the reciprocal lattice near the center. The distortion increases both for angle and for scale toward the edges with increasing rapidity. This is also true for equi-inclination Sauter photographs.

For numerous reasons, the ideal way to take Schiebold photographs is with the equi-inclination method. An equi-inclination net is provided in Fig. 176. As in Weissenberg photographs, the family of curves is the same for all layers, although the d labeling varies with cos μ.

LITERATURE

The Sauter method

E. SCHIEBOLD. *Ergebnisse der technische Röntgenkunde.* II. (Leipzig, 1931.) Pages 86–87.

ERWIN SAUTER. Zur Kenntnis des Rotations-Röntgengoniometer-diagrams. *Z. Krist.* (*A*), **84** (1933), 461–467.

ERWIN SAUTER. Das Rotationsgoniometerdiagramm und das reziproke Gitter. *Naturwissenschaften,* **20** (1932), 889–890.

ERWIN SAUTER. Eine einfache Universalkamera für Röntgen-Kristallstruktur-analysen. *Z. Krist.* (*A*), **85** (1933), 156–159.

E. SCHIEBOLD. Über ein neues Röntgengoniometer. Gleichzeitig Bemerkung zu der Arbeit von E. Sauter: " Eine einfache Universalkamera für Röntgenkris-tallstrukturanalysen." *Z. Krist.* (*A*), **86** (1933), 370–377.

ERWIN SAUTER. Universalkamera und " selbstindizierende " Drehkristallkamera. *Z. physik. Chem.* (*B*), **23** (1933), 370–378.

ERWIN SAUTER. Über die Herstellung von vollständigen Faserdiagrammen. *Z. Krist.* (*A*), **93** (1936), 93–106.

ROBERT B. HULL and VICTOR HICKS. A universal x-ray photogoniometer. *Z. Krist.* (*A*), **96** (1937), 311–321.

Schiebold method

E. SCHIEBOLD. Über ein neues Röntgengoniometer. *Z. Krist.* (*A*), **86** (1933), especially 377–383.

CHAPTER 17

THE DE JONG AND BOUMAN METHOD

INTRODUCTION

Each of the moving-film methods discussed so far provides a projection of the reciprocal lattice. These projections may be regarded as distorted pictures of the reciprocal lattice. It remained for de Jong and Bouman to discover the conditions under which the reciprocal lattice might be photographed without any geometrical distortion.

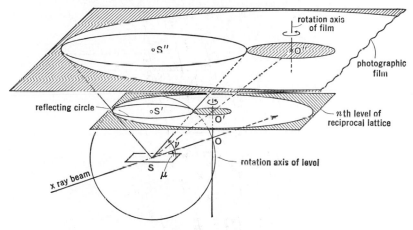

FIG. 177.

Figure 177 shows the scheme of the de Jong and Bouman method of photographing the reciprocal lattice. The x-ray beam, partly enveloped by its sphere of reflection, makes, in general, an angle μ with the levels of the reciprocal lattice. Attention is directed to some particular n-level, which alone is shown. This level rotates about its level origin, and, as each of its lattice points passes through the reflecting circle, a reflection develops along the generator of a cone of opening angle, $180° - 2\nu$. The generators of this cone are, in effect, rays of projection emanating from the center of the sphere. To make a projection geometrically similar to the reciprocal lattice level, it is only necessary to place the plane of projection (the photographic film) parallel with the plane level to be projected. For any given static position of the reciprocal lattice level, then, a perfect undistorted pro-

331

Fig. 178*A* (above). Apparatus for making de Jong and Bouman photographs, in position ready for recording.

Fig. 178*B* (below). Apparatus for making de Jong and Bouman photographs, with layer line screen removed.

FIG. 178C. Instrument for making de Jong and Bouman photographs, with layer line screen removed and μ setting returned to zero, ready for adjusting crystal to center of x-ray beam.

jection of the locations of its lattice points which are making contact with the reflecting circle is obtained on the surface of the film. In order to maintain this projection relation undisturbed during the rotation of the crystal, it is only necessary to have the film rotate about a point in its plane which is the projection of the point about which the reciprocal lattice rotates in its own plane, i. e., about the level origin, and to observe that the two rates of rotation are identical and in the same direction. Under these conditions, as each point in the reciprocal lattice passes through the reflecting circle (which occurs, in general, twice during a rotation), it records itself in projection on the appropriate spot on the film. Figure 177 shows that, in general, the reflecting circle sweeps through an area having the shape of an annulus. Within a circular area, therefore, the developed film is an undistorted projection of the level of the reciprocal lattice under observation, except that the photographic record in the center of the film is lacking because the corresponding circle in the reciprocal lattice level never passes through the sphere of reflection. The only exceptions to this are when equi-inclination technique is followed, i. e., $\mu = -\nu$, or when anti-equi-inclination technique is followed, i. e., $\mu = +\nu$.

Figure 178 shows an instrument designed by the author to make de Jong and Bouman photographs, and Fig. 179 shows several photographs made with this instrument.

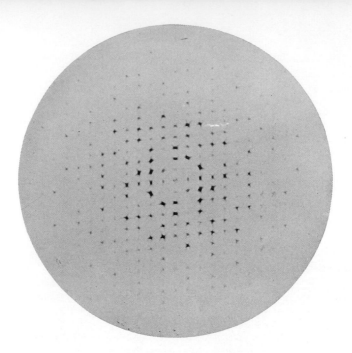

FIG. 179A. de Jong and Bouman photograph, symmetry C_2, parallelogram reciprocal cell. Level origin is marked by a short exposure to direct beam. (Pectolite, $Ca_2NaSi_3O_8(OH)$, monoclinic; b-axis rotation, zero level; Mo$K\alpha$ radiation monochromatized by reflection from NaCl crystal.)

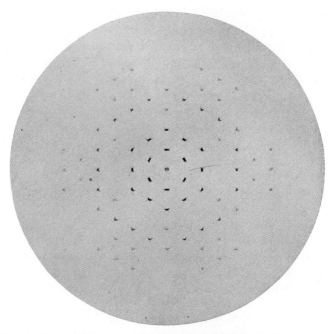

FIG. 179B. de Jong and Bouman photograph, symmetry C_{6l}, hexagonal reciprocal cell. Level origin is marked by a short exposure to direct beam. (Low-quartz, SiO_2, hexagonal; c-axis rotation, zero level; Mo$K\alpha$ radiation monochromatized by reflection from NaCl crystal.)

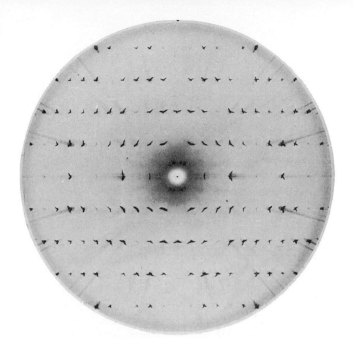

FIG. 179C. de Jong and Bouman photograph, symmetry C_{2l}, rectangular reciprocal cell. Level origin is marked by uncovering a pinhole on the rotation axis and producing a dot on the film center by a short exposure to light. (Claudetite, As_2O_3, monoclinic; c-axis rotation, zero level; $CuK\alpha$ radiation from hot-cathode tube, filtered through nickel foil.)

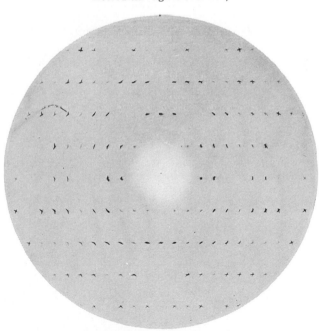

FIG. 179D. de Jong and Bouman photograph, symmetry C_l, rectangular reciprocal cell. (Same data as above, but 2nd level.)

Note: (a) There is a blind area in the center of the photograph.
(b) The horizontal row of spots, corresponding with the central horizontal row in Fig. 179C, is displaced downward from the level origin (dot). From the magnitude of this displacement and the level coordinate, ζ, the interaxial angle, β, of this monoclinic crystal can be computed.

The location of the film rotation axis. It is important to investigate the proper position for the rotation axis of the film. This may be derived from Fig. 180, which is a cross section of Fig. 177 taken through the center of the sphere and the axis of rotation. It will be observed that it is necessary to move the center of rotation of the

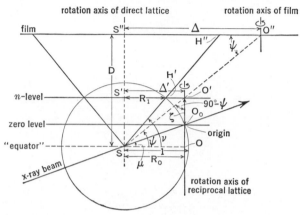

Fig. 180.

film, O'', away from the center of the diffraction cone, S'' (i. e., away from the axis of rotation of the direct lattice), and opposite in direction to the source of x-radiation, by an amount Δ. This can be computed provided that the angle ψ is first found. ψ can be involved with quantities already known, as follows: Applying the law of sines to triangle $SO_0 O'$:

$$\frac{\sin (90° - \psi)}{1} = \frac{\sin (\psi - \mu)}{\zeta}. \tag{1}$$

This may be reduced in the following stages:

$$\zeta \sin (90 - \psi) = \sin (\psi - \mu), \tag{2}$$

$$\zeta \cos \psi = \sin \psi \cos \mu - \cos \psi \sin \mu, \tag{3}$$

$$\zeta = \tan \psi \cos \mu - \sin \mu, \tag{4}$$

from which

$$\tan \psi = \frac{\zeta + \sin \mu}{\cos \mu}, \tag{†5}$$

$$\tan \psi = \frac{\zeta}{\cos \mu} + \tan \mu,$$

and

$$\cot \psi = \frac{\cos \mu}{\zeta + \sin \mu}. \tag{6}$$

† An alternative derivation is $\tan \psi = \dfrac{O'O}{SO} = \dfrac{O'O_0 + O_0 O}{SO} = \dfrac{\zeta + \cos \mu}{\cos \nu}. \tag{5}$

This can be put also in other forms. Thus, it was shown in (5), Chapter 15, that

$$\zeta = \sin \nu - \sin \mu, \tag{7}$$

whence

$$\zeta + \sin \mu = \sin \nu. \tag{8}$$

Substituting this for $\zeta + \sin \mu$ in (6) then gives

$$\cot \psi = \frac{\cos \mu}{\sin \nu}. \tag{†(9)}$$

Alternative forms result from the following reduction:

$$\cot \psi = \frac{\sqrt{1 - \sin^2 \mu}}{\sin \nu}$$

$$= \frac{\sqrt{1 - (\sin \nu - \zeta)^2}}{\sin \nu} \tag{10}$$

Figure 180 shows that the required displacement, Δ, of the film is

$$\Delta - D \cot \psi. \tag{11}$$

Substituting in this relation the values of $\cot \psi$ given in (6), (9), and (10) yields the required expression

$$\Delta = D \frac{\cos \mu}{\zeta + \sin \mu}, \tag{12}$$

or, alternatively,

$$\Delta = D \frac{\cos \mu}{\sin \nu}. \tag{12'}$$

$$= D \frac{\sqrt{1 - (\sin \nu - \zeta)^2}}{\sin \nu}. \tag{13}$$

Solutions of (13) are given in Table 19. Figure 181A presents the same data graphically for the convenient crystal-to-film distance, $D = 3$ cm.

The magnification factor. In order to give proper interpretation to a de Jong and Bouman photograph, it is necessary to know how much the reciprocal lattice appears enlarged on the photograph. This magnification factor, m, of the reciprocal lattice as it appears projected along the diffraction cone to the film is easily obtained by proportionality in the similar triangles $SS''H''$ and $SS'H'$:

$$m = \frac{S''H''}{S'H'}. \tag{14}$$

† An alternative derivation is $\cot \psi = \dfrac{SO}{O'O} = \dfrac{\cos \mu}{\sin \nu}.$ $\qquad\qquad$ (9)

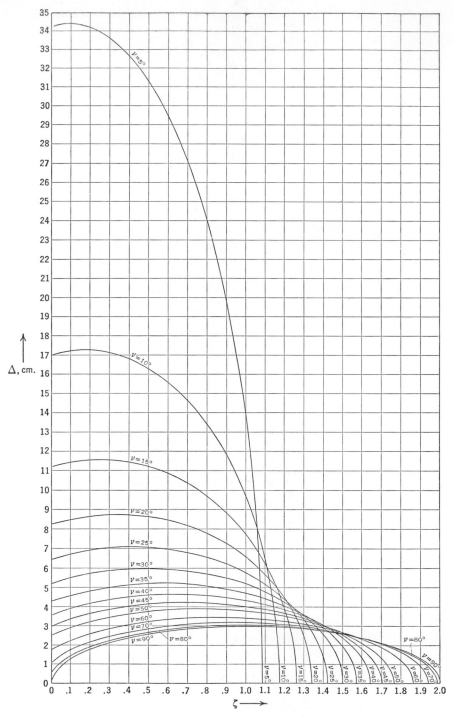

FIG. 181*A*. Variation of the film-rotation-axis displacement, Δ, as a function of the level coordinate, ζ, for various cone angles, ν. (Crystal-to-film distance, $D = 3$ cm.)

338

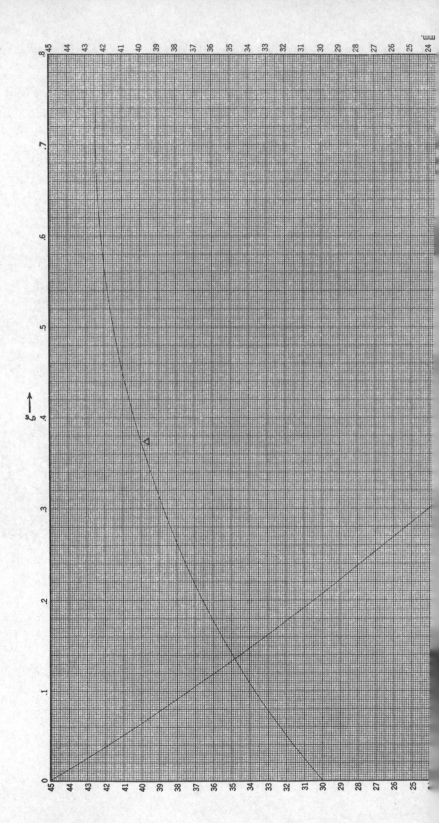

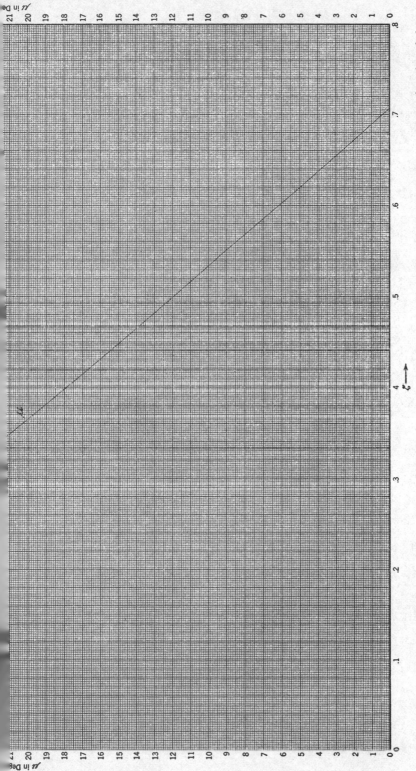

FIG. 181B. Chart for graphically determining the two instrumental settings, μ and Δ, of the de Jong and Bouman apparatus, as a function of the level coordinate ζ. (Cone angle, $\nu = 45°$, crystal-to-film distance, $D = 3$ cm.)

TABLE 19

DATA FOR SETTING THE POSITION OF THE FILM ROTATION AXIS FOR THE DE JONG AND BOUMAN INSTRUMENT

Solution of $\Delta = D \dfrac{\sqrt{1 - (\sin \nu - \zeta)^2}}{\sin \nu}$

Crystal-to-film distance, D, taken as unity. For any other distance, multiply Δ by D.

Δ

ζ	$\nu = 5°$	$\nu = 10°$	$\nu = 15°$	$\nu = 20°$	$\nu = 25°$	$\nu = 30°$	$\nu = 35°$	$\nu = 40°$	$\nu = 45°$	$\nu = 50°$	$\nu = 60°$	$\nu = 70°$	$\nu = 80°$	$\nu = 90°$
0.0	11.424	5.673	3.732	2.748	2.145	1.732	1.428	1.192	1.000	0.839	0.577	0.353	0.177	0.000
0.1	11.467	5.745	3.815	2.835	2.240	1.833	1.535	1.307	1.124	0.974	0.742	0.578	0.473	0.436
0.2	11.395	5.758	3.857	2.894	2.307	1.908	1.617	1.395	1.219	1.076	0.861	0.716	0.629	0.600
0.3	11.205	5.714	3.861	2.921	2.349	1.560	1.677	1.461	1.291	1.155	0.952	0.818	0.740	0.714
0.4	10.892	5.611	3.825	2.919	2.366	1.590	1.717	1.509	1.346	1.215	1.022	0.896	0.824	0.800
0.5	10.445	5.445	3.750	2.887	2.359	2.000	1.739	1.540	1.384	1.259	1.075	0.956	0.888	0.866
0.6	9.845	5.210	3.632	2.825	2.329	1.930	1.743	1.554	1.406	1.287	1.113	1.001	0.937	0.917
0.7	9.062	4.898	3.468	2.730	2.274	1.960	1.729	1.553	1.414	1.303	1.139	1.033	0.973	0.954
0.8	8.042	4.490	3.249	2.599	2.191	1.908	1.698	1.536	1.408	1.305	1.152	1.054	0.998	0.980
0.9	6.681	3.959	2.965	2.426	2.079	1.853	1.648	1.503	1.388	1.294	1.154	1.063	1.012	0.995
1.0	4.684	3.244	2.594	2.202	1.932	1.732	1.577	1.453	1.352	1.269	1.144	1.062	1.015	1.000
1.1		2.169	2.089	1.907	1.741	1.600	1.482	1.384	1.301	1.230	1.123	1.050	1.009	0.995
1.2			1.306	1.501	1.488	1.423	1.359	1.292	1.231	1.176	1.088	1.028	0.992	0.980
1.3				0.838	1.135	1.200	1.198	1.173	1.139	1.104	1.040	0.993	0.964	0.954
1.4					0.500	0.872	0.982	1.016	1.020	1.010	0.976	0.945	0.924	0.917
1.5							0.657	0.801	0.862	0.887	0.893	0.882	0.870	0.866
1.6								0.450	0.637	0.720	0.784	0.799	0.801	0.800
1.7									0.169	0.466	0.637	0.691	0.710	0.714
1.8											0.413	0.543	0.588	0.600
1.9												0.297	0.409	0.436
2.0														0.000
ζ for $\Delta = 0$	1.087	1.174	1.259	1.342	1.423	1.500	1.574	1.643	1.707	1.766	1.866	1.940	1.985	2.000

Since, in the similar triangles $SS''H''$ and $SS'H'$,

$$\frac{S''H''}{S'H'} = \frac{SS''}{SS'},\qquad(15)$$

the right member of (15) may be substituted in (14) for its left, giving, in place of (14),

$$m = \frac{SS''}{SS'}$$

$$= \frac{D}{\sin\,\nu}.\qquad(16)$$

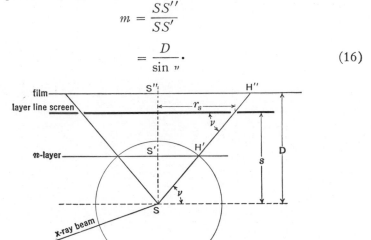

Fig. 182.

Thus, a measurement, x'', made on the photograph, corresponds with a reciprocal lattice distance

$$x' = \frac{1}{m}\,x'' = \left(\frac{\sin\,\nu}{D}\right)x''.\qquad(17)$$

This result is expressed in wavelength units, where λ is taken as unity. In absolute units, any measurement, x'', made on the photograph, represents an absolute reciprocal lattice distance of

$$x' = \lambda\frac{1}{m}\,x'' = \left(\lambda\frac{\sin\,\nu}{D}\right)x''.\qquad(18)$$

The layer line screen setting.   In order to prevent more than one layer line from reaching the film, it is necessary to provide the usual layer line screen.   This takes the form of a metal sheet, Fig. 182, from which a narrow circular slit is removed.   Figure 182 shows that the dimensions and placing of this slit are related by

$$s = r_s\,\tan\,\nu.\qquad(19)$$

SOME CHARACTERISTICS OF THE DE JONG AND BOUMAN METHOD

De Jong and Bouman's method of recording the reciprocal lattice is unique in affording an undistorted picture of the reciprocal lattice. It must not be supposed, however, that it is a method of unalloyed advantages; it has some serious disadvantages:

Normal-beam method. Figure 183A illustrates the taking of normal-beam photographs by the de Jong and Bouman method. Each level is photographed at a different cone angle, ν. If the film were left in the same plane for all the photographs, this would give rise to different magnifications, according to (16). In order to compare and interpret the photographs easily, it is desirable to have the magnifications of all levels the same. This requires that the film occupy different positions, as shown in Fig. 183A. The several proper film distances, D, for this condition can easily be derived by transposing (16):

$$D = m \sin \nu. \tag{20}$$

Note that, when ν is zero, D is zero, regardless of the magnification, m. Simultaneously the magnification, m, becomes indeterminate. Figure 183B, which illustrates the situation clearly, indicates that under these conditions it is geometrically impossible to record the usual photograph of points. In other words, it is impossible, by the normal-beam method, to record the zero layer. It is also extremely difficult for both mechanical and optical reasons to record layers of small ζ coordinate.

The disadvantages of recording de Jong and Bouman photographs by the normal-beam method, accordingly, are:

1. It is impossible to record the zero layer. Since the determination of space group depends on this layer, this disadvantage is serious.

2. It is extremely difficult to record layers of small ζ coordinate.

3. If the film position is maintained constant, the magnifications of the several layers are different, being proportional to $\dfrac{1}{\sin \nu}$.

In order to keep the magnifications the same, the film distances must be made proportional to $\sin \nu$. This weakens the intensities of the spots on the lower level photographs owing to inclined x-ray path.

4. There is a blind spot in the center of the film, where lattice points of comparatively low index would occur (Fig. 177).

Equi-inclination method. Figure 183C illustrates the taking of an equi-inclination photograph by the de Jong and Bouman method. As in the normal-beam method, each level must be photographed at a

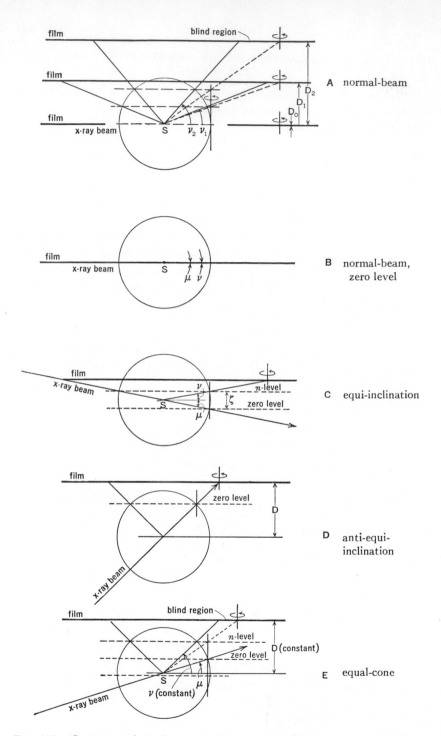

FIG. 183. Geometry of de Jong and Bouman recording by various techniques.

different cone angle, ν, and consequently, if the film is left in the same position, the various levels are recorded at different magnifications, given by (16). They can be recorded at the same magnification by changing the film distance, D, for each level, in accordance with relation (20).

The equi-inclination and normal-beam methods reduce to the same situation for the zero layer. The equi-inclination situation for the zero layer is thus also illustrated by Fig. 183B; consequently it is impossible to record the zero layer by equi-inclination methods in de Jong and Bouman photographs. As in the normal-beam method, it is also physically very difficult to record n-levels of small ζ coordinate. This difficulty is greatly exaggerated over the normal-beam method because in the equi-inclination method the several cone angles, ν, are much smaller than in the normal-beam case, for a given set of level coordinates, ζ. Consequently, the equi-inclination method offers more difficulties in the taking of de Jong and Bouman photographs than the normal-beam method, or, indeed, than any of the several methods.

The equi-inclination method has the further mechanical disadvantage that the x-ray beam must be introduced into the physical set-up by means of a pinhole system. This traverses a region which might otherwise be photographic film, and consequently a considerable area of potentially recordable lattice points is unrecorded.

The disadvantages of recording de Jong and Bouman photographs by the equi-inclination method, accordingly, are:

1. It is impossible to record the zero layer (see, however, the mitigating scheme below, under " anti-equi-inclination condition ").

2. It is even more difficult than in the normal-beam method to record n-layers of small ζ coordinate.

3. If the film position is maintained constant, the magnifications of the several layers are different, being proportional to $\dfrac{1}{\sin \nu}$. In order to keep the magnifications the same, the film distances must be made proportional to $\sin \nu$. This weakens the intensities of the spots on the lower level photographs due to inclined x-ray path.

4. For physical reasons, it is necessary to sacrifice some potentially recordable reciprocal lattice points of high ξ value, owing to the necessity of introducing the x-ray beam.

The anti-equi-inclination condition. The case $\mu = +\nu$ (rather than $\mu = -\nu$ of the equi-inclination method) is illustrated in Fig. 183D and has been discussed on page 301. It has already been pointed out that, under these conditions, it is possible to record only the zero layer. This condition is very closely related to the equi-inclination condition. It therefore provides a way of making the equivalent of an equi-inclination photograph of the zero layer. For any given n-layer, whose equi-inclination recording requires a cone angle, ν, a zero-layer photograph can be taken at the same cone angle, ν, by inclining the direct beam by an angle, $\mu = +\nu$. This n-layer and zero-layer pair will have the same magnification. The only difficulty with this scheme as a routine method of pattern analysis is that this identical magnification holds only for this particular pair; for any other pair of n- and zero-levels, the angle ν is different, the magnification is consequently different, and the two pairs cannot be directly compared.

The equal-cone method. The equal-cone method, illustrated in Fig. 183E, is the best adapted for the taking of de Jong and Bouman photographs. It has the following advantages:

1. It provides for the recording of the zero layer. It consequently provides a method of making a complete space group study (see, however, disadvantage 1, below).

2. It automatically provides for recording all levels at the same magnification, with the same crystal-to-film distance. This brings with it the further advantage:

3. Since all layers are recorded with diffraction beams whose paths are constant (i. e., the beams are all generators of identical cones), the intensities of the spots are comparable in respect to air absorption and in respect to decrease of intensity with distance. This advantage is unique to the equal-cone method.

The equal-cone method of recording de Jong and Bouman photographs has the following disadvantage:

1. All n-layer photographs have a central circular blind area. This area increases from a minimum at $\zeta = 0$ (i. e., the zero layer) to a maximum when $\mu = 0$ [this occurs when (7) reduces to $\zeta = \sin \nu$], and then decreases again until the equi-inclination condition [i. e., when (7) reduces to $\zeta = 2 \sin \nu$]. Beyond this value of ζ, the blind area increases until it entirely eliminates the record.

In the actual use of the de Jong and Bouman method, one first decides what cone angle, ν, will be employed. Then, making use of (19), the layer line screen is permanently set to record this cone. Next a decision is made as to what crystal to film distance, D, will be employed, and this distance is set for the instrument. Since the resulting film is in reciprocal lattice form, and easy to interpret no matter how small, it is ordinarily desirable to make D as small as possible in order to reduce the time of exposure to as little as possible. For mechanical reasons, it is not easy to reduce D below about 2 cm. A value of 3 cm. is probably an optimum and Fig. 181 is accordingly drawn for this scale.

The cone angle having been decided upon, and the level spacing of the reciprocal lattice being known by the aid of some sort of rotating-crystal photograph, it is necessary to solve (7) for each level setting. This can be done graphically with sufficient accuracy for the purpose. Graphical solutions for various cone angles, ν, differing by 5° intervals can be obtained directly from Fig. 162. The value of the film rotation axis displacement, Δ, is obtained directly from Fig. 181 or its equivalent. For instruments set at $\nu = 45°$, $D = 3$ cm., all settings can be obtained from Fig. 181B.

The films resulting from the preparations made as discussed above present one with a partially complete reciprocal lattice, level by level. They can be indexed, but can best be interpreted by inspection according to the theory developed in Chapter 22.

In the dimensional measurement of the reciprocal lattice constants from de Jong and Bouman films, it is only necessary to make the necessary linear measurements, x'', and find the true lengths these represent by multiplying them by the instrumental constant in the parentheses of (18):

$$x' = \left(\lambda \frac{\sin \nu}{D} \right) x''. \tag{18}$$

It might be pointed out in this place that unfortunately the precision of such lattice constants cannot be highly refined as can lattice constants derived from equatorial Weissenberg photographs (see Chapter 20).

Since each photograph is simply a true enlargement of the reciprocal lattice itself, any angular measurements made entirely within the plane of one photograph are angle-true in the reciprocal lattice. Angles involving lattice points from different levels may be computed by superposing the several films in question, making appropriate linear measures, and computing the angles from these lengths. These lengths, it should be remembered, are the projections of the true lengths on the plane of the reciprocal lattice level.

LITERATURE

W. F. DE JONG and J. BOUMAN. Das Photographieren von reziproken Kristall-netzen mittels Röntgenstrahlen. *Z. Krist.* (*A*), **98** (1938), 456–459.

W. F. DE JONG, J. BOUMAN, and J. J. DE LANGE. X-ray photography of zero-order reciprocal net planes of a crystal. *Physica*, **5** (1938), 188–192.

W. F. DE JONG and J. BOUMAN. Das Photographieren von reziproken Netzebenen eines Kristalles mittels Röntgenstrahlen. *Physica*, **5** (1938), 220–224.

W. F. DE JONG. Axinit. Das reziproke und das Bravaissche Gitter. *Z. Krist.* (*A*), **99** (1938), 326–335.

W. F. DE JONG and J. BOUMAN. Das reziproke und das Bravaissche Gitter von Gips. *Z. Krist.* (*A*), **100** (1938), 275–276.

J. BOUMAN and W. F. DE JONG. Die Intensitäten der Punkte einer photographier-ten reziproke Netzebene. *Physica*, **5** (1938), 817–832.

W. F. DE JONG and J. BOUMAN. Over het reciproke tralie van kristallen. Natuur-wetenschappelijk Tijdschrift, **21** (1939), 291–303.

CHAPTER 18

THE GEOMETRY OF OBLIQUE CELLS AND THEIR RECIPROCALS

In Chapter 6, the fundamental properties of reciprocal lattices were developed. In this chapter are discussed some properties of oblique cells and their reciprocals which are of aid in the experimental determination of the lattice constants of crystals belonging to the oblique crystal systems.

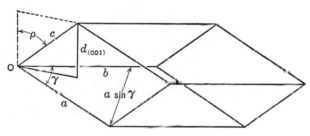

FIG. 184.

THE VOLUME OF A TRICLINIC CELL IN TERMS OF ITS OWN ELEMENTS

Analytical geometric derivation. The volume of a triclinic cell in terms of its elements, a, b, c, α, β, γ, can be derived by analytical geometric means as follows: Fig. 184 shows a triclinic parallelepiped. Its volume is

$$V = \text{Base} \times \text{Altitude}$$

$$= (a \sin \gamma)b \times d_{(001)}. \tag{1}$$

If the angle between the vertical and c is designated by ρ, Fig. 184, then,

$$d_{(001)} = c \cos \rho. \tag{2}$$

Substituting this in (1), the latter becomes

$$V = abc \, \sin \gamma \cos \rho. \tag{3}$$

It now remains to evaluate ρ in terms of α, β, and γ. To simplify the discussion, the cell is given a specialized orientation with respect to

347

three mutually orthogonal reference axes, OX, OY, and OZ, Fig. 185, namely, b is placed along OY, and the basal plane, (001), is placed in the plane OXY. The direction cosines of c with respect to OX, OY, and OZ are then

$$\cos \psi, \ \cos \alpha, \ \cos \rho.$$

The usual relation between these three direction cosines obtains, namely,

$$\cos^2 \psi + \cos^2 \alpha + \cos^2 \rho = 1, \tag{4}$$

from which

$$\cos^2 \rho = 1 - \cos^2 \alpha - \cos^2 \psi. \tag{5}$$

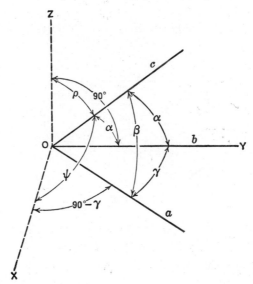

FIG. 185.

The only unknown quantity in this is $\cos \psi$. This can be involved in the standard expression for the cosine between the two directed lines a and c, i. e., the expression for $\cos \beta$, thus:

$$\cos \beta = \cos (90° - \gamma) \cos \psi + \cos \gamma \cos \alpha + \cos 90° \cos \rho. \tag{6}$$

This may be rearranged to give

$$\cos \psi = \frac{\cos \beta - \cos \gamma \cos \alpha - 0}{\cos (90° - \gamma)}, \tag{7}$$

which reduces to

$$\cos \psi = \frac{\cos \beta - \cos \gamma \cos \alpha}{\sin \gamma}. \tag{8}$$

Equation (5) may now be evaluated by substituting in it the value $\cos \psi$ given by (8). This results in

$$\cos^2 \rho = 1 - \cos^2 \alpha - \left(\frac{\cos \beta - \cos \gamma \cos \alpha}{\sin \gamma}\right)^2. \tag{9}$$

This can be reduced to a more symmetrical form by the following steps:

$$\cos^2 \rho = 1 - \cos^2 \alpha - \frac{\cos^2 \beta - 2 \cos \alpha \cos \beta \cos \gamma + \cos^2 \alpha \cos^2 \gamma}{\sin^2 \gamma}$$

$$= \frac{\sin^2 \gamma - \cos^2 \alpha \sin^2 \gamma - \cos^2 \beta + 2 \cos \alpha \cos \beta \cos \gamma - \cos^2 \alpha \cos^2 \gamma}{\sin^2 \gamma}$$

$$= \frac{\sin^2 \gamma - (\cos^2 \alpha \sin^2 \gamma + \cos^2 \alpha \cos^2 \gamma) - \cos^2 \beta + 2 \cos \alpha \cos \beta \cos \gamma}{\sin^2 \gamma}$$

$$= \frac{\sin^2 \gamma - \cos^2 \alpha (\sin^2 \gamma + \cos^2 \gamma) - \cos^2 \beta + 2 \cos \alpha \cos \beta \cos \gamma}{\sin^2 \gamma}$$

$$= \frac{\sin^2 \gamma - \cos^2 \alpha - \cos^2 \beta + 2 \cos \alpha \cos \beta \cos \gamma}{\sin^2 \gamma}$$

$$= \frac{1 - \cos^2 \gamma - \cos^2 \alpha - \cos^2 \beta + 2 \cos \alpha \cos \beta \cos \gamma}{\sin^2 \gamma}, \tag{10}$$

from which

$$\cos \rho = \frac{1}{\sin \gamma} \sqrt{1 - \cos^2 \alpha - \cos^2 \beta - \cos^2 \gamma + 2 \cos \alpha \cos \beta \cos \gamma}. \tag{11}$$

The value of $\cos \rho$ given by (11) is expressed in terms of the inter-axial cell angles α, β, and γ, and it can be substituted directly into (3), which then becomes

$$V = abc \sqrt{1 - \cos^2 \alpha - \cos^2 \beta - \cos^2 \gamma + 2 \cos \alpha \cos \beta \cos \gamma}. \tag{12}$$

Vector algebraic derivation. The derivation of the volume of a tri-clinic cell by vector algebraic methods is exceedingly neat. To derive the volume by this means, each of the oblique axes, **a**, **b**, and **c**, is first resolved into its components along any three mutually orthogonal axes OX, OY, and OZ. If the components of the length of a along the X, Y and Z axes are a_x, a_y, and a_z, and the unit vectors along X, Y, and Z are **i**, **j** and **k**, then

$$\mathbf{a} = a_x \mathbf{i} + a_y \mathbf{j} + a_z \mathbf{k},$$

and similarly,

$$\mathbf{b} = b_x \mathbf{i} + b_y \mathbf{j} + b_z \mathbf{k}, \tag{13}$$

and

$$\mathbf{c} = c_x \mathbf{i} + c_y \mathbf{j} + c_z \mathbf{k}.$$

The volume of the triclinic parallelepiped is

$$V = [\mathbf{a\ b\ c}], \qquad (14)$$

and substituting for $\mathbf{a}$, $\mathbf{b}$, and $\mathbf{c}$ the components given by (13), it is well known that (14) becomes [and this can also be derived from (55), Chapter 2]

$$V = \begin{vmatrix} a_x & a_y & a_z \\ b_x & b_y & b_z \\ c_x & c_y & c_z \end{vmatrix} [\mathbf{i\ j\ k}] = \begin{vmatrix} a_x & a_y & a_z \\ b_x & b_y & b_z \\ c_x & c_y & c_z \end{vmatrix} \cdot 1. \qquad (15)$$

In order to evaluate this, two mathematical tricks are resorted to. The first of these is squaring the right and left members of (15):

$$V^2 = \begin{vmatrix} a_x & a_y & a_z \\ b_x & b_y & b_z \\ c_x & c_y & c_z \end{vmatrix} \cdot \begin{vmatrix} a_x & a_y & a_z \\ b_x & b_y & b_z \\ c_x & c_y & c_z \end{vmatrix} . \qquad (16)$$

The second trick depends upon the well-known property of a determinant that its value is not altered by interchanging columns and rows. This interchange is applied to the second determinant of (16), giving

$$V^2 = \begin{vmatrix} a_x & a_y & a_z \\ b_x & b_y & b_z \\ c_x & c_y & c_z \end{vmatrix} \cdot \begin{vmatrix} a_x & b_x & c_x \\ a_y & b_y & c_y \\ a_z & b_z & c_z \end{vmatrix} . \qquad (17)$$

The multiplication of the two determinants is carried out according to the same rules as given for the multiplication of matrices, pages 17–18. Following these rules, (17) expands to

$$V^2 = \begin{vmatrix} a_x a_x + a_y a_y + a_z a_z & a_x b_x + a_y b_y + a_z b_z & a_x c_x + a_y c_y + a_z c_z \\ b_x a_x + b_y a_y + b_z a_z & b_x b_x + b_y b_y + b_z b_z & b_x c_x + b_y c_y + b_z c_z \\ c_x a_x + c_y a_y + c_z a_z & c_x b_x + c_y b_y + c_z b_z & c_x c_x + c_y c_y + c_z c_z \end{vmatrix} . \qquad (18)$$

Each element of the determinant in the right member of (18) is the expression in component form for the scalar product of the two vectors, for example,

$$b_x a_x + b_y a_y + b_z a_z = \mathbf{b \cdot a}. \qquad (19)$$

Converting the elements of the right member of (18) from component form into vector form by substitutions similar to (19), (18) reduces to

$$V^2 = \begin{vmatrix} \mathbf{a \cdot a} & \mathbf{a \cdot b} & \mathbf{a \cdot c} \\ \mathbf{b \cdot a} & \mathbf{b \cdot b} & \mathbf{b \cdot c} \\ \mathbf{c \cdot a} & \mathbf{c \cdot b} & \mathbf{c \cdot c} \end{vmatrix} . \qquad (20)$$

If the absolute values of each of these scalar products are now substituted, (20) becomes

$$V^2 = \begin{vmatrix} a^2 & ab \cos \gamma & ac \cos \beta \\ ab \cos \gamma & b^2 & bc \cos \alpha \\ ac \cos \beta & bc \cos \alpha & c^2 \end{vmatrix}. \qquad (21)$$

The expansion of this determinant is

$$V^2 = a^2(b^2c^2 - b^2c^2 \cos^2 \alpha) - ab \cos \gamma \, (abc^2 \cos \gamma - abc^2 \cos \alpha \cos \beta)$$
$$+ ac \cos \beta \, (ab^2c \cos \alpha \cos \gamma - ab^2c \cos \beta)$$

$$= a^2b^2c^2 \{ (1 - \cos^2 \alpha) - (\cos^2 \gamma - \cos \alpha \cos \beta \cos \gamma)$$
$$+ (\cos \alpha \cos \beta \cos \gamma - \cos^2 \beta) \}$$

$$= a^2b^2c^2(1 - \cos^2 \alpha - \cos^2 \beta - \cos^2 \gamma + 2 \cos \alpha \cos \beta \cos \gamma). \quad (22)$$

Therefore, the volume of the triclinic cell is evidently

$$V = abc\sqrt{1 - \cos^2 \alpha - \cos^2 \beta - \cos^2 \gamma + 2 \cos \alpha \cos \beta \cos \gamma}. \qquad (23)$$

Volumes of specialized oblique cells. The cells of isometric, tetragonal, and orthorhombic crystals have their axes at right angles to one another. For this reason they may be called orthogonal cells. The cells of hexagonal, monoclinic, rhombohedral, and triclinic crystals are non-orthogonal; they may be termed oblique cells.

The expressions for the volumes of the orthogonal cells are very simple and have the general form abc. The expressions for the volumes of the oblique cells are more complicated, but only in the triclinic case are they as complicated as (12). The next most complicated case is the rhombohedral one, in which $a = b = c$ and $\alpha = \beta = \gamma$. In this case, (12) reduces to

$$V = a^3\sqrt{1 - 3 \cos^2 \alpha + 2 \cos^3 \alpha}. \qquad (24)$$

This may also be trigonometrically transformed into an alternative product form:

$$V = 2a^3 \sin \frac{\alpha}{2} \sqrt{\sin \frac{\alpha}{2} \sin \frac{3\alpha}{2}}. \qquad (25)$$

In the monoclinic case, $\alpha = \gamma = 90°$ and $\cos \alpha = \cos \gamma = 0$, so (12) reduces to

$$V = abc\sqrt{1 - \cos^2 \beta}$$
$$= abc\sqrt{\sin^2 \beta}$$
$$= abc \sin \beta. \qquad (26)$$

This can easily be derived directly from a sketch like Fig. 184 by noting that V = Base × Altitude, where the base = $ac \sin \beta$ and the altitude = b.

The hexagonal cell is a special case of the monoclinic one in which $a = b$ and β has the specialized value of 120°. In this case (16) reduces to

$$V = a^2c \sin 120°$$

$$= a^2c \frac{\sqrt{3}}{2}. \tag{27}$$

These several volume expressions are listed in Table 10, page 159.

THE LENGTHS OF THE RECIPROCAL CELL AXES

Relations (36)–(38) and (56)–(58) of Chapter 6 are expressions for the lengths of the reciprocal cell axes in terms of the elements of the direct cell and the volume of the direct cell. With the volumes of the several kinds of cells evaluated, relations (36)–(38) and (56)–(58) can be specifically evaluated for the several special kinds of cells. These evaluations for all except the general, triclinic case are given in Table 10, page 159.

RELATIONS BETWEEN THE AXIAL ANGLES OF DIRECT AND RECIPROCAL CELLS

Triclinic case. The angles of the reciprocal cell are functions of the angles of the direct cell. The interrelations between these angles are shown in Fig. 186A. This diagram is constructed as follows: The directions of the direct cell axes are given by the lines a, b, and c. The planes through each pair of these axes are pinacoids of the direct cell:

Plane ab = (001),

Plane ac = (010),

Plane bc = (100).

Through any point, O', three planes are passed, each perpendicular to a direct cell axis. These three planes intersect each other in three lines, $O'P$, $O'Q$, and $O'R$. Each of these lines is perpendicular to two direct cell axes, and hence perpendicular to the pinacoid containing them. Each of these lines is therefore a reciprocal cell axis. This reasoning may be tabulated thus:

$O'P \perp b, c, \therefore \perp$ plane bc, (100), $\therefore O'P = a^*$.

$O'Q \perp a, c, \therefore \perp$ plane ac, (010), $\therefore O'Q = b^*$.

$O'R \perp a, b, \therefore \perp$ plane ab, (001), $\therefore O'R = c^*$.

It should be observed that not only is a reciprocal cell axis perpendicular to a direct cell pinacoid, for example, $a^* \perp (100)$, but also a direct cell axis is perpendicular to a reciprocal cell pinacoid by construction, for example, $a \perp A\,R\,O'Q = b^*c^* = (100)^*$.

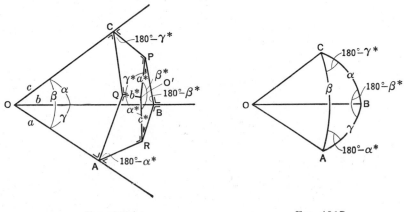

FIG. 186A.   FIG. 186B.

In the above construction, each reciprocal cell pinacoid intersects two direct cell pinacoids to outline a quadrilateral on the reciprocal pinacoid. For example, the reciprocal pinacoid b^*c^* intersects ac in AQ and intersects ab in AR to form quadrilateral $O'QAR$. In this particular quadrilateral, $\angle Q$ and $\angle R$ are right angles. Since the sum of the angles of a quadrilateral equals four right angles,

$$A + O' = 2 \text{ right angles};$$

therefore

$$A = 180° - O',$$

or

$$A = 180° - \alpha^*,$$

similarly,

$$B = 180° - \beta^*, \qquad (28)$$

and

$$C = 180° - \gamma^*.$$

A, B, and C are evidently the dihedral angles between the direct cell pinacoids.

Now, construct a sphere about O as a center. This intersects the direct cell pinacoids in arcs α, β, and γ. The angles between the arcs are A, B, and C. This important segment of the sphere is shown in Fig. 186B. Evidently ABC is a spherical triangle, and within it the angles and arcs are related by several well-known relations.

Assuming that the direct cell angles are known, and it is required to find the reciprocal cell angles, then the fundamental relation between A, B, C, α, β, and γ is expressed by the law of cosines which has the form

$$\cos A = \frac{\cos \alpha - \cos \beta \cos \gamma}{\sin \beta \sin \gamma}. \tag{29}$$

Since, from (28), $A = 180° - \alpha^*$,

$$\cos (180° - \alpha^*) = \frac{\cos \alpha - \cos \beta \cos \gamma}{\sin \beta \sin \gamma}, \tag{30}$$

from which

$$\cos \alpha^* = \frac{\cos \beta \cos \gamma - \cos \alpha}{\sin \beta \sin \gamma}, \tag{31}$$

and similarly,

$$\cos \beta^* = \frac{\cos \alpha \cos \gamma - \cos \beta}{\sin \alpha \sin \gamma}, \tag{32}$$

and

$$\cos \gamma^* = \frac{\cos \alpha \cos \beta - \cos \gamma}{\sin \alpha \sin \beta}. \tag{33}$$

Unfortunately, these forms involve subtractions, and hence are not well adapted to computation. Several functions of the half-angles $\frac{A}{2}$, $\frac{B}{2}$, and $\frac{C}{2}$ give rise to good computing forms. If, for compactness, the abbreviation

$$\sigma = \frac{\alpha + \beta + \gamma}{2} \tag{34}$$

is adopted, then the most convenient relation for computation is:

$$\tan \tfrac{1}{2}A = \sqrt{\frac{\sin (\sigma - \beta) \sin (\sigma - \gamma)}{\sin \sigma \sin (\sigma - \alpha)}}. \tag{35}$$

According to (28),

$$\tan \tfrac{1}{2}A = \tan [\tfrac{1}{2}(180° - \alpha^*)]$$

$$= \tan \left(90° - \frac{\alpha^*}{2} \right)$$

$$= \cot \frac{\alpha^*}{2}. \tag{36}$$

Substituting this in left member of (35), that relation becomes the required

$$\cot \frac{\alpha^*}{2} = \sqrt{\frac{\sin (\sigma - \beta) \sin (\sigma - \gamma)}{\sin \sigma \sin (\sigma - \alpha)}}; \qquad (37)$$

similarly,

$$\cot \frac{\beta^*}{2} = \sqrt{\frac{\sin (\sigma - \gamma) \sin (\sigma - \alpha)}{\sin \sigma \sin (\sigma - \beta)}} \qquad (38)$$

and

$$\cot \frac{\gamma^*}{2} = \sqrt{\frac{\sin (\sigma - \alpha) \sin (\sigma - \beta)}{\sin \sigma \sin (\sigma - \gamma)}}. \qquad (39)$$

These formulas have the advantage over corresponding sine and cosine formulas (given below) in that, for the actual computation of the entire set of angles, only four quantities need be looked up in tables, namely: $\sin \sigma$, $\sin (\sigma - \alpha)$, $\sin (\sigma - \beta)$, and $\sin (\sigma - \gamma)$, or their logarithms.

The three corresponding cosine formulas are of the form

$$\cos \frac{\alpha^*}{2} = \sqrt{\frac{\sin (\sigma - \beta) \sin (\sigma - \gamma)}{\sin \beta \sin \gamma}}. \qquad (40)$$

The solution of equations of this form for α^*, β^*, and γ^* require looking up six quantities in tables, namely: $\sin (\sigma - \alpha)$, $\sin (\sigma - \beta)$, $\sin (\sigma - \gamma)$, $\sin \alpha$, $\sin \beta$, and $\sin \gamma$, or their logarithms.

The three corresponding sine formulas are of the form

$$\sin \frac{\alpha^*}{2} = \sqrt{\frac{\sin \sigma \sin (\sigma - \alpha)}{\sin \beta \sin \gamma}}. \qquad (41)$$

The solution of equations of this form for α^*, β^*, and γ^* requires looking up seven quantities in tables, namely, those just listed for the cosine formula plus $\sin \sigma$ or its logarithm.

The sine formulas are highly insensitive in the region $\frac{\alpha^*}{2} = 90°$, $\alpha^* = 180°$; and the cosine formulas are highly insensitive in the region $\frac{\alpha^*}{2} = 0°$, $\alpha^* = 0°$. Fortunately such conditions are rarely encountered for interaxial angles of cells. The cotangent formulas are sensitive in all regions.

Non-triclinic cases. Only triclinic crystals have angular relations as complicated as those just given. All others except the rhombohedral case have very simple angular relations between direct and reciprocal cell.

In the rhombohedral case, $\alpha = \beta = \gamma$, and (31) reduces to

$$\cos \alpha^* = \frac{\cos \alpha \cos \alpha - \cos \alpha}{\sin \alpha \sin \alpha}$$

$$= \frac{\cos \alpha \, (\cos \alpha - 1)}{\sin^2 \alpha}$$

$$= \frac{\cos \alpha \, (\cos \alpha - 1)}{1 - \cos^2 \alpha}$$

$$= \frac{\cos \alpha \, (\cos \alpha - 1)}{(1 - \cos \alpha)(1 + \cos \alpha)}$$

$$= - \frac{\cos \alpha}{1 + \cos \alpha} . \tag{42}$$

In the only other general oblique cell, the monoclinic, the direct and reciprocal cell general angles are supplementary,

$$\beta^* = 180° - \beta. \tag{43}$$

Table 10, page 159, contains a tabulation of reciprocal cell angles in terms of direct angles.

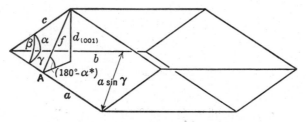

FIG. 187.

THE VOLUME OF A TRICLINIC CELL IN TERMS OF DIRECT AND RECIPROCAL ELEMENTS

The expression for the volume of a triclinic cell can be much simplified if it is permitted to contain both direct and reciprocal cell angles. The resulting volume expression is convenient for certain computations and is also useful in the proof of other relations.

In Fig. 187 the volume of the triclinic cell is

$$V = \text{Base} \times \text{Altitude}$$

$$= (b {\cdot} a \sin \gamma) \times d_{(001)}$$

$$= abd_{(001)} \sin \gamma. \tag{44}$$

In order to evaluate d, pass a plane through the end of the vector c, and normal to a. In the discussion of Fig. 186A, it was shown that this plane (Q A R, Fig. 186A) was a pinacoid, and that it intersected the two direct lattice pinacoids in two lines which met at an angle A, Figs. 186A and 187. It was also shown, (28), that $A = 180° - \alpha^*$. Adopting this value, it is evident from Fig. 187 that

$$\sin A = \frac{d_{(001)}}{f}, \qquad (45)$$

and that

$$\sin \beta = \frac{f}{c}. \qquad (46)$$

Combining (45) and (46) gives

$$d_{(001)} = f \sin A$$
$$= (c \sin \beta) \sin A$$
$$= c \sin \beta \sin (180° - \alpha^*)$$
$$= c \sin \beta \sin \alpha^*. \qquad (47)$$

Subotituting this value of $d_{(001)}$ in (44) gives

$$V = abc \sin \alpha^* \sin \beta \sin \gamma. \qquad (48)$$

In a similar way it can be shown that

$$V = abc \sin \alpha \sin \beta^* \sin \gamma \qquad (49)$$

and also

$$V = abc \sin \alpha \sin \beta \sin \gamma^*. \qquad (50)$$

These three relations are identical except for the position of the star. Each involves three direct lattice lengths and three angles, any one of which may be a reciprocal lattice angle.

LENGTHS OF CELL AXES IN TERMS OF DIRECT AND RECIPROCAL ELEMENTS

The fundamental relations between the reciprocal and direct cell axes were shown in Chapter 6 to be

$$a^* = \frac{bc \sin \alpha}{V}, \qquad (51)$$

$$b^* = \frac{ac \sin \beta}{V}, \qquad (52)$$

$$c^* = \frac{ab \sin \gamma}{V}. \qquad (53)$$

Substituting the expressions for the volume supplied by (49) and (50) in (51), also by (48) and (50) in (52), also by (48) and (49) in (53), gives the following relations, respectively:

$$a^* = \frac{1}{a \sin \beta^* \sin \gamma} = \frac{1}{a \sin \beta \sin \gamma^*}, \tag{54}$$

$$b^* = \frac{1}{b \sin \gamma^* \sin \alpha} = \frac{1}{b \sin \gamma \sin \alpha^*}, \tag{55}$$

$$c^* = \frac{1}{c \sin \alpha^* \sin \beta} = \frac{1}{c \sin \alpha \sin \beta^*}. \tag{56}$$

These relations are of considerable importance in the experimental determination of the lattice constants of oblique cells. It is of interest, in this connection, to see a geometrical interpretation of these relations.

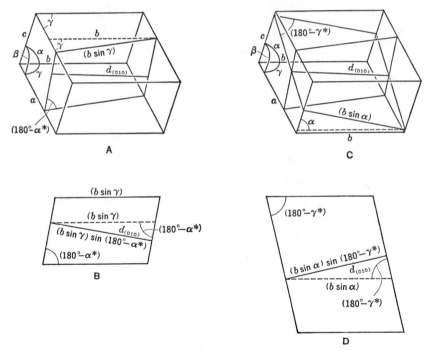

FIG. 188.

The geometrical proof of (55) is illustrated in Fig. 188. Figure 188A shows a triclinic direct cell. If a plane is passed perpendicular to edge a, it intersects the cell sides (010) and (001), the intersections outlining a parallelogram. Since this plane is $\perp a$, it is also $\perp$ (010).

and consequently contains the spacing $d_{(010)}$, which is shown as a full line in the parallelogram. For clearness the parallelogram is also shown separately in Fig. 188B. From Fig. 188A, it is evident that the upper side of the parallelogram is $(b \sin \gamma)$. Figure 188B shows that the spacing $d_{(010)}$ is $(b \sin \gamma) \sin (180° - \alpha^*)$, which reduces directly to $(b \sin \gamma) \sin \alpha^*$. According to the fundamentals of the reciprocal lattice theory (page 119), $d_{(010)}$ and b^* are reciprocals, hence

$$b^* = \frac{1}{d_{(010)}} = \frac{1}{b \sin \gamma^* \sin \alpha}.$$

This is the geometrical proof of the first half of (55). In Fig. 188C, it is evident that the spacing $d_{(010)}$ is not only contained in a plane $\perp a$, but also in a plane $\perp c$. The trace of this plane on the pinacoids (100) and (010) also outlines a parallelogram, shown separately in Fig. 188D. The upper side of this parallelogram is evidently $(b \sin \alpha)$, and the spacing $d_{(010)}$, according to Fig. 188D, is therefore $(b \sin \alpha) \sin (180° - \gamma^*)$, or $(b \sin \alpha) \sin \gamma^*$. Consequently,

$$b^* = \frac{1}{d_{(010)}} = \frac{1}{b \sin \alpha \sin \gamma^*}.$$

This is the second half of (55).

RECIPROCAL VOLUME RELATIONS

It can be shown by either algebraic or vector algebraic methods that the volumes of the direct and reciprocal cells are reciprocal.

Algebraic proof. Starting with the fundamental reciprocal relation [(40), Chapter 6]:

$$a = \frac{b^* c^* \sin \alpha^*}{V^*}, \tag{57}$$

the reciprocal cell volume is

$$V^* = \frac{b^* c^* \sin \alpha^*}{a}. \tag{58}$$

The starred forms of b^* and c^* may be eliminated by substituting their values given in (52) and (53):

$$V^* = \left(\frac{ac \sin \beta}{V}\right)\left(\frac{ab \sin \gamma}{V}\right)\frac{\sin \alpha^*}{a}. \tag{59}$$

This reduces to

$$V^* = \frac{abc \sin \alpha^* \sin \beta \sin \gamma}{V^2}. \tag{60}$$

TABLE 20

COLLECTION OF RELATIONS FOR TRICLINIC RECIPROCAL CELL ELEMENTS

$$a^* = \frac{bc \sin \alpha}{V}. \tag{51}$$

$$b^* = \frac{ac \sin \beta}{V}. \tag{52}$$

$$c^* = \frac{ab \sin \gamma}{V}. \tag{53}$$

$$a^* = \frac{1}{a \sin \beta^* \sin \gamma} = \frac{1}{a \sin \beta \sin \gamma^*}. \tag{54}$$

$$b^* = \frac{1}{b \sin \alpha^* \sin \gamma} = \frac{1}{b \sin \alpha \sin \gamma^*}. \tag{55}$$

$$c^* = \frac{1}{c \sin \alpha^* \sin \beta} = \frac{1}{c \sin \alpha \sin \beta^*}. \tag{56}$$

$$a^* : b^* : c^* = \frac{\sin \alpha}{a} : \frac{\sin \beta}{b} : \frac{\sin \gamma}{c}. \tag{68}$$

$$V^* = a^*b^*c^*\sqrt{1 - \cos^2 \alpha^* - \cos^2 \beta^* - \cos^2 \gamma^* + 2 \cos \alpha^* \cos \beta^* \cos \gamma^*}. \tag{12}, (23)$$
$$V^* = a^*b^*c^* \sin \alpha^* \sin \beta^* \sin \gamma. \tag{48}$$
$$= a^*b^*c^* \sin \alpha^* \sin \beta \sin \gamma^* \tag{49}$$
$$= a^*b^*c^* \sin \alpha \sin \beta^* \sin \gamma^*. \tag{50}$$

$$V^* = \frac{1}{V}. \tag{61}, (64)$$

$$\cos \alpha^* = \frac{\cos \beta \cos \gamma - \cos \alpha}{\sin \beta \sin \gamma}. \tag{31}$$

$$\cos \beta^* = \frac{\cos \alpha \cos \gamma - \cos \beta}{\sin \alpha \sin \gamma}. \tag{32}$$

$$\cos \gamma^* = \frac{\cos \alpha \cos \beta - \cos \gamma}{\sin \alpha \sin \beta}. \tag{33}$$

$$\cot \frac{\alpha^*}{2} = \sqrt{\frac{\sin (\sigma - \beta) \sin (\sigma - \gamma)}{\sin \sigma \sin (\sigma - \alpha)}}. \tag{37}$$

$$\cot \frac{\beta^*}{2} = \sqrt{\frac{\sin (\sigma - \alpha) \sin (\sigma - \gamma)}{\sin \sigma \sin (\sigma - \beta)}}. \tag{38}$$

$$\cot \frac{\gamma^*}{2} = \sqrt{\frac{\sin (\sigma - \alpha) \sin (\sigma - \beta)}{\sin \sigma \sin (\sigma - \gamma)}}, \tag{39}$$

$$\text{where } \sigma = \frac{\alpha + \beta + \gamma}{2}.$$

$$\frac{\sin \alpha^*}{\sin \alpha} = \frac{\sin \beta^*}{\sin \beta} = \frac{\sin \gamma^*}{\sin \gamma}. \tag{70}$$

TABLE 21
Collection of Relations for Triclinic Direct Cell Elements

$$a = \frac{b^* c^* \sin \alpha^*}{V^*} . \qquad\qquad (51')$$

$$b = \frac{a^* c^* \sin \beta^*}{V^*} . \qquad\qquad (52')$$

$$c = \frac{a^* b^* \sin \gamma^*}{V^*} . \qquad\qquad (53')$$

$$a = \frac{1}{a^* \sin \beta \sin \gamma^*} = \frac{1}{a^* \sin \beta^* \sin \gamma} . \qquad\qquad (54')$$

$$b = \frac{1}{b^* \sin \alpha \sin \gamma^*} = \frac{1}{b^* \sin \alpha^* \sin \gamma} . \qquad\qquad (55')$$

$$c = \frac{1}{c^* \sin \alpha \sin \beta^*} = \frac{1}{c^* \sin \alpha^* \sin \beta} . \qquad\qquad (56')$$

$$a : b : c = \frac{\sin \alpha^*}{a^*} : \frac{\sin \beta^*}{b^*} : \frac{\sin \gamma^*}{c^*} . \qquad\qquad (68')$$

$$V = abc \sqrt{1 - \cos^2 \alpha - \cos^2 \beta - \cos^2 \gamma + 2 \cos \alpha \cos \beta \cos \gamma.} \qquad (12'), (23')$$

$$V = abc \sin \alpha \sin \beta \sin \gamma^* \qquad\qquad (48')$$

$$= abc \sin \alpha \sin \beta^* \sin \gamma \qquad\qquad (49')$$

$$= abc \sin \alpha^* \sin \beta \sin \gamma. \qquad\qquad (50')$$

$$V = \frac{1}{V^*} \qquad\qquad (61'), (64')$$

$$\cos \alpha = \frac{\cos \beta^* \cos \gamma^* - \cos \alpha^*}{\sin \beta^* \sin \gamma^*} . \qquad\qquad (31')$$

$$\cos \beta = \frac{\cos \alpha^* \cos \gamma^* - \cos \beta^*}{\sin \alpha^* \sin \gamma^*} . \qquad\qquad (32')$$

$$\cos \gamma = \frac{\cos \alpha^* \cos \beta^* - \cos \gamma^*}{\sin \alpha^* \sin \beta^*} . \qquad\qquad (33')$$

$$\cot \frac{\alpha}{2} = \sqrt{\frac{\sin (\sigma^* - \beta^*) \sin (\sigma^* - \gamma^*)}{\sin \sigma^* \sin (\sigma^* - \alpha^*)}} . \qquad\qquad (37')$$

$$\cot \frac{\beta}{2} = \sqrt{\frac{\sin (\sigma^* - \alpha^*) \sin (\sigma^* - \gamma^*)}{\sin \sigma^* \sin (\sigma^* - \beta^*)}} . \qquad\qquad (38')$$

$$\cot \frac{\gamma}{2} = \sqrt{\frac{\sin (\sigma^* - \alpha^*) \sin (\sigma^* - \beta^*)}{\sin \sigma^* \sin (\sigma^* - \gamma^*)}} , \qquad\qquad (39')$$

$$\text{where } \sigma^* = \frac{\alpha^* + \beta^* + \gamma^*}{2} .$$

$$\frac{\sin \alpha}{\sin \alpha^*} = \frac{\sin \beta}{\sin \beta^*} = \frac{\sin \gamma}{\sin \gamma^*} . \qquad\qquad (70')$$

The numerator of the right member of this is V, according to (48). Relation (60) thus reduces to

$$V^* = \frac{1}{V}. \tag{61}$$

Therefore, the volume of the reciprocal cell is the reciprocal of the volume of the direct cell.

Vector algebraic proof. Starting with the fundamental volume relation

$$V^* = \mathbf{a}^* \cdot \mathbf{b}^* \times \mathbf{c}^*, \tag{62}$$

substitute the fundamental vector expression, given in (53), Chapter 6, for each part of the right member:

$$V^* = \frac{\mathbf{b} \times \mathbf{c}}{V} \cdot \frac{\mathbf{c} \times \mathbf{a}}{V} \times \frac{\mathbf{a} \times \mathbf{b}}{V}. \tag{63}$$

This reduces in the following steps:

$$V^* = \frac{\mathbf{b} \times \mathbf{c} \cdot \mathbf{c} \times \mathbf{a} \times \mathbf{a} \times \mathbf{b}}{V^3}$$

$$= \frac{\mathbf{b} \times \mathbf{c} \cdot ([\mathbf{cab}]\mathbf{a} - [\mathbf{caa}]\mathbf{b})}{V^3}$$

$$= \frac{\mathbf{b} \times \mathbf{c} \cdot (V \cdot \mathbf{a} - 0)}{V^3} = \frac{\mathbf{a} \cdot \mathbf{b} \times \mathbf{c} \cdot V}{V^3}$$

$$= \frac{V^2}{V^3};$$

therefore

$$V^* = \frac{1}{V}. \tag{64}$$

RATIOS OF DIRECT AND RECIPROCAL ELEMENTS

The Goldschmidt polarform ratio. An interesting ratio discovered by Goldschmidt by comparatively heavy, straightforward methods may easily be proved with the relations now at hand: Dividing (51) by (52), there results

$$\frac{a^*}{b^*} = \frac{b}{a} \frac{\sin \alpha}{\sin \beta}$$

$$= \frac{\dfrac{\sin \alpha}{a}}{\dfrac{\sin \beta}{b}}, \tag{65}$$

or

$$a^* : b^* = \frac{\sin \alpha}{a} : \frac{\sin \beta}{b}. \tag{66}$$

This ratio can be extended to all axes:

$$a^* : b^* : c^* = \frac{\sin \alpha}{a} : \frac{\sin \beta}{b} : \frac{\sin \gamma}{c}. \tag{67}$$

Similarly, the reciprocal ratio can be proved:

$$a : b : c = \frac{\sin \alpha^*}{a^*} : \frac{\sin \beta^*}{b^*} : \frac{\sin \gamma^*}{c^*}. \tag{68}$$

The modified law of sines. A relation which ranks with Goldschmidt's polarform ratio is a modified law of sines. In the spherical triangle ABC, Fig. 186B, the well-known law of sines of spherical trigonometry supplies the relation

$$\frac{\sin A}{\sin \alpha} = \frac{\sin B}{\sin \beta} = \frac{\sin C}{\sin \gamma}. \tag{69}$$

Since, according to (28), $A = 180° - \alpha^*$, and since $\sin (180° - \alpha^*) = \sin \alpha^*$, (69) can be easily transformed into

$$\frac{\sin \alpha^*}{\sin \alpha} = \frac{\sin \beta^*}{\sin \beta} = \frac{\sin \gamma^*}{\sin \gamma}. \tag{70}$$

LITERATURE

A. BRAVAIS. Abhandlung über die Systeme von regelmässig auf einer Ebene oder in Raum vertheilten Punkten (1848). (Translated by C. and E. Blasius, and appearing as No. 90 of Ostwald's *Klassiker der exakten Wissenschaften;* Wilhelm Engelmann, Leipzig, 1897.)

VICTOR GOLDSCHMIDT. *Index der Krystallform der Mineralien.* Vol. I. (Julius Springer, Berlin, 1886.) See especially pages 5–19, 78–79, 83.

E. H. KRAUS and G. MEZ. Ueber topische Axenverhältnisse. *Z. Krist.* (*A*), **34** (1901), especially 390–391.

D. CROWFOOT. The interpretation of Weissenberg photographs in relation to crystal symmetry. *Z. Krist.* (*A*), **90** (1935), especially 218–222.

M. J. BUERGER. The x-ray determination of lattice constants and axial ratios of crystals belonging to the oblique systems. *Am. Mineral.,* **22** (1937), especially 418–420.

W. F. DE JONG and J. BOUMAN. Kristallographische Berechnungen und Konstruktionen mittels des reziproken Gitters. *Z. Krist.* (*A*), **101** (1939), 317–336.

W. F. DE JONG and J. BOUMAN. Over het reciproke tralie van kristallen. *Natuurwetenschappelijk Tijdschrift,* **21** (1939), 291–303.

CHAPTER 19

THE EXPERIMENTAL DETERMINATION OF THE LATTICE CONSTANTS OF THE CRYSTALS BELONGING TO THE OBLIQUE SYSTEMS

THE CHOICE OF ELEMENTS AND SETTING OF A TRICLINIC CRYSTAL

Introduction. For a given crystal, x-ray diffraction methods provide a picture of the reciprocal lattice (for example, the method of de Jong and Bouman gives an undistorted image of the reciprocal lattice). To this unique reciprocal lattice there corresponds a unique direct lattice. Unfortunately there are an infinite number of ways of defining this lattice because there are an infinite number of ways in which cells can be chosen from it. This circumstance requires the adoption of certain conventional rules so that all investigators may arrive at the same cell for a given lattice and define the same lattice in the same way. Such rules must provide not only for the selection of an appropriate cell but also for its orientation and labeling.

Choice of cell. In crystal systems other than triclinic, it is often necessary to choose a non-primitive cell in order to gain the advantages of orthogonality, or in order to have a cell whose outlines correspond with the symmetry of the crystal. Such alternative choices gain nothing in the triclinic crystal, so a primitive cell should always be chosen.

There are an infinite number of ways of choosing primitive cells in a triclinic lattice. It is standard practice to choose for descriptive purposes the *reduced cell*, i. e., that particular cell which has as cell edges the three shortest translations. This cell can usually be selected by inspection of a set of elevations of the lattice. If any doubt arises as to its right to be called a reduced cell, it may be tested by showing that (vectorially)

$$\text{Conditions A} \begin{cases} |a| < |a+b|, & |b| < |a+b|, & |c| < |a+c|, \\ |a| < |a-b|, & |b| < |a-b|, & |c| < |a-c|, \\ |a| < |a+c|, & |b| < |b+c|, & |c| < |b+c|, \\ |a| < |a-c|, & |b| < |b-c|, & |c| < |b-c|. \end{cases}$$

This set of requirements may be epitomized by saying that each cell edge must be shorter than the diagonals of the faces bordering it.

This test indicates that a particular axial translation is less than any combination of axial translation, and consequently is one of the set of three shortest translations.   The first rule in defining a triclinic lattice is therefore:

Rule 1.   Select the reduced primitive cell.

The reciprocal of a primitive cell is a primitive cell of the reciprocal lattice.   The reciprocal of the primitive reduced cell is that particular primitive cell of the reciprocal lattice whose three pinacoidal spacings, $d^*_{(100)}$, $d^*_{(010)}$, and $d^*_{(001)}$, are greater than any of the co-zonal prismatic spacings, i. e.,

$$\text{Conditions B} \begin{cases} d^*_{(100)} > d^*_{(110)}, & d^*_{(010)} > d^*_{(110)}, & d^*_{(001)} > d^*_{(101)}, \\ d^*_{(100)} > d^*_{(1\bar{1}0)}, & d^*_{(010)} > d^*_{(110)}, & d^*_{(001)} > d^*_{(10\bar{1})}, \\ d^*_{(100)} > d^*_{(101)}, & d^*_{(010)} > d^*_{(011)}, & d^*_{(001)} > d^*_{(011)}, \\ d^*_{(100)} > d^*_{(10\bar{1})}, & d^*_{(0\bar{1}0)} > d^*_{(01\bar{1})}, & d^*_{(001)} > d^*_{(01\bar{1})}. \end{cases}$$

Contrary to the opinions expressed in certain publications, these conditions are not equivalent to

$$\text{Conditions C} \begin{cases} |a^*| < |a^*+b^*|, & |b^*| < |a^*+b^*|, & |c^*| < |a^*+c^*|, \\ |a^*| < |a^*-b^*|, & |b^*| < |a^*-b^*|, & |c^*| < |a^*-c^*|, \\ |a^*| < |a^*+c^*|, & |b^*| < |b^*+c^*|, & |c^*| < |b^*+c^*|, \\ |a^*| < |a^*-c^*|, & |b^*| < |b^*-c^*|, & |c^*| < |b^*-c^*|. \end{cases}$$

It is therefore not necessarily true that the reciprocal of the direct reduced primitive cell is the reduced primitive cell of the reciprocal lattice.   This is unfortunate, because, had this been the case, an easy way to select the direct reduced primitive cell would have been to pick out the reduced primitive cell of the reciprocal lattice, and its reciprocal would have been the required cell.   Although, as mentioned above, Conditions C do not necessarily imply Conditions B, they frequently do.   In attempting to select the correct cell, it is therefore strategic to select the reduced reciprocal primitive cell according to Conditions C, and then test for Conditions B.   If Conditions B are met, the correct direct cell is the reciprocal of the one chosen; if not, then the correct reciprocal cell is one very closely related to the one first selected.

For all except triclinic crystals, Conditions C imply Conditions B, and the direct reduced primitive cell is the reciprocal of the reduced primitive cell of the reciprocal lattice.

Choice of axial label.   Given the three translations of the primitive reduced cell edges, t_1, t_2, and t_3, and the three labels a, b, and c, there are forty-eight different ways of applying the labels to the translations. For, the label a can be applied to t_1, $-t_1$, t_2, $-t_2$, t_3, or $-t_3$ (i. e., there are six ways of placing the label a). Once the position of a is decided upon, there are four different ways of applying the label b. For example, suppose that the label a is applied to t_1, then $-a$ automatically becomes $-t_1$, and b can be applied to any one of the remaining four t's. If consideration is limited to right-handed coordinate systems, which is customary in crystallography, then, when a and b are fixed, c is automatically fixed. There are six ways of fixing a, and to each of these there are four ways of fixing b, making twenty-four possible ways of fixing labels to the translations. If left-handed coordinate systems are also included, this number is doubled.

From this discussion, it is evident that, unless some simple conventional rules are adopted, a crystal lattice might be defined in twenty-four different ways by different crystallographers, and, to say the least, confusion would be the result. A simple way to prevent this confusion is with the additional aid of two arbitrary but simple rules. The first of these new rules may be made a consequence of a definition: Let a be, by definition, the shortest axis; let b be the intermediate axis; and let c be the longest axis; i.e., by definition, $a < b < c$. As a result of this definition, one rule obviously is:

Rule 2. †‡   *Label the shortest translation a, the intermediate translation b, and the longest translation c.* This rule fixes the directions, but not the senses ($+$ and $-$) of a, b, and c. These alternatives are taken care of by applying an additional rule:

Rule 3.   Take $+a$, $+b$, and $+c$ in such directions that the interaxial angles α, β, and γ are all obtuse. (Obtuse angles rather than acute ones are specified because this is customary crystallographic usage.)

This last rule has the interesting corollary that the interaxial angles, $α^*$, $β^*$, and $γ^*$, of the reciprocal cell are acute. This can be proved with the aid of relations (31), (32), and (33), Table 20. These relations

---

† In certain instances it is desirable to bring out the similarity in structure between a triclinic crystal and some other crystal, possibly of more specialized symmetry. To facilitate such comparison it may be desirable to label the axes in some order other than $a < b < c$, and it may even be desirable to choose a non-primitive or non-reduced cell. The rules suggested here are not intended to apply to such cases, but rather to the general systematic listing or other systematic geometrical treatment of crystal cells.

‡ An ill-defined custom prevails of labeling the three axes so that their lengths are in the order $c < a < b$. Though it is true that the choice of labels is arbitrary, the same arguments can be advanced against the choice of this order as against the choice of the order $γ < α < β$ for refractive indices.

have the form:

$$\cos \alpha^* = \frac{\cos \beta \cos \gamma - \cos \alpha}{\sin \beta \sin \gamma}.$$

If α, β, and γ are obtuse, their sines are $(+)$ and their cosines are $(-)$. The sign scheme is then

$$\cos \alpha^* = \frac{(-)(-) - (-)}{(+)(+)}$$

$$= \frac{(+) + (+)}{(+)} = \frac{(+)}{(+)} = (+).$$

Since the sign of $\cos \alpha^*$ is $(+)$, α^* is acute. This consequence of Rule 3 can similarly be shown to hold for β^* and γ^*.

These three rules, although arbitrary, are geometrical; i. e., they depend only on the geometry of the pattern, and no two investigators can apply them differently.

Unfortunately, those who have considered the matter of the choice of axes for triclinic crystals up to the present have given unwarranted attention to the external form of the crystal and have been strongly influenced by certain precedents in crystallographic practice. Attempts to set up rules hampered by this background always stumble over the conception of a " chief zone," and it is not uncommon to find rules such as the following in a paper in which a " unique orientation is the main objective ":

The proper vertical axis is properly selected from a consideration of the normal crystal habit. If a crystal species is habitually acicular or columnar, the axis of the acicular or columnar zone is almost certainly an edge of the properly chosen lattice cell; and most morphologists will set this axis vertical for the sake of appearance alone. But aside from the matter of appearance, it is highly desirable for practical reasons that the axis of habitual morphological elongation be taken as the vertical axis, since this axis is the most convenient, sometimes the only practical axis of adjustment on the Goldschmidt reflecting goniometer and the Weissenberg x-ray goniometer . . . .
In the case of crystals which are habitually elongated and tabular to a plane in the elongated zone, morphologists are less consistent in setting the edge of the elongated zone vertical. But here the practical consideration mentioned above is equally valid and it is equally desirable that the edge of an elongated table be set upright.

Quite apart from the fact that such rules do not lead to a " unique orientation," they are based upon the fallacious assumption that a given crystal species has a habit characteristic of its lattice. The habit of a crystal is known to depend upon the packing and bonding of its atoms, plus the environment of the crystal when it was grown, and not upon the lengths of its identity periods. An interesting sidelight on the fictitious nature of these requirements is that the elongated habit axis, which is the " only practical axis of adjustment " on the Goldschmidt and Weissenberg goniometers, for which reason it is allotted the sacred office of " vertical " (i. e., c-axis), is actually *horizontal* on these goniometers.

In some quarters a variant of Rule 3 is given which is equivalent to the following form:

> *Rule 3A.   Take $+a$, $+b$, and $+c$ in such directions that the two interaxial angles β and γ are obtuse, and ϕ_{001} lies between $0°$ and $90°$.*

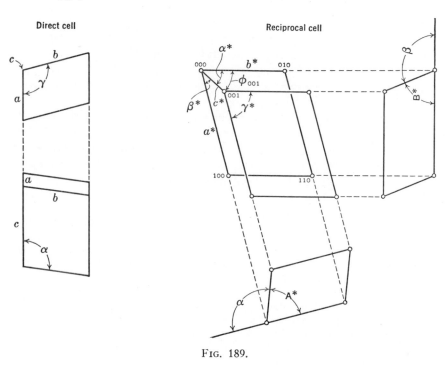

Fig. 189.

The right-hand side of Fig. 189, which shows a plan and two elevations of the reciprocal cell, illustrates the significance of this alternative rule. The requirement that ϕ_{001} lies between $0°$ and $90°$ permits the dihedral angle, A^*, to be acute within the range $\phi_{001} = 0° - \phi_{100}$, but A^* is obtuse when ϕ lies beyond this range.   It was shown in the last chapter that A^* is the supplement of α.   Hence, in the range $\phi_{001} = 0° - 90°$, α changes from obtuse to acute as ϕ_{001} passes the value $\phi_{001} = \phi_{100} = \gamma^*$.   This is a most arbitrary rule.   The angles β and γ share the favored role of being always obtuse, while α is either obtuse or acute, a capricious condition controlled by the size of γ^*. As a consequence of this, the sign analysis given above may be used to prove that the reciprocal angles α^*, β^*, and γ^* may be either acute or obtuse in an unpredictable fashion. Certainly, not only is Rule $3A$ arbitrary, but also its consequences are chaotic.

Conventional orientation. It is conventional to orient the axes of a crystal so that the c-axis is vertical, the b-axis runs roughly left to right, and the a-axis runs back to front. This orientation can be defined precisely by requiring that c be vertical and that the ϕ coordinate of (010) be zero. The conventional plan and front elevation of the direct cell are then as shown at the left of Fig. 189. The conventional orientation of the reciprocal lattice is then as shown at the right of Fig. 189.

THE TECHNIQUE OF DETERMINING LATTICE ELEMENTS

The determination of the lattice constants of a crystal resolves itself into a problem of measuring the appropriate linear and angular positions of the points of the reciprocal lattice of the crystal. In one particular x-ray method — the de Jong and Bouman method — the x-ray photographs provide an undistorted image of the reciprocal lattice, and these constants, therefore, may be measured directly on the film. In all other methods, however, the x-ray photographs constitute distorted projections of the levels of the reciprocal lattice. In such cases, the positions of the diffraction spots on the films are but distorted representations of the true positions of the corresponding reciprocal lattice points, and the film spot positions must undergo interpretation before the desired lattice point positions become available. This process of interpretation can take two forms:

(a) Charts, such as Figs. 79, 142, 144, and 148, which map out a film in reciprocal lattice coordinates, can be laid on the film, and then the film spot positions can be read directly in terms of reciprocal lattice coordinates; or,

(b) the film coordinates, x and z, may be accurately determined by linear scale measurements, and from these the appropriate reciprocal lattice coordinates can be computed with the aid of transformation equations.

For rough or preliminary values, the chart method of interpretation is to be recommended because it is ordinarily much more rapid. On the other hand, the chart method is useless for more accurate values, for two reasons:

(*A*) The charts contain whatever lack of precision was inherent in the original chart computations, and they also contain the inaccuracies of the original chart drawing and the inaccuracies of the subsequent reproduction.

(*B*) The successive graduations on charts are not uniform, and hence it is impossible to use verniers with charts to determine the precise positions of spots falling between graduations.

On the other hand, the measurements of the relative positions of spots on the film in terms of film coordinates x and z can be accomplished by means of steel scale and vernier devices.   Such measurements are consequently capable of giving film position values of extreme precision. The reciprocal lattice positions derived from these measurements have the same high precision, subject to the provisos discussed in the next chapter.   For these reasons it is evident that, if any precision in the values of the lattice constants is desired, measurements should be made in terms of the film coordinates by means of steel scales, and not in terms of reciprocal lattice coordinates by means of charts.

Regardless of the method of measurement employed — chart or scale — it is possible to make measurements of the reciprocal cell which will lead to either the lattice constants of the reciprocal cell or to the lattice constants of the direct cell.   Since the lattice constants of the direct cell are the ones ultimately desired, the latter scheme of measurement would seem to be the obvious one to use.   The decision between these, however, is not so simple, for the alternative method of measuring the lattice constants of the reciprocal cell and then converting them to the lattice constants of the direct cell is capable of providing extreme precision if handled in accordance with the practice recommended in Chapter 21.   The last section of the present chapter gives an introduction to the reasons for this possible precision.

DETERMINATION OF THE ELEMENTS OF THE RECIPROCAL CELL

General Remarks

The measurement of the reciprocal cell constants is much more easily accomplished on the reciprocal lattice than the measurement of the direct cell constants.   Furthermore, as already pointed out, these measurements are capable of being refined by a systematic treatment detailed in Chapter 21, and thus extremely accurate cell constants may be arrived at.

Each moving-film photograph represents a two-dimensional level of the three-dimensional reciprocal lattice.   Consequently any dimension in the level is determinable by measuring the corresponding dimension of the film.

Linear Constants of the Reciprocal Cell

The constants of the reciprocal cell are most easily determined from photographs made by the de Jong and Bouman method. In this case any photograph is a direct image of the corresponding reciprocal lattice level, i. e., it is an image of the plane lattice of that level. To determine the linear lattice constants lying in that level, it is only necessary to decide upon an appropriate cell, Fig. 189, measure the cell edges, and divide by the appropriate magnification factor, page 337. Unfortunately, no means of refining such lattice constants has yet been devised.

The corresponding measurements may also be made on the Weissenberg film, Fig. 190B. Such measurements can be made with the aid of the $\xi_{\bar{z}}$ scale, Fig. 144, in which case the computation of the absolute reciprocal lattice constant, say a^*, takes the form

$$|a^*| = \lambda\xi, \tag{1}$$

if the position of the first lattice point is measured. More generally, if the position of the nth lattice point from the origin is measured,

$$|a^*| = \lambda\frac{\xi}{n}. \tag{2}$$

Since this method may be made to yield lattice constants of extreme precision when refined by the method discussed in Chapter 21, it is often desirable to determine ξ with greater precision than possible with the chart of Fig. 228. This is possible by measuring the film distance x, of the spot, converting to Υ, and then to ξ, thus: From (1) and (3), Chapter 12,

$$\Upsilon = C_1x, \tag{3}$$

where the instrumental constant,

$$C_1 = \left[\frac{360°}{2\pi r_F}\right],$$

and, for the zero layer, (9), Chapter 14, reduces to

$$\xi = 2\sin\frac{\Upsilon}{2}. \tag{4}$$

Combining (3) and (4) gives

$$\xi = 2\sin\frac{C_1x}{2}. \tag{5}$$

This is expressed in λ units. In order to reduce to absolute units, relation (2) above may be used.

This method is capable of refinement to extreme precision, as discussed in Chapter 21. One of the advantages of the Weissenberg method is the possibility of making these refinements.

TABLE 22

DATA FOR SUBDIVISION OF ZERO-LAYER ξ SCALE

Solution of $\Upsilon = 2 \sin^{-1} \dfrac{\xi}{2}$

To transform Υ to x for cylindrical film, multiply by $\dfrac{1}{C_1} = \dfrac{2\pi r_F}{360°}$.

$\downarrow \xi \rightarrow$	.00	.01	.02	.03	.04	.05	.06	.07	.08	.09
0.0	0.00	0.57	1.15	1.72	2.29	2.87	3.44	4.01	4.58	5.16
0.1	5.73	6.31	6.88	7.45	8.03	8.60	9.18	9.75	10.33	10.90
0.2	11.48	12.05	12.63	13.21	13.78	14.36	14.94	15.52	16.10	16.67
0.3	17.25	17.82	18.41	18.99	19.58	20.16	20.74	21.32	21.91	22.49
0.4	23.07	23.66	24.24	24.83	25.42	26.01	26.59	27.18	27.76	28.36
0.5	28.96	29.55	30.14	30.73	31.33	31.92	32.52	33.12	33.72	34.32
0.6	34.92	35.52	36.12	36.72	37.33	37.93	38.54	39.15	39.75	40.36
0.7	40.97	41.59	42.20	42.81	43.43	44.05	44.67	45.29	45.91	46.53
0.8	47.16	47.78	48.41	49.04	49.67	50.30	50.94	51.57	52.21	52.85
0.9	53.49	54.13	54.77	55.42	56.07	56.72	57.37	58.02	58.68	59.34
1.0	60.00	60.66	61.33	62.00	62.67	63.34	64.01	64.69	65.37	66.05
1.1	66.73	67.42	68.13	68.81	69.50	70.20	70.90	71.61	72.31	73.03
1.2	73.74	74.46	75.18	75.90	76.63	77.36	78.10	78.84	79.58	80.33
1.3	81.08	81.84	82.60	83.36	84.13	84.91	85.69	86.47	87.26	88.06
1.4	88.85	89.66	90.47	91.29	92.11	92.94	93.77	94.61	95.46	96.32
1.5	97.18	98.05	98.93	99.81	100.71	101.61	102.52	103.44	104.37	105.31
1.6	106.26	107.22	108.19	109.17	110.17	111.18	112.20	113.23	114.28	115.34
1.7	116.42	117.52	118.63	119.77	120.92	122.09	123.28	124.50	125.75	127.02
1.8	128.32	129.65	131.01	132.41	133.85	135.34	136.87	138.46	140.10	141.82
1.9	143.61	145.49	147.48	149.59	151.86	154.32	157.04	160.13	163.78	168.54
2.0	180.00									

Angular Constants of the Reciprocal Cell

The determination of the angular reciprocal lattice constant in the plane of a de Jong and Bouman photograph is comparatively simple. It is only necessary to lay a protractor on the angle between the two axial lattice lines, Fig. 189, in order to determine the interaxial angle.

There are several methods of determining the angular reciprocal lattice constants from Weissenberg photographs, and these are capable of achieving different degrees of precision:

Method of ω separations. The axial lattice rows of Fig. 190A are represented on zero layer Weissenberg film by the two sloping straight lines of 0k0 and h00 reflections. The angle γ^* is represented by the z component between these two sloping lines. Fortunately these are straight lines, and their distance apart can be measured accurately by placing the film at a slant in a coordinate-measuring device such as shown in Fig. 229. From the resulting measurement, ∂, the film equivalent, z, of the angular separation of the lines can be computed by

$$\sin Я = \frac{\partial}{z}, \tag{6}$$

where Я, according to (15), Chapter 14, is an instrumental constant given by

$$\tan Я = \frac{x}{z} = \frac{\Upsilon/C_1}{\omega/C_2} = \frac{\Upsilon}{\omega} \cdot \frac{C_2}{C_1} = 2\frac{C_2}{C_1}.$$ (7)

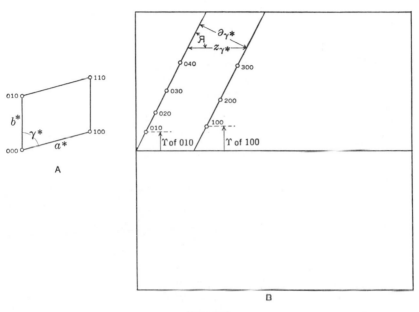

FIG. 190.

For Weissenberg photographs of undistorted scale, $\tan Я = 2$. If there is an error in the orientation of the crystal axis to the rotation axis, then the sloping lines do not necessarily make an angle, Я, with the center line of the Weissenberg film. When such deviation is detected, then the distance is inaccurately given, but the best may be made of a bad situation by using not the entire sloping lines of spots, but only the spots nearest to the center lines, where the z error is least developed, or, better, by the extrapolation of the sloping line of spots to the center line of the film.

Method of triangulation. In Fig. 191A, it will be observed that the reciprocal lattice vectors a^*, $-b^*$, and $a^* - b^*$ (i. e., $[1\bar{1}0]^*$) form a triangle which includes the reciprocal interaxial angle γ^*. These three linear elements can be accurately measured, in Weissenberg projection, as indicated in Fig. 191B, and they may also be further refined by the methods discussed in Chapter 21; hence it is possible to com-

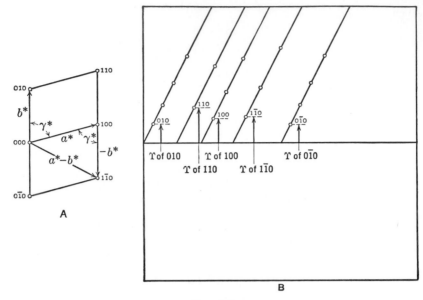

Fig. 191.

pute γ^* to the same precision. Any of the several formulas for computing an angle of the triangle, given its three sides, may be used, namely:

$$\cos \frac{\gamma^*}{2} = \sqrt{\frac{\sigma^*(\sigma^* - [1\bar{1}0]^*)}{a^*b^*}}, \tag{8}$$

$$\sin \frac{\gamma^*}{2} = \sqrt{\frac{(\sigma^* - a^*)\,(\sigma^* - b^*)}{a^*b^*}}, \tag{9}$$

$$\tan \frac{\gamma^*}{2} = \sqrt{\frac{(\sigma^* - a^*)\,(\sigma^* - [1\bar{1}0]^*)}{\sigma(\sigma - b^*)}}, \tag{10}$$

where

$$\sigma^* = \frac{a^* + b^* + [1\bar{1}0]^*}{2}. \tag{11}$$

The cosine and sine formulas are insensitive in the regions of $\gamma^* = 0°$ and $\gamma^* = 180°$, respectively; the tangent formula is sensitive over the entire range. The tangent formula, however, does not give rise to the usual minimum of labor (see page 355), but the maximum, because σ is different for α^*, β^*, and γ^*. The cosine formula gives rise to the all-around minimum of computation.

DETERMINATION OF THE DIRECT CELL CONSTANTS

General Remarks

The lattice constants of the direct cell can be determined by measuring certain components of the lattice constants of the reciprocal cell. These components are easily amenable to measurement on rotating-crystal photographs and on photographs taken by the method of de Jong and Bouman, but are, in general, less easily, and sometimes considerably less accurately, measurable on sets of Weissenberg photographs.

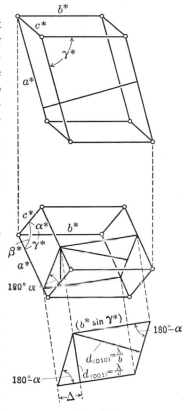

The important aspects of the geometry of the desired components are shown in Fig. 192. This figure is essentially the reciprocal equivalent to Fig. 188. A set of moving-film photographs taken for rotations about an axis normal to one of the sides of the cell in Fig. 192 gives pictures of the base of the cell. Consequently one can use the level photographs to make measurements of that part of the geometry of the cell which exists in these bases. This permits making certain measurements in the cross sectioning plane, but measurements cannot be made in the alternative cross-sectioning plane shown in Fig. 188D.

Fig. 192.

Unfortunately, methods have not been perfected for refining the accuracy of direct cell constants, directly measured.

Angular Constants of Direct Cells

Method of level offsets. The upper left part of Fig. 193 shows a reciprocal cell projected on one of its pinacoids. Such a projection

(a) is directly visible by superposing two de Jong and Bouman photographs of adjacent levels, and

(b) may be graphically constructed from two adjacent level photographs made by any other moving-film method.

The direct lattice interaxial angles are the obtuse dihedral angles between the several pinacoidal planes of the reciprocal cell. For any

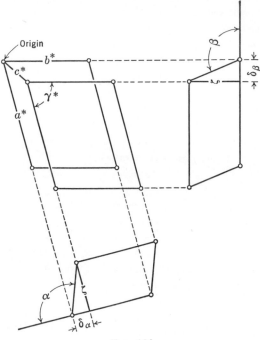

F<small>IG</small>. 193.

given level pair, two such direct angles can be determined by the amount of displacement between levels. If the level spacing is ζ, then, evidently, according to Fig. 193,

$$\tan \alpha = -\frac{\zeta}{\delta_\alpha},$$

and (12)

$$\tan \beta = -\frac{\zeta}{\delta_\beta}.$$

The level offsets, δ, can be determined in the graphical reconstruction of the reciprocal lattice by measuring the distance between the lattice rows. When using de Jong and Bouman films, the two adjacent level photographs may be clipped together in correct registry, and the pair attached to the film coordinate-measuring device, Fig 229. This permits an accurate measure of the distance between corresponding lattice rows of adjacent levels. The coordinate ζ is known from the rotating-crystal photograph for that axis of rotation.

The level offset, δ, can be measured directly in Weissenberg photographs by considering the zero- and first-level photographs. For the particular cell having its origin at the reciprocal lattice origin, the two cell edges on the zero level, Fig. 193, include the origin and hence are parts of central lattice lines. They therefore record as straight lines on the Weissenberg film, Fig. 196B. The offset cell edges on the first level, however, do not include the origin and are hence parts of non-central lattice lines of spacing δ (see page 269). This spacing can be measured on the Weissenberg first-level film as the height of the lattice line festoon above the film center line. Unfortunately, the entire line is not present but only some of its lattice points. The chart of Fig. 148 can be used to outline the rest of the curve, and the arrangement of Fig. 145 can be used to measure the ξ coordinate of the lowest point of the curve. This ξ value is δ. For monoclinic crystals, Fig. 194B, the symmetry of the crystal provides a reflection, 001, at this minimum point. In this case the chart can be dispensed with, and either

(a) the height of the spot measured roughly with triangle scale, or

(b) the height of the spot measured accurately in terms of film coordinate x and then converted to ξ with the aid of (5). Remembering that the measurement is made on the first layer photograph, for which the equi-inclination setting is μ, then [see (10), Chapter 14],

$$\delta = \cos \mu \left(2 \sin \frac{C_1 x}{2} \right). \tag{13}$$

The direct angles are then computed with the aid of (12).

Method of angular lag. The methods of determining δ from a Weissenberg film, just discussed, are not very accurate. A quite accurate method of deriving δ, and hence the direct interaxial angles, is by measuring the angular lag of the reflections of the first layer behind those of the zero layer. Since not only the understanding of the methods but also the technique of making the required measurements are much easier in the monoclinic case, owing to symmetry, this special case will be discussed first:

Monoclinic case. The following discussion applies to a monoclinic crystal rotated about the c-axis, but, by interchange of notation, applies also to rotation about the a-axis. The upper part of Fig. 194A shows the important aspects of the plan of the first (or any n-)

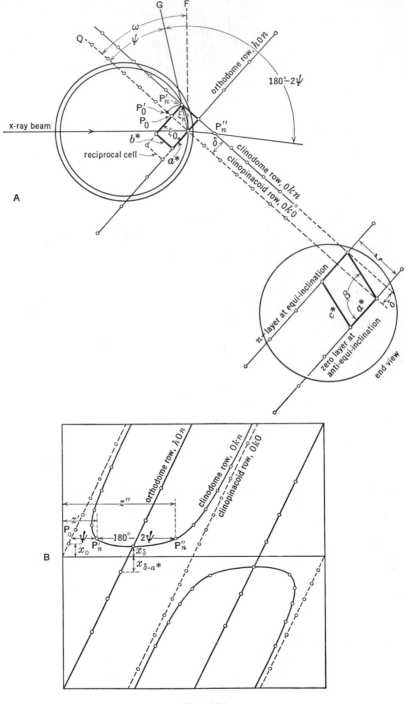

378

level of the reciprocal lattice in equi-inclination position with respect to the sphere of reflection. The n-level is recording in accordance with the equi-inclination $\mu = -\nu$ scheme. If the zero level were also recording simultaneously it would be recording according to the anti-equi-inclination scheme, $\mu = +\nu$. The full beaded lines are the orthodome and clinodome series of reciprocal lattice points on the n-level. The latter is displaced a distance, δ, behind the corresponding clinopinacoid line (dashed) on the zero level. As a consequence of this, the n-level line reaches its reflecting circle after the zero-level line and consequently the n-level reflects later than the zero level. In order to get a quantitative measure of this angular lag in the reflection of the n-level, consider two points, P_0' and P_n', of the same radial coordinate (i. e., the same distance from the origin) ξ_n, one point on the zero level and another on the n-level. When the zero-level line has reached the position indicated by OG in the figure, the imaginary point P_0' strikes its circle and reflects. It is necessary to rotate the crystal through an angle, ψ, beyond this position before the point P_n' on the n-level reflects. This angle is evidently defined by

$$\tan \psi = \frac{\delta}{\xi_0}, \tag{14}$$

where ξ_0 is the radial coordinate of P_0, the pinacoid reflection on the zero level corresponding with the dome reflection P_n' on the n-level. Combining (12) and (14) gives

$$\beta = \tan^{-1}\left(-\frac{\zeta}{\xi_0 \tan \psi}\right). \tag{15}$$

Now, the point P_n occurs in pairs symmetrically located on each side of the orthodome series, namely, P_n' and P_n''. The angle subtended by these two points from the origin, Fig. 194A, is $180° - 2\psi$. When P_n' is just touching the circle in reflecting position, the crystal must be rotated through this angle before P_n'' reflects. If z' and z'' are the film coordinates of the reflections of P_n' and P_n'', then $z'' - z'$ is the film representation of this angular difference. Applying the coupling constant,

$$C_2(z'' - z') = 180° - 2\psi, \tag{16}$$

from which

$$\psi = 90° - \frac{C_2(z'' - z')}{2}. \tag{17}$$

Finally, substituting the value of ψ given by (17) into (15) gives

$$\beta = \tan^{-1}\left(-\frac{\zeta}{\xi_0 \tan\left\{90° - \dfrac{C_2(z'' - z')]}{2}\right\}}\right)$$

$$= \tan^{-1}\left(-\frac{\zeta}{\xi_0}\tan\left\{\frac{C_2(z'' - z')]}{2}\right\}\right). \tag{18}$$

In this relation,

> ζ is the height of the reciprocal lattice level, as determined, for example, from a layer line photograph, or from ξ measurements on films made for another rotation axis.

> ξ_0 is the cylindrical coordinate of the spot P_n, which may be accurately determined according to (5) from an ordinary normal-beam, zero layer film.

> $(z'' - z')$ is the distance between spots P_n'' and P_n', which may be accurately determined with the aid of micrometer calipers, or with the aid of the device shown in Fig. 229.

It should be observed that the precision of ξ_0 can be considerably improved by the methods discussed in the next chapter, for example by the method of Bradley and Jay. If this refinement is resorted to, then a refinement is made in it in the form of the spacing $d_{(010)}$, from which ξ_0 can be computed by the reciprocal relation (see Fig. 70):

$$\xi_0 = \frac{\lambda}{d_{(010)}}. \tag{19}$$

Returning to relation (18), it may be said that, of the several variables, ξ_0 can be determined with relatively great precision; ζ can be determined with the same great precision by the same method, provided that a zero-layer photograph for rotation about another axis is measured. The angular difference represented by $z'' - z'$ is subject to the main lack of precision. An error in the orientation of the crystal axis to the rotation axis gives rise to a considerable error in this factor.

The determination of β for a monoclinic crystal by the method of angular lag may be illustrated for the crystal realgar, AsS. The values of ξ_0 and ζ used in (18) have not been refined by the method of Bradley and Jay, but represent single determinations, and are therefore not very precise. The other necessary measurement, $z'' - z'$, was made with a vernier caliper on the c-axis, first-layer photograph, Fig. 195. The values for substitution in equation (18) are:

Quantity	For reduced cell	For cell of Goldschmidt's classical orientation
ζ	0.2346	0.2346
ξ_0	0.1140	0.1140
$z'' - z'$	58.7 mm.	47.8 mm.
C_2	2°/mm.	2°/mm.

The reflection spots, P'_n and P''_n, for the reduced cell are the two strong spots near the center on the upper half of the photograph, while the corresponding ones for Goldschmidt's less simple, alternative

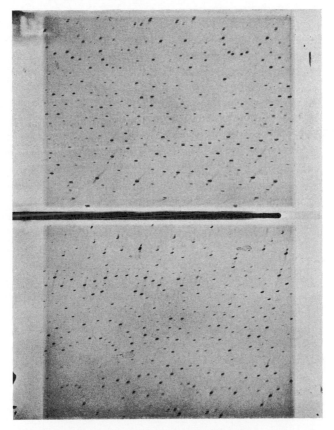

Fig. 195. n-Level, equi-inclination Weissenberg photograph, illustrating the determination of the interaxial angle, β, for monoclinic crystals. The two spots nearest the center line on the upper half of the film correspond with P'_n and P''_n of Fig. 194. (Realgar, AsS, monoclinic; c-axis rotation, 1st level; unfiltered CuK radiation from gas x-ray tube.)

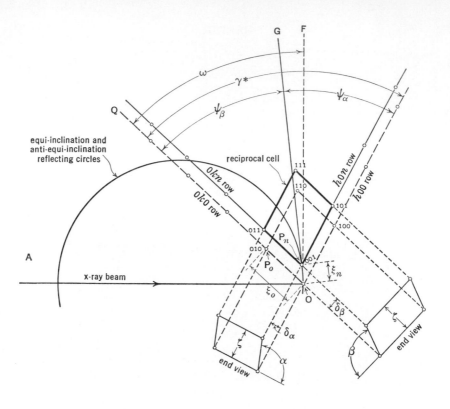

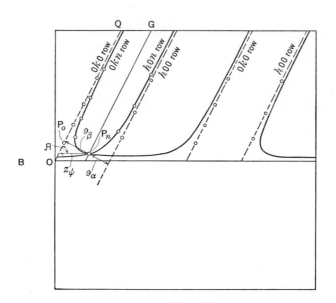

Fig. 196.

382

orientation are the two nearest the center on the lower half of the photograph. Substitution, into formula (18), of the measured values above, gives the following values of β:

	For reduced cell	For cell of Goldschmidt's classical orientation
β from x-ray film measurements	106° 28′	113° 47′
β from reflection goniometer on crystal surfaces	106° 33′	113° 45′
Difference	−0° 05′	+0° 02′

Triclinic case. In Figs. 196A and 196B is shown the triclinic case corresponding with the monoclinic case of Figs. 194A and 194B. The triclinic case offers a more complicated situation, and the lack of symmetry makes new measurement strategies necessary.

The value of ξ_0 can no longer be used in computing δ; instead the relation

$$\sin \psi = \frac{\delta}{\xi_n} \qquad (20)$$

must be used. Furthermore, the reflection P_n no longer has a symmetrical companion, so some other method must be devised for the measurement of the film equivalent of ψ. The absence of a natural zero mark on the n-layer film puts an obstacle in the way of measuring ψ intervals. The difficulty can be overcome if some equivalent fiducial mark can be made on both the zero- and n-layer films. This can be accomplished in either of two ways:

(a) Both the zero-layer and the n-layer photographs may be recorded in succession on the same film with the same angle, ν. This gives rise to the composite picture shown in Fig. 196B. The relations between the recording positions of the two layers along the axis of the film cylinder is shown in Fig. 197. The direct beam for the zero layer records (for $\mu = +\nu$) at z_0; the direct beam for the n-layer subsequently would record (for equi-inclination, $\mu = -\nu$) at z_n.

(b) The zero-layer and n-layer photographs may be recorded on separate films, and, in order to make a mark of known position on each film, the direct beam may be allowed to fall directly on each film at the same rotation setting of the crystal spindle. Reference to Fig. 197 shows that the direct beam strikes the zero-layer film at z_0 but it strikes the n-layer film at z_n. The direct beam mark on the n-layer film is thus at $+2s$ with respect to the mark on the zero-layer film, where

$$s = r_F \tan \nu. \tag{21}$$

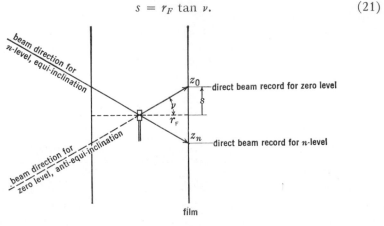

FIG. 197.

Suppose that both the zero- and n-layer photographs are recorded on the same film.   Note in Fig. 196A that OQ is a central lattice line on the zero level, and OG is a central line also, although it contains only one lattice point, P_n, i. e., it is irrational.   The angle QOG is ψ. It appears on the Weissenberg photograph, Fig. 196B, as the z component between the two sloping central lines OQ and OG.   In order to measure this separation, it is only necessary to lay the film on the coordinate-measuring device, Fig. 229, align the spots of the central lattice line, OQ, parallel with the hair line, and measure the normal distance, ∂, of P_n from this line.   According to (6),

$$z = \frac{\partial}{\sin Я}. \tag{22}$$

The angle

$$\psi = C_2 z_\psi$$

$$= C_2 \frac{\partial}{\sin Я}. \tag{23}$$

Equation (20) may be rewritten

$$\delta = \xi_n \sin \psi; \tag{24}$$

and, substituting from (23), this becomes

$$\delta = \xi_n \sin \left(\frac{C_2}{\sin Я} \partial \right). \tag{25}$$

This makes possible evaluation of (12), which then takes the form:

$$\tan \begin{vmatrix} \alpha, \\ \beta, \\ \gamma, \end{vmatrix} = - \frac{\zeta}{\xi_n \sin \left(\dfrac{C_2}{\sin \mathcal{A}} \partial \right)}. \tag{26}$$

Note that the term $C_2/\sin \mathcal{A}$ is an instrumental constant. ξ_n is measured on an n-layer film; hence the corrected form of (5) for this measurement is

$$\xi_n = 2 \sin \left(\frac{C_1 x}{2} \right) \cos \nu. \tag{27}$$

Method of triangulation. The identity periods of a crystal along a, c, and $[10\bar{1}]$ may be obtained from rotation photographs about these three directions. These three identity periods form a triangle whose obtuse angle is the crystallographic angle β. Since the three sides of the triangle are known, the angle may be calculated. All three interaxial angles of a triclinic crystal may be calculated from a knowledge of the six identity periods a, $[\bar{1}10]$, b, $[0\bar{1}1]$, c, and $[10\bar{1}]$, and these data may be obtained from the six corresponding rotation photographs.

This method of calculating interaxial angles is not ordinarily of much practical importance since, if the crystal is sufficiently well developed to permit orientations of the rotation axes parallel with these zones, the reciprocal interaxial angles may be measured with an optical goniometer and the direct interaxial angles calculated from them.

Linear Constants of the Direct Cell

There are two categories of components in Fig. 192 which permit measurements leading to direct cell axial lengths, namely, layer line spacings and plane-lattice row spacings:

Layer line spacings. In order to set the layer line screen for any kind of moving-film photograph, it is first necessary to have made a rotating-crystal photograph. The positions of the layer lines of the rotating-crystal photograph are functions of the level spacing of the reciprocal lattice. In the particular orientation shown in Fig. 192, this corresponds with the measurement of the component $d^*_{(001)}$ from which it is possible to compute c by the relation

$$c = \frac{\lambda}{d^*_{(001)}}. \tag{28}$$

The actual measurement on the rotating-crystal photograph consists of measuring, with either chart or scale, the position of an nth layer line.   If a chart is used, the height of the layer line appears directly as the height, ζ_n, of the nth level of the reciprocal lattice, and the computation takes the form given by (39), Chapter 8, namely:

$$c = \frac{n\lambda}{\zeta_n}. \tag{29}$$

The values computed for the identity period along the rotation axis may be made considerably more accurate by measuring the actual film coordinate, y, of the nth layer by means of a steel scale and vernier device such as shown in Fig. 229.   In this case, the computation takes the form given by (6), Chapter 5, namely:

$$c = \frac{n\lambda}{\sin \tan^{-1}(y/r_F)}, \tag{30}$$

where r_F is the radius of the film cylinder.

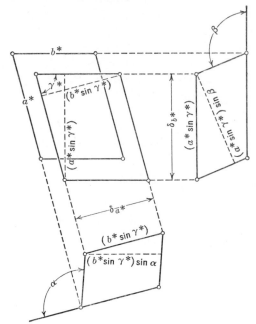

Fig. 198.

Plane-lattice row spacings.   If the spacings of the axial rows of the reciprocal lattice can be measured, then it is possible to compute direct lattice constants.   Figure 198 illustrates the situation for rotation about the c-axis.   In this case the reciprocal lattice provides plans of the $(001)^*$ levels of the reciprocal lattice, and consequently spacings δ_{a*} and δ_{b*} of the $a*$ and $b*$ axial rows. If α and β of the direct lattice have already been determined, as discussed in the last section, then a and b can be computed by the reciprocal equivalents of (54) and (55), Chapter 18. The several relations which can be utilized for the several axes of rotation, and their reduction to measurements of axial row spacings, δ, are:

Rotation about a
$$b = \frac{1}{(b* \sin \alpha*) \sin \gamma} = \frac{1}{\delta_{c*} \sin \gamma}, \tag{31}$$

$$c = \frac{1}{(c* \sin \alpha*) \sin \beta} = \frac{1}{\delta_{b*} \sin \beta}, \tag{32}$$

Rotation about b
$$a = \frac{1}{(a* \sin \beta*) \sin \gamma} = \frac{1}{\delta_{c*} \sin \gamma}, \tag{33}$$

$$c = \frac{1}{(c* \sin \beta*) \sin \alpha} = \frac{1}{\delta_{a*} \sin \alpha}, \tag{34}$$

Rotation about c
(Fig. 198)
$$a = \frac{1}{(a* \sin \gamma*) \sin \beta} = \frac{1}{\delta_{b*} \sin \beta}, \tag{35}$$

$$b = \frac{1}{(b* \sin \gamma*) \sin \alpha} = \frac{1}{\delta_{a*} \sin \alpha}. \tag{36}$$

The axial row spacings, δ, can be easily measured on a de Jong and Bouman photograph. A single photograph representing any level is placed in the coordinate-measuring device, Fig. 229, with the required axial row parallel with the hairline. The row spacing, $\delta_{a*,b*,c*}$, is then determined directly, and need only be multiplied by the magnification factor (page 337) to give the absolute row spacing. From the combined photographs of two adjacent levels, the direct cell angles are determined according to (12) with the aid of the same instrument. All data are now available for computing one of the direct cell constants with the appropriate relation from (31) to (36).

The same data may be derived from Weissenberg photographs, but with less precision, because the lattice lines of non-central lines are not straight lines on a Weissenberg photograph. The required relationships are shown in Fig. 199. The required measurements for the case illustrated are δ_{a*} and δ_{b*}, Fig. 199A. These represent the shortest distances from the origin to the first $a*$ and $b*$ lattice rows, respectively. In the Weissenberg projection, the reciprocal lattice origin is drawn out to become the center line of the photograph, and the minimum distance to the first lattice line is represented by the minimum ξ or

x distance to the corresponding lattice row festoon.   Note that these required minima are just ahead of reflection 010 and just behind 100, respectively.   Note also that the required 010 and 100 pair are those with an acute angle between them (because reciprocal γ^* is acute if direct α, β, and γ are obtuse, page 367).   This represents selection

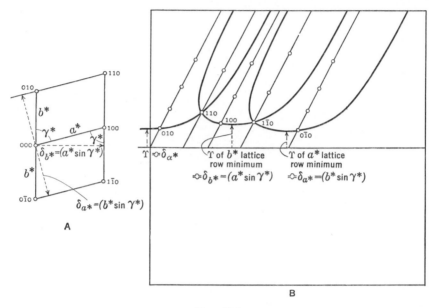

Fig. 199.

of the $0\overline{1}0$ and 100 pair closest to one another on the Weissenberg photograph.   (This precaution eliminates selection of a pair like 010 and 100, for which the separation is greater than $0\overline{1}0$ and 100, because it represents $180° - \gamma^*$, which is obtuse.)   It is difficult to measure to the minimum of a festoon on a Weissenberg photograph, because the festoon is imaginary.   It can, however, be sketched in with the aid of the equi-inclination lattice row template, Fig. 148.   The ξ of the spacing δ can then be measured on the coordinate-measuring instrument, Fig. 229, or, more crudely, with the aid of the ξ_{Ξ} scale, Fig. 144.

THE DETERMINATION OF ALL LATTICE CONSTANTS FROM ONE CRYSTAL SETTING

An important problem which occasionally arises is the determination of all the lattice constants of a crystal from a single setting of the crystal. This problem occurs, for example, when the crystal is only sufficiently well developed to permit orientation about one zone; a small acicular crystal without termination presents a case in point. It should be observed that optical goniometric measurements can never lead to sets of crystal elements in such a case. The problem also arises, even more acutely, when the crystal has surfaces too poorly developed to permit optical orientation with a goniometer. The crystal may then be oriented to a rational axis by trial oscillations.

This problem may be solved with the aid of one rotation photograph and two moving-film photographs, preferably a zero- and first-level photograph, for rotation about the one rational axis. Suppose that the rational axis is the c-axis, for sake of definiteness. Suppose, furthermore, that the moving-film method is the de Jong and Bouman method. Then, according to the preceding discussion, data leading directly to the following lattice constants can be determined:

	From moving-film photographs	From rotation photograph
Reciprocal elements	$a^*, b^*, —$; $— — \gamma^*$	
Direct elements	$a,\ b,\ —$; $\alpha\ \beta\ —$	c

The only direct element missing from this list is γ, and this can be quickly computed from such relations as (31) and (33).

If any other moving-film method is employed, then a straightforward duplication of the above scheme may be had by reconstructing the reciprocal lattice graphically and making appropriate measurements on it. This, unfortunately, carries with it the lack of precision inherent in the graphical reconstruction and in any graphical solutions attempted.

A scheme leading to constants of superior accuracy is to make measurements on the film leading directly to the following elements:

	From moving-film photographs	From rotation photograph
Reciprocal elements	$a^*\ b^* —$; $— — \gamma^*$	
Direct elements	$— — —$; $\alpha\ \beta\ —$	c

By recasting (54) and (55), Chapter 18, in the following form:

from (54):
$$a = \frac{1}{a^* \sin \beta \sin \gamma^*}, \tag{37}$$

from (55):
$$b = \frac{1}{b^* \sin \alpha \sin \gamma^*}, \tag{38}$$

a and b may be computed.

The following elements are now known:

$$a^*, b^*, —; \quad — — \gamma^*$$
$$a, \quad b, \quad c; \quad \alpha \quad \beta —.$$

In order to compute the remaining element γ, it should be noted that the cell volume is determined in several ways by different groups of elements. One volume expression is given by (50), Chapter 18:

$$V = abc \sin \alpha \sin \beta \sin \gamma^*. \tag{39}$$

Another volume expression is given by (12), Chapter 18:

$$V = abc\sqrt{1 - \cos^2 \alpha - \cos^2 \beta - \cos^2 \gamma + 2 \cos \alpha \cos \beta \cos \gamma}. \tag{40}$$

Squaring and equating the right member of (39) and (40), there results

$$\sin^2 \alpha \sin^2 \beta \sin^2 \gamma^* = 1 - \cos^2 \alpha -$$
$$\cos^2 \beta - \cos^2 \gamma + 2 \cos \alpha \cos \beta \cos \gamma. \tag{41}$$

This can be rearranged in the form

$$\cos^2 \gamma - (2 \cos \alpha \cos \beta) \cos \gamma +$$
$$(1 - \cos^2 \alpha - \cos^2 \beta - \sin^2 \alpha \sin^2 \beta \sin^2 \gamma^*) = 0. \tag{42}$$

The terms in parentheses are known. This quadratic may be solved for the unknown element γ.

PROCEDURE LEADING TO MAXIMUM ACCURACY IN LATTICE CONSTANTS

It has already been intimated that, of all the possible cell elements, the most precisely determinable are those of the reciprocal lattice. This is especially true when measurements are made on Weissenberg zero-level photographs with the aid of a steel scale and vernier. There are numerous reasons for the possibility of precision in this case:

(a) A distance which is a multiple of the lattice row spacing, for example, 10 a^*, can be measured. This permits a greater distance to be measured and thus increases the precision by the factor of the multiple.

(b) If the multiple distance is measured to a spot in the region of $\Upsilon = 180°$, the radial coordinate scale approaches great accuracy. This is because, according to (4), $\xi = 2 \sin \dfrac{\Upsilon}{2}$. Figure 200$A$ shows

that, in the region $\Upsilon = 180°$, i. e., $\dfrac{\Upsilon}{2} = 90°$, a scale division of Υ, which is the quantity measured (in terms of the film distance x), determines a minute change in ξ. This means that a normally accurate measurement of Υ gives a supernormally accurate determination of ξ.

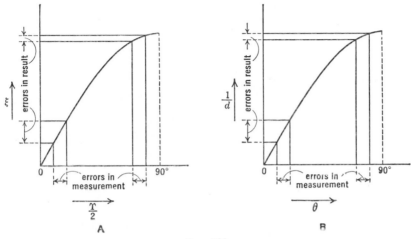

FIG. 200.

(c) The spots become sharpened in the region of $\Upsilon = 180°$, because of focusing, as explained in the next chapter. This permits the hair line of the measuring instrument to be precisely set on the spot.

(d) The errors inherent in recording tend to become eliminated in the region of $\Upsilon = 180°$, as explained in the next chapter.

(e) Finally, there are methods of eliminating all inaccuracies by measuring the positions of several spots in a lattice row in the region $\Upsilon = 180°$. This makes the Weissenberg instrument an absolute measuring instrument of precision surpassing any other measuring device.

Only in a zero-layer Weissenberg photograph is it possible to take advantage of the precision gains in (b), (c), (d), and (e). This gain in precision is unavailable to measurements taken from de Jong and Bouman and also Sauter photographs, because the precision requires measurements of reflection in which 2θ approaches $180°$; such reflections cannot be recorded in de Jong and Bouman or in ordinary Sauter photographs.

Evidently lattice constants expressible as functions of x (i. e., Υ or ξ) measurements, and taken from Weissenberg equatorial photographs, are alone capable of being determined with high precision. Fortunately, all lattice constants can be so expressed. From a practical point of view, it is only necessary that the crystal being investigated have a sufficiently well-developed surface to permit orientation about any three non-coplanar rational axes. Suppose that these three axes turn out to be desirable crystallographic axes (if they do not, it is merely necessary to transform them to desirable ones). Then, with the aid of three zero-layer Weissenberg photographs, the data given in column II of the following table can be determined.

I	II	III
Rotation axis	Precisely determinable constants	Additional constants precisely computable from values in column II
a	b^*, $[0\bar{1}1]^*$, c^*	α^*
b	c^*, $[10\bar{1}]^*$, a^*	β^*
c	a^*, $[\bar{1}10]^*$, b^*	γ^*

From the six distinct linear reciprocal values in column II, the reciprocal lattice angles in column III can be computed with equal precision. From these reciprocal cell angles, the direct cell angles can be computed with equal precision with the aid of the reciprocal equivalents of (37), (38), and (39), Chapter 18. Finally, with the values of a^*, b^*, c^*; α^*, β^*, γ^*; α, β, and γ, which are then available, the values of a, b, and c may be precisely computed with the aid of (54), (55), and (56), Chapter 18. In Chapter 21, further details on the precision determination of lattice constants will be given.

LITERATURE

Tom. F. W. Barth and George Tunell. The space lattice and optical orientation of chalcanthite ($CuSO_4 \cdot 5H_2O$): an illustration of the use of the Weissenberg x-ray goniometer in the triclinic system. *Am. Mineral.*, **18** (1933), 187–194.

D. Crowfoot. The interpretation of Weissenberg photographs in relation to crystal symmetry. *Z. Krist. (A)*, **90** (1935), 215–236.

M. J. Buerger. Crystals of the realgar type: the symmetry, unit cell, and space group of nitrogen sulfide. *Am. Mineral.*, **21** (1936), especially 580.

M. J. Buerger. The x-ray determination of lattice constants and axial ratios of crystals belonging to the oblique systems. *Am. Mineral.*, **22** (1937), 416–435.

W. F. de Jong and J. Bouman. Kristallographische Berechnungen und Konstruktionen mittels des reziproken Gitters. *Z. Krist. (A)*, **101** (1939), 317–336.

W. F. de Jong and J. Bouman. Over het reciproke tralie van kristallen. *Natuurwetenschappelijk Tijdschrift*, **21** (1939), 291–303.

CHAPTER 20

THE THEORY OF ATTAINING PRECISION IN THE DETERMINATION OF LATTICE CONSTANTS

Methods for the achievement of precision in the determination of lattice constants have been in the process of development since the very beginning of x-ray diffraction investigations. Almost the entire development has been made in connection with the derivation of lattice constants from powder photographs, but the theories involved are equally applicable to the derivation of lattice constants from photographs made with single crystals.

The attainment of precision in the determination of lattice constants depends upon two factors:

1. the utilization of d's in the sensitive region; and

2. the elimination of systematic errors inherent in the recording of reflections and the measurement of reflection angles. These errors are due to eccentricity of the specimen, to lack of knowledge of the radius of the camera, to shrinkage of the film in development, and to absorption by the specimen.

THE UTILIZATION OF d's IN THE SENSITIVE REGION

Factor 1, above, is most easily discussed. It should be stated at this point, however, that factor 1 alone gives great sensitivity, but this sensitivity gives a false impression of the accuracy attained unless the systematic errors, 2, are also eliminated.

Factor 1 is the same as that discussed near the close of the last chapter and illustrated in Fig. 200A. In that place, the coordinates of the reciprocal lattice were used. It is desirable, now, to translate this material into the language of reflection measurement. Determinations of lattice constants are experimentally made by determining the interplanar spacings, d, of appropriate planes. These spacings are controlled by the Bragg relation

$$n\lambda = 2d \sin \theta. \tag{1}$$

For the present purposes, this is most conveniently thrown in the form

$$\frac{n}{d_{(hkl)}} = \frac{1}{d_{(nh\ nk\ nl)}} = \left(\frac{2}{\lambda}\right) \sin \theta. \tag{2}$$

Figure 200B shows a plot of this relationship. It is evident that if measurements are made of reflections whose θ's fall in the 90° region, the d's of the planes corresponding with these reflections can be determined with much greater sensitivity than those whose θ's occur at the lower angles.   For this reason, lattice constants based upon measurements of reflections in the region of $\theta = 90°$ are comparatively precise, other things being equal, and they are the more precise the nearer the θ's are to 90°.

It will become evident later that there are other advantages in selecting for measurement reflections whose θ's lie in the 90° region.

THE METHODS OF ATTAINING PRECISION

General remarks.   A publication† by Hadding[1] in 1921, on the effect of absorption on the apparent lattice constant as computed from powder photographs, was the first systematic attempt to make allowances for errors inherent in computing lattice constants from film measurements. Since then, a number of major schools of strategy have arisen in an attempt to eliminate such errors.   Before discussing the errors themselves, a brief outline is offered of the attempts to eliminate errors by these several schools of strategy:

Error elimination by comparison.[3]   One of the earliest attempts to eliminate lattice constant errors consisted of making a powder photograph, not of a single crystalline substance but of a mixture of this substance with another whose lattice constant was accurately known. The position of a reflection from the substance of unknown lattice constants was then, not measured with respect to the film origin, but compared with the positions of the reflections, in that region, of the substance of known lattice constant.   This method therefore ignores the sources of errors and attempts to eliminate them by comparison. Quite apart from the absolute merits or demerits of this method, it will not be discussed further here because it would be difficult to apply properly to single-crystal methods.

Error elimination by careful experimental technique.[4-18]   Straumanis and his associates have very recently started a school of strategy which attempts to eliminate the errors at their source by very careful experimental technique.   The strategy was first devised for the powder method but has been subsequently applied to the rotating-crystal method.   The main features of the experimental procedure are as follows:

† The literature pertaining to the subject matter of this chapter is of such recent interest and so voluminous that it has been thought advisable to make specific references, by means of numerical superscripts, to the list of publications collected at the end of the chapter.

1. A specimen so tiny is used that any corrections due to absorption by the specimen are reduced to a vanishing amount. In the case of powder photographs, the powder is stuck to the outside of a piece of practically non-absorbing Lindemann glass rod having a diameter as small as 0.05 to 0.08 mm. The finished powder mount has a diameter in the neighborhood of 0.15 mm.

2. The specimen is accurately centered to the rotation axis with the aid of a microscope. This reduces the volume swept out by the speci-

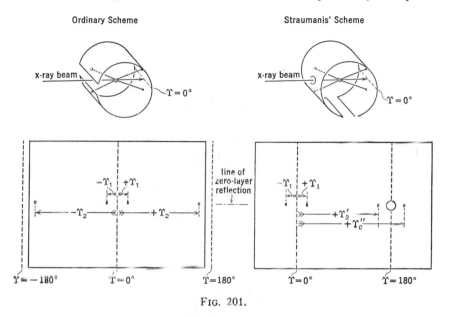

FIG. 201.

men in rotating, and consequently diminishes the width of a powder-photograph line.

3. The specimen axis is accurately centered to the film cylinder axis, thus eliminating errors due to lack of centering.

4. Straumanis' major contribution lies in a strategy for determining the exact effective camera radius, and thus eliminating radius and shrinkage errors. The idea behind this is illustrated, for rotating-crystal photographs, in Fig. 201. The ordinary scheme is shown on the left, and Straumanis' scheme, applied to rotating-crystal films, on the right. The upper part of the figure shows the films as they take the form of cylinders in the camera; the lower part shows them unrolled as flat sheets.

In both the ordinary scheme and Straumanis' scheme, the zero for Υ can be easily located because it is bracketed halfway between pairs of identical spots of coordinates $+\Upsilon_1$ and $-\Upsilon_1$. In the ordinary scheme, the direct beam enters at the position $\Upsilon = 180°$, and the film ends are placed here to permit the entrance of the pinhole system. For this reason, the position $\Upsilon = 180°$ is not on the film. In Straumanis' scheme the pinhole system enters through a hole in the film. Although the $\Upsilon = 180°$ position lies within this hole, it is accurately bracketed halfway between equivalent reflections of coordinates $+\Upsilon_2'$ and $+\Upsilon_2''$. This permits the 0° and 180° position for Υ to be accurately located. The actual distances between these positions may be accurately measured on a device such as shown in Fig. 229. This measure is half the effective camera circumference *after* development. From this the effective camera radius or diameter is easily computed.

Straumanis' method of eliminating errors is subject to certain drawbacks when applied to single crystals. In the first place, the crystal must be exceptionally tiny in order to eliminate the absorption correction. It is easy enough to make a powder mount of tiny diameter, but it is another matter to select a single crystal of a diameter as small as 0.1 mm. in every case. Even should a crystal be found having this cross section, then (a) it is usually too tiny to orient conveniently by optical means, and (b) its third dimension is probably considerably larger, so that rotations about other than the needle axis cannot be treated by the Straumanis method.

Error elimination by systematic allowance.[2, 19-23] A third school has made a very careful study of the effect of divergence, preparation thickness, etc., on the apparent lattice constant. This technique is passed by here without further comment other than that it is quite complicated.

Error elimination by utilizing high-angle reflections and by " back-reflection " cameras. About 1927 a development began which eventually led to the several extrapolation techniques now in common use for eliminating systematic errors in the determination of lattice constants. This had its inception when Dehlinger[24] observed that all errors tended to vanish in the region of $\theta = 90°$, $2\theta = 180°$. To take advantage of this situation Dehlinger devised a back-reflection camera, in which the direct beam enters at the center of the film; he made measurements only of reflections which appeared in the immediate region of $2\theta = 180°$ Van Arkel,[25] von Göler and Sachs,[26] Sachs and Weerts,[27] and Braekken[28] took advantage of a similar strategy, the last applying it to rotating-crystal photographs.

Error elimination by graphical extrapolation. Although the investigators of the foregoing group realized that the errors were eliminated at $\theta = 90°$, $2\theta = 180°$, none of them apparently used the idea of extrapolating the lattice constant until the correction actually vanished. It remained for Kettmann[29] to do this. Kettmann made no particular analysis of the errors themselves, but was content to show that, since errors vanish at $\theta = 90°$, the calculated values of a lattice constant approach the true value at $\theta = 90°$, and therefore the true value can be ascertained by simply plotting the calculated value against θ, drawing a smooth curve, and extrapolating the curve to $\theta = 90°$. The main drawback of the method is that the form of this curve is unknown, and therefore its extension to $\theta = 90°$ is uncertain and cannot be carried out with great precision.

Bradley and Jay[30] developed a different extrapolation method by examining the manner in which the individual errors themselves vanished at $\theta = 90°$. They were able to show that the errors vanished very approximately linearly with $\cos^2 \theta$. By plotting the lattice constants, as computed from reflections in various parts of the film, against $\cos^2 \theta$ (instead of against θ, as Kettmann did), the computed value approaches the true value along a straight line, and attains the true value as extrapolated along this line at $\theta = 90°$, $\cos^2 \theta = 0$. Bradley and Jay used a front-reflection method, and were apparently unaware of the fact that radius and shrinkage errors could be eliminated by the back-reflection method. As a consequence, their method was weakened by the necessity of calibrating the radius and film shrinkage.

Up to this time extrapolation had been accomplished graphically. Cohen[31-37] made the contribution of showing how the extrapolation could be carried out analytically with the aid of the method of least squares.

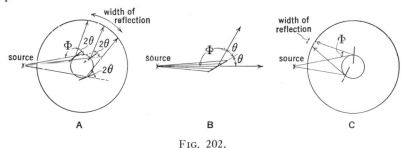

FIG. 202.

DIVERGENCE OF THE X-RAY BEAM

Before entering into a discussion of the sources of error in the recording of x-ray reflections, a feature of the x-rays themselves must be mentioned which, though not in itself leading to errors, causes complications in the errors arising from other sources. This is the property

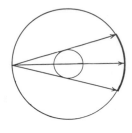

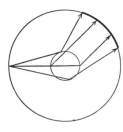

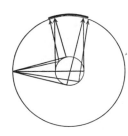

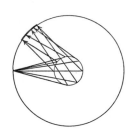

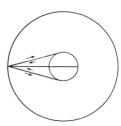

of divergence. It is impossible to control x-rays by means of lenses in the way light is controlled. For this reason it is impossible to collimate a point source of x-radiation into a parallel beam. At best, x-rays can be limited only to a beam. This beam may be controlled so as to appear to come from a point source, but it will be a diverging beam, Fig. 202, because it must bathe a crystal specimen (shown as a circle in all illustrations) of finite size.

This divergence of the x-ray beam gives rise to a distortion of the size of the x-ray reflection which reaches the film, and the amount of distortion is a function of the position of the reflection on the film. This situation and the way it arises are illustrated diagrammatically in Fig. 202. The x-ray beam is shown in Fig. 202A diverging from a source at the left and impinging on a single crystal of circular shape, located in the center of a cylindrical camera. The entire crystal does not reflect at once, because the divergence of the beam causes its rays to make varying angles with the crystal plane, Fig. 202B. Only one ray of the diverging beam can make the correct glancing angle, θ, with the planes of the crystal. Therefore, the crystal continues to give rise to reflections from various parts of its volume while it is rotating through a small angle equal to the angle which the crystal subtends from the point of divergence. The size of the reflection recorded on the film can be graphically constructed, Fig. 202A, by drawing rays from the point of divergence to all points within the volume of the crystal, and at each point laying off the angle

$$\Phi = 180° - 2\theta. \qquad (3)$$

(Φ must not be confused with the reciprocal lattice cylindrical coordinate φ.) The second leg of the angle so constructed is a ray from the point in the crystal to the film. An easy way to estimate the size of the reflection quickly is to lay off Φ for a number of points on the outline of the

398

crystal. This construction demonstrates that, in the region of small Φ angles, Fig. 202C, the width of the reflection is smaller than the crystal, and in regions of large Φ angles, Fig. 202A, the width of the reflection is larger than the crystal.

A particular arrangement of instrumental dimensions is of considerable importance, Fig. 203. This arrangement is the particular one in which the beam diverges from a point on the circumference of the cylindrical film. In this case, the width of the reflection is:

Position of reflection		Width of reflection
2θ	Φ	
0°	180°	$2t$ (where t = thickness of crystal)
90°	90°	$\approx t$
180°	0°	0

This arrangement of instruments is of importance because, in the region $\Phi = 0$, $2\theta = 180°$, the line width approaches zero, and the width of the reflection is therefore an image of the width of the aperture of divergence, i. e., of the limiting pinhole of the pinhole system. This condition is often referred to as *Bragg focusing*. It gives rise to sharp reflections which permit accurate measurements of the reflection position. Doublets are clearly resolved in this region.

Within the region $\Phi = 0° - 90°$, the width of a reflection is approximately $t \sin \Phi$. In actual practice, the reflection width appears considerably broader than this, in part because the wavelength itself is not strictly monochromatic, but covers a small range, and in part because temperature changes in the specimen during exposure cause lattice changes which make the reflection sweep out a range.

THE SOURCES OF SYSTEMATIC ERROR

As already briefly noted, there are several situations which cause a shift in the center of an x-ray reflection from the position it ideally should occupy. These give rise to errors which are systematic functions of the position of the reflection. The chief of these errors is due to:

Geometrical errors:
- Eccentricity of the specimen with respect to the axis of the film cylinder.
- Lack of knowledge of the exact film radius.
- Change of film radius due to shrinkage in the developing process.

Physical errors: Absorption of the specimen.

Measurement of deviation angles. When some degree of precision in lattice constants is desired, it is customary to measure, not the film equivalent, x, of the deviation angle 2θ, Fig. 204, but rather the film equivalent, $2x$, of the double deviation angle, 4θ. There are three reasons for this:

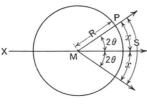

FIG. 204.

(a) There is then no necessity for recording the zero for θ;

(b) the measurement of the double distance increases precision; and

(c) this automatically eliminates part of the eccentricity correction and makes the other part subject to a systematic treatment, as demonstrated below.

The doubled distance, i. e., the distance between a reflection and its mate on the other side of the beam exit position, is customarily designated by S. S corresponds with 4θ. In radian measure,

$$4\theta = \frac{S}{R},\qquad(4)$$

from which

$$\theta = \frac{S}{4R}.\qquad(5)$$

Eccentricity error. If, owing to an error in the construction or assembly of the apparatus, the crystal axis is eccentric to the cylinder axis of the camera, an error in the position of the reflections results. The following analysis of this error is due to Bradley and Jay.[30]

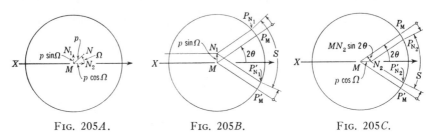

FIG. 205A.   FIG. 205B.   FIG. 205C.

Suppose that the crystal is eccentric with respect to the axis of the camera cylinder, and stands at N, not M, Fig. 205A. The displacement can be expressed in terms of a distance, p, and a direction, Ω. For the present purposes, it is convenient to resolve the displacement into its components perpendicular to, and parallel with, the x-ray beam direction. These are equal to $p \sin \Omega$ and $p \cos \Omega$, respectively. The effect of the first component, Fig. 205B, is to shift both reflections, P_N, in the same direction by very approximately the same amount. This leaves the measured distance, S, unaffected and without error. The effect of the second component, Fig. 205C, is to shift both reflections toward (or away) from one another, and so introduces a total error of

$$\Delta S = 2P_M P_{N_2}$$
$$= -2P_{N_2}P_M$$
$$= -2MN_2 \sin 2\theta$$
$$= -2(p \cos \Omega) \sin 2\theta. \qquad (6)$$

The angular error (expressed in radians) corresponding with this linear error, according to (5), is

$$\Delta\theta = -\frac{[2(p \cos \Omega) \sin 2\theta]}{4R}$$

$$= -\frac{p \cos \Omega}{2R} \sin 2\theta \qquad (7)$$

$$= -\frac{p \cos \Omega}{R} \sin \theta \cos \theta. \qquad (7A)$$

Radius and shrinkage errors. It is not always possible to know precisely the radius of the film as it lies rolled in the form of a cylinder in the cylindrical camera. The error in the assumed radius is known as the *radius error*. Another error of very similar characteristics is caused by the change in length of the film due to the wetting and drying processes of the photographic development. The net change in length of the film (known as a *film shrinkage*) introduces a shrinkage error. If the film shrinks, for example, the entire film and its record are uniformly reduced in size, and the result is the same as if the film had been made in a proportionally smaller camera. Both radius error and shrinkage error are thus inherently the same, and can be treated at one time.

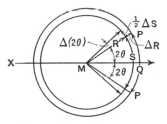

FIG. 206.

The characteristics of the radius and shrinkage errors can be derived from Fig. 206. For simplicity, suppose that the radius error is due to expansion of the film in development. The film, originally of radius R, has expanded so that its radius is $R + \Delta R$. The reflection positions, P, are measured, and from this measurement, θ is computed on the assumption that the radius is R. The measured arc between reflections is Q, while the true arc for the assumed radius is S. For a positive radius error, ΔR, Q is greater than S; consequently the angle θ computed from (5) is too large. The excess of the measured arc, Q, over the true arc, S, for the assumed radius, is the resulting measurement error,

$$\Delta S = Q - S. \tag{8}$$

In radian measure, from Fig. 206,

$$Q = 4\theta \cdot (R + \Delta R) \tag{9}$$

and

$$S = 4\theta \cdot R. \tag{10}$$

Substituting these in (8),

$$\Delta S = 4\theta (R + \Delta R) - 4\theta \cdot R$$
$$= 4\theta \Delta R. \tag{11}$$

This error in length measurement gives rise to a computed error in θ of an amount (in radian measure) equal to

$$\Delta \theta = \frac{\Delta(S/4)}{R} = \frac{\Delta S}{4R} \tag{12}$$

A slightly different derivation of (12) is as follows: The error in θ is

$$\Delta \theta = \theta_{\text{measured}} - \theta_{\text{true}}$$

$$= \frac{Q}{4R} - \frac{S}{4R}$$

$$= \frac{Q - S}{4R}$$

$$= \frac{\Delta S}{4R}. \tag{12}$$

Making substitution from (11), (12) becomes

$$\Delta \theta = \frac{4\theta \Delta R}{4R} = \frac{\Delta R}{R} \theta. \tag{13}$$

Absorption error for parallel radiation. The general effect of high absorption in a crystal upon its apparent lattice constant, as inferred from the position of a reflection, was first recognized by Hadding[1] in 1921. Hadding assumed a non-divergent beam of x-rays. Under these conditions the effect of absorption is shown in Figs. 207A, B, and C. In Fig. 207A it is assumed that the crystal is non-absorbing. The full sample therefore diffracts radiation equally well in all directions.   If the crystal is highly absorbing, i. e., opaque to x-radiation, Fig. 207C, then only reflections from the surface can form complete paths and all reflection paths involving some traversing of the crystal are suppressed.   It should be observed that the width of the unsup-

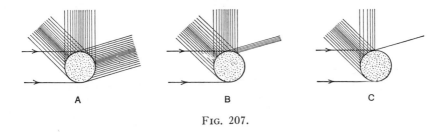

FIG. 207.

pressed reflected radiation beam decreases with decreasing θ.   Figure 207B shows the case of intermediate absorption, which is very similar in its characteristics to that of high absorption.

In the case of complete absorption, Fig. 208A, the surface reflections include those rays which fall on that arc of the specimen's circumference which is limited by the two tangents to the crystal cylinder

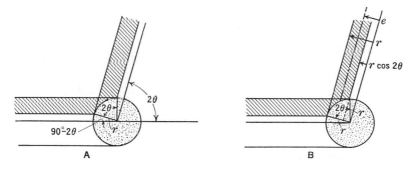

FIG. 208.

parallel with the incoming and outgoing beam directions, respectively. Only these rays are included in the reflected beam.   The radii to these tangents meet at an angle equal to 2θ, Fig. 208A.   In Fig. 208B, it is evident that the inner and outer sides of the reflected bundle of rays are displaced away from the true center of the ideal reflection by amounts:

$$\text{Inner displacement} = r \cos 2\theta, \qquad (14)$$

$$\text{Outer displacement} = r. \qquad (15)$$

The center of the reflected bundle is displaced by an absolute distance on the film which is the mean of these two, namely,

$$e = \frac{r + r \cos 2\theta}{2} = \frac{r(1 + \cos 2\theta)}{2}. \tag{16}$$

The entire error in the measurement of S, therefore, is

$$\Delta S = 2e = r(1 + \cos 2\theta) \tag{17}$$

$$= 2r \cos^2 \theta. \tag{17A}$$

The angular error (expressed in radians) corresponding with this measurement error is, according to (5),

$$\Delta\theta = \frac{\Delta S}{4R} = \frac{r}{4R}(1 + \cos 2\theta) \tag{18}$$

$$= \frac{r}{2R}(\cos^2 \theta). \tag{18A}$$

The width of the reflection increases with θ. Figure 208B shows that the reflection width in a non-divergent beam is

$$W = r - r \cos 2\theta$$

$$= r(1 - \cos 2\theta) \tag{19}$$

$$= 2r \sin^2 \theta. \tag{19A}$$

Note that both these results are for the ideal case of non-divergent beams.

Absorption error for divergent radiation. When not only absorption but also divergence is present, which are the usual conditions, then the situation is more complex. The exact treatment for this case is quite complicated, but an exceedingly close approximation, followed here, was devised by Bradley and Jay.[30] The geometry involved for complete absorption and any divergence is shown in Fig. 209A. The radius of the specimen is r, and the radius of the film is R. Part of the approximation is due to the fact that r is very small compared with R.

If the x-ray beam were parallel, the reflected rays would be included in the shaded path of Fig. 209A. The x-ray beam, however, actually diverges from point X (which might be the limiting pinhole, for example). Owing to this divergence, the x-ray beam does not strike the film over the range LK but over the range $L'K'$. The shift of $L \rightarrow L'$ is due to the fact that the x-ray beam to the point D' is not parallel with the axis XM, but makes an angle MXD' with the axis. The shift $K \rightarrow K'$

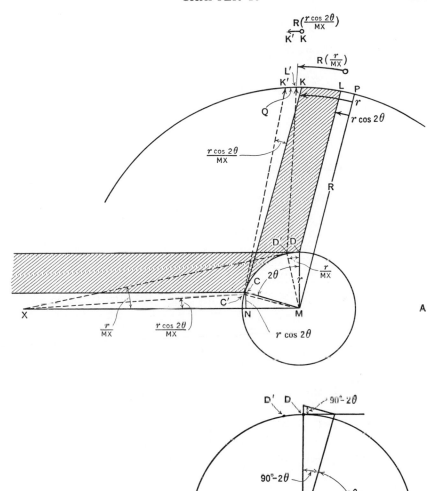

F<small>IG</small>. 209.

occurs for a similar reason. To compute these shifts, it is necessary
to know the sizes of the angles mentioned. In radian measure, very
approximately,

$$\angle MXD' = \frac{MD'}{MX} = \frac{r}{MX},\qquad(20)$$

and

$$\angle MXC' = \frac{NC'}{NX}.\qquad(21)$$

In this last case, a further approximation is

$$NC' \approx NC = r \sin (90° - 2\theta) = r \cos 2\theta, \tag{22}$$

whence

$$\angle MXC' = \frac{r \cos 2\theta}{MX}. \tag{23}$$

It should now be observed that, since r is very small, D' practically coincides with D. Nevertheless, the ray $D'L'$ diverges from DL by the angle given in (20). In radian measure, the displacement

$$LL' = R \left(\angle LDL' \right)$$
$$= R \left(\angle MXD' \right), \tag{24}$$

and according to (20), this becomes

$$LL' = R \frac{r}{MX}. \tag{25}$$

Similarly, since r is small, C' practically coincides with C. The angle KCK' is equal to $\angle MXC'$ given by (21). In radian measure, the displacement

$$KK' = R \left(\angle KCK' \right)$$
$$= R \left(\angle MXC' \right), \tag{26}$$

and according to (23), this becomes

$$KK' = R \frac{r \cos 2\theta}{MX}. \tag{27}$$

The total displacements of the edges of the reflections from the ideal center point, P, are

$$PL' = PL + LL' \tag{28}$$

and

$$PK' = PK + KK'. \tag{29}$$

In (14) and (15) PL and PK were shown to be

$$PL = r \cos 2\theta, \tag{30}$$

$$PK = r. \tag{31}$$

Substituting from (30) and (25) into (28), and from (31) and (27) into (29), (28) and (29) expand into

$$PL' = r \cos 2\theta + R \frac{r}{MX} = r \left(\cos 2\theta + \frac{R}{MX} \right) \tag{32}$$

and

$$PK' = r + R \frac{r \cos 2\theta}{MX} = r\left(1 + \frac{R}{MX} \cos 2\theta\right). \tag{33}$$

The displacement PQ of the midpoint of the reflection is

$$PQ = \frac{PL' + PK'}{2} \tag{34}$$

$$= \frac{1}{2}\left\{r\left(\cos 2\theta + \frac{R}{MX}\right) + r\left(1 + \frac{R}{MX} \cos 2\theta\right)\right\}. \tag{35}$$

This can be rearranged as follows:

$$PQ = \frac{r}{2}\left\{\cos 2\theta\left(1 + \frac{R}{MX}\right) + \left(1 + \frac{R}{MX}\right)\right\}$$

$$= \frac{r}{2}(\cos 2\theta + 1)\left(1 + \frac{R}{MX}\right)$$

$$= \frac{r}{2}\left(1 + \frac{R}{MX}\right)(1 + \cos 2\theta) \tag{36}$$

$$= r\left(1 + \frac{R}{MX}\right)\cos^2 \theta. \tag{36A}$$

The underscored terms are constants of the experiment and involve the specimen radius, camera radius, and limiting pinhole distance.

In (36), PQ represents $\frac{1}{2}\Delta S$. Substituting $2PQ$ for ΔS in (5) gives the error which results in θ in cases of complete absorption:

$$\Delta\theta = \frac{\Delta S}{4R}$$

$$= \frac{r}{4R}\left(1 + \frac{R}{MX}\right)(1 + \cos 2\theta) \tag{37}$$

$$= \frac{r}{2R}\left(1 + \frac{R}{MX}\right)\cos^2 \theta. \tag{37A}$$

Note that the trigonometric part of (37) is the same as that in (18) for parallel radiation, which means that the error varies in the same systematic way. Note also that, for parallel radiation, $MX = \infty$, and (37) reduces to (18), the parallel radiation case.

The width of the x-ray line in the divergent radiation case is given by

$$W = L'K' = PK' - PL'. \tag{38}$$

Making substitution from (33) and (32), this becomes

$$W = r\left(1 + \frac{R}{MX}\cos 2\theta - \cos 2\theta - \frac{R}{MX}\right). \tag{39}$$

This reduces to

$$W = r\left\{\cos 2\theta\left(\frac{R}{MX} - 1\right) - \left(\frac{R}{MX} - 1\right)\right\}$$

$$= r\left(\cos 2\theta - 1\right)\left(\frac{R}{MX} - 1\right)$$

$$= r\left(1 - \frac{R}{MX}\right)(1 - \cos 2\theta) \tag{40}$$

$$= 2r\left(1 - \frac{R}{MX}\right)\sin^2\theta. \tag{40A}$$

Note that the trigonometric form of this is the same as that given in (19) for parallel radiation, which means that the width varies according to the same general scheme. For parallel radiation, $MX = \infty$ and (40) reduces to (19), the parallel radiation case.

Note also that when $MX = R$, i. e., when the limiting pinhole is on the camera circumference, $\dfrac{R}{MX} = 1$ and the second term of (40) vanishes. The width of a line then becomes zero. This means that, if the limiting pinhole is placed on the camera circumference, perfect focusing is produced by a crystal of very high absorption. This perfection, however, is subject to the limitations of the approximations used in the above derivation. The chief error in this approximation is due to the assumption that D and D' coincide and that C and C' coincide. The width of the reflection is actually decreased by the amounts of the projection of DD' in the direction MP plus the projection of CC' in the direction MP. Since CC' is always very small it can be neglected. The arc DD' is

$$DD' = r(\angle DMD') = r(\angle MXD') = r\left(\frac{r}{MX}\right) = \frac{r^2}{MX}. \tag{41}$$

Figure 209B shows that the projection of DD' in the reflection direction makes an angle $(90° - 2\theta)$ with DD':

$$\text{Proj. } DD' = DD'\cos(90° - 2\theta)$$

$$= DD'\sin 2\theta$$

$$= \frac{r^2}{MX}\sin 2\theta. \tag{42}$$

The true line width is then approximately the difference between (40) and (42), namely,

$$W = r\left(1 - \frac{R}{MX}\right)(1 - \cos 2\theta) - \frac{r^2}{MX}\sin 2\theta. \tag{43}$$

This also reduces to (19) for the parallel radiation condition, $MX = \infty$. When the limiting pinhole is on the camera circumference, $MX = R$, and (43) reduces to

$$W = \frac{r^2}{R}\sin 2\theta. \tag{44}$$

If r is small compared with R, this is a very small value.

KETTMANN'S EXTRAPOLATION

The method of extrapolation devised by Kettmann[29] to eliminate systematic errors, though no longer used, is a direct prelude to the method of Bradley and Jay, which follows. Kettmann's method is distinguished by not inquiring into the nature of the variation of the individual errors; rather it lumps them together. Kettmann's treatment, slightly modified for present purposes, is as follows:

A measurement of θ determines a spacing, d, according to the Bragg formula,

$$n\lambda = 2d \sin \theta. \tag{45}$$

Consequently, any d depends on θ in the following manner:

$$d = \frac{n\lambda}{2 \sin \theta}. \tag{46}$$

If, instead of the true θ, an erroneous Θ is measured, an erroneous D is computed according to

$$D = \frac{n\lambda}{2 \sin \Theta}. \tag{47}$$

Dividing (47) by (46),

$$\frac{D}{d} = \frac{\sin \theta}{\sin \Theta}, \tag{48}$$

from which

$$D = d\frac{\sin \theta}{\sin \Theta}. \tag{49}$$

The true angle θ and the uncorrected angle Θ are related by

$$\theta = \Theta - \Delta\theta, \tag{50}$$

where $\Delta\theta$ is the error in determining θ. Making this substitution, (49) becomes

$$D = d \, \frac{\sin \, (\Theta - \Delta\theta)}{\sin \, \Theta}, \tag{51}$$

which may be expanded to

$$D = d \left(\frac{\sin \, \Theta \cos \, \Delta\theta - \cos \, \Theta \sin \, \Delta\theta}{\sin \, \Theta} \right)$$

$$= d(\cos \, \Delta\theta - \cot \, \Theta \sin \, \Delta\theta). \tag{52}$$

Now, as the measured angle, Θ, approaches 90°, cot Θ approaches zero, and (52) approaches

$$\lim_{\Theta \to 90°} D = d \cos \, \Delta\theta. \tag{53}$$

Since the error, $\Delta\theta$, is a small angle, cos $\Delta\theta$ is very nearly unity, and (53) may be evaluated as

$$\lim_{\Theta \to 90°} D = d. \tag{54}$$

This result simply means that, if spacing values, D, are computed from various reflections of different measured Θ value on the film, and

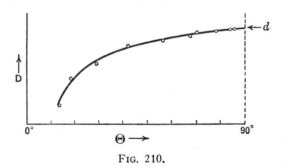

Fig. 210.

plotted against the measured Θ, the values of D, so computed, approach the true value of d when Θ is 90°. Because of the extremely small size of the correction angle, $\Delta\theta$, cos $\Delta\theta$ is substantially unity, and sin $\Delta\theta$ may be replaced by $\Delta\theta$. Relation (52) may therefore be simplified to

$$D = d(1 - \cot \, \Theta \, \Delta\theta). \tag{55}$$

This provides the general form of the extrapolation curve, an example of which is shown in Fig. 210.

The major difficulty with Kettmann's extrapolation method is that $\Delta\theta$ is a function of Θ, and consequently the curve shown in Fig. 210 is not that of the curve $(1 - \cot \, \Theta)$, but also depends on the unknown

form of the variation of $\Delta\theta$ with Θ. The character of the error variation for any particular case can be experimentally derived from the
data by a manipulation of (55):

$$D = d - d \cot \Theta \; \Delta\theta,$$

$$D - d = -d \cot \Theta \; \Delta\theta,$$

$$\frac{D - d}{d} = -\cot \Theta \; \Delta\theta, \qquad (56)$$

and

$$\Delta\theta = -\frac{D - d}{d} \tan \Theta. \qquad (56')$$

BRADLEY AND JAY'S EXTRAPOLATION

General remarks. Bradley and Jay[30] further developed the extrapolation method by examining the systematic variation of $\Delta\theta$ with Θ
for eccentricity and absorption errors. This extrapolation method is
exactly suited to the refining of lattice constants as computed from the
zero-layer lines of rotating-crystal photographs and from Weissenberg
photographs made in the ordinary manner (" front-reflection," i. e.,
the center of the film is at $\Upsilon = 0$, as contrasted with the " back-reflection " method described in the next chapter, for which the x-ray beam
enters a hole in the film at $\Upsilon = 180°$). Radius and shrinkage errors
cannot be eliminated by this kind of
extrapolation (see next section), and
so it is necessary to calibrate the
film.

Calibration. The film may be
calibrated by laying against it, in
the camera, a metal piece which will
cast a shadow (with respect to scattered radiation which normally
reaches and darkens the film) over
a definite angular range. Any other

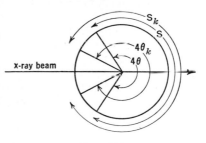

FIG. 211.

angular range may then be compared with this range. This is illustrated in Fig. 211. From the arc-angle proportionality,

$$\frac{S}{S_k} = \frac{4\theta}{4\theta_k} = \frac{\theta}{\theta_k}. \qquad (57)$$

Once the angle θ_k for the metal piece is known, any arc measurement,
S, can be converted directly into the corresponding angle, θ, by measur-

ing both S and S_k on the same film and computing θ from

$$\theta = S\left(\frac{\theta_k}{S_k}\right),\qquad(58)$$

where the term in parentheses is a constant for any film. S_k varies from film to film on account of shrinkage, but θ_k is a constant for the camera. It remains to determine θ_k. This can be done in several ways:

1. Compare θ_k with $180°$ by the Straumanis method, on the same camera.

2. Compare θ_k with the known range, 4θ, of the reflection of some crystal of well-known lattice constant, taking great care to minimize absorption by using a tiny specimen of low absorbing material, and making mechanically certain that no eccentricity errors are introduced.

3. Instead of a metal piece, slip into the camera a ring with 36 equally spaced teeth, and expand the teeth, collet-fashion, equally against the film. Each space between teeth represents $\dfrac{360°}{36} = 10°$.

Fundamental spacing-error equation. Starting with the Bragg law,

$$n\lambda = 2d \sin \theta,\qquad(59)$$

the spacing, d, is evidently computed from a measurement of θ, with the aid of

$$d = \left(\frac{n\lambda}{2}\right)\frac{1}{\sin \theta} = \frac{n\lambda}{2} \csc \theta.\qquad(60)$$

Using Δ as the symbol for the differential to avoid confusion with the spacing symbol, d, and differentiating (60) with respect to θ,

$$\frac{\Delta d}{\Delta \theta} = \left(\frac{n\lambda}{2}\right)(-\csc \theta \cot \theta),\qquad(61)$$

from which

$$\Delta d = -\frac{n\lambda}{2} \csc \theta \cot \theta\ \Delta\theta.\qquad(62)$$

This gives the absolute error, Δd, as the result of an error in θ, namely, $\Delta\theta$. The fractional (percentage) error is $\dfrac{\Delta d}{d}$, and can be obtained by dividing (62) by (60):

$$\frac{\Delta d}{d} = -\frac{\dfrac{n\lambda}{2}\csc\theta\cot\theta\,\Delta\theta}{\dfrac{n\lambda}{2}\csc\theta}$$

$$= -\cot\theta\,\Delta\theta. \tag{63}$$

This fundamental fractional error equation is substantially the equivalent of equation (56), derived by Kettmann. Up to this point Bradley and Jay differ from Kettmann only in the method of arriving at (63).

In order to evaluate $\Delta\theta$ for eccentricity and absorption errors, Bradley and Jay derived (7A) and (37A):

$$\text{Eccentricity error:}\quad \Delta\theta = -\frac{p\cos\Omega}{R}\sin\theta\cos\theta \tag{7A}$$

$$\text{Absorption error}\dagger:\quad \Delta\theta = \frac{r}{2R}\left(1 + \frac{R}{MX}\right)\cos^2\theta. \tag{37A}$$

Fractional error due to eccentricity. If the value of $\Delta\theta$ given by (7A) is substituted into (63), there results

$$\frac{\Delta d}{d} = -\cot\theta\,\Delta\theta$$

$$= -\cot\theta\left(-\frac{p\cos\Omega}{R}\sin\theta\cos\theta\right)$$

$$= \left[\frac{p\cos\Omega}{R}\right]\cos^2\theta. \tag{64}$$

This is the fractional (percentage) error in the spacing resulting from an eccentricity error of magnitude and direction p and Ω.

† For reasons unknown, Bradley and Jay[30] used an " approximate " form of (37) namely:

$$\Delta\theta \approx \frac{r}{2R}\left(1 + \frac{R}{MX}\right)\frac{\sin 2\theta}{2\theta} \tag{37'}$$

$$\approx \frac{r}{2R}\left(1 + \frac{R}{MX}\right)\frac{\sin\theta\cos\theta}{\theta}. \tag{37A'}$$

Fractional error due to absorption. If the value of $\Delta\theta$ given by $(37A)$ is substituted into (63), there results

$$\frac{\Delta d}{d} = -\cot\theta \; \Delta\theta$$

$$= -\cot\theta \cdot \frac{r}{2R}\left(1 + \frac{R}{MX}\right)\cos^2\theta$$

$$= \left[-\frac{r}{2R}\left(1 + \frac{R}{MX}\right)\right]\cot\theta\cos^2\theta. \qquad (65)\ddagger$$

This gives the fractional (percentage) error in the spacing resulting from complete absorption by a specimen of radius r immersed within a beam diverging from a point distant MX from the specimen.

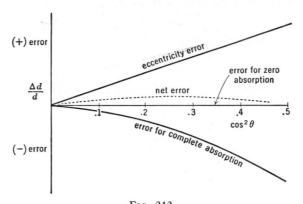

FIG. 212.

Vanishing of errors at $\theta = 90°$. When $\theta = 90°$, $\cos^2\theta = 0$, and the right members of both (64) and (65) vanish, i. e., both eccentricity and absorption errors vanish, and the value of d is the true value if the measurement of θ is made at 90°. This is better brought out graphically. Since (64) and (65) contain the common variable, $\cos^2\theta$,

‡ By substituting the " approximate " value given by $(37A')$, Bradley and Jay[30] get, instead,

$$\frac{\Delta d}{d} = -\cot\theta \; \Delta\theta$$

$$= -\cot\theta \cdot \frac{r}{2R}\left(1 + \frac{R}{MX}\right)\frac{\sin\theta\cos\theta}{\theta}$$

$$= \left[-\frac{r}{2R}\left(1 + \frac{R}{MX}\right)\right]\frac{\cos^2\theta}{\theta}. \qquad (65')$$

the form of the curves of (64) and (65) are both simplified by plotting against $\cos^2 \theta$. Figure 212 shows such a plot. Equation (64) is a straight line, and (65) is a curve.

Note that whereas the absorption error, (65), is always negative, the eccentricity error, (64), may be either positive or negative depending

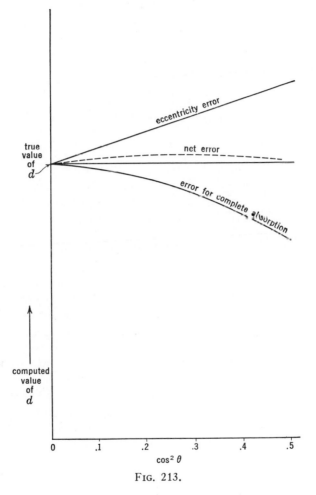

Fig. 213.

upon Ω, the direction of the eccentricity. If the specimen axis is off center toward the beam exit, (64) is $(+)$; if it is off center toward the beam entrance, (64) is $(-)$. Note also that the size of each error depends upon the coefficients in square brackets, which are constants for the experiment. Note further that the absorption-error expression has been computed for complete absorption. For zero absorption,

the error is zero, and the error curve is a horizontal line. For intermediate absorption, the absorption curve lies between the curve shown
and the horizontal line.

Each of the two full error lines in Fig. 212 represents a single source
of error. The total error is the algebraic sum of these, shown as a
dotted line in Fig. 212. The location and shape of this net-error
curve depend upon the experimental coefficients in square brackets.
If the absorption of the specimen is not too great, and if the radius of
the specimen is small, the experimental coefficient of the absorption
error is small, and the absorption-error curve approaches a horizontal
line. Under these conditions the total-error curve is dominated by
the eccentricity curve and approaches a straight line.

The computed value of the spacing is the true value plus the error,
i. e., $d + \Delta d = d + d\left(\dfrac{\Delta d}{d}\right)$. Curves of the computed spacing, plotted
against $\cos^2 \theta$, are shown in Fig. 213. The curve of the computed
spacing containing an eccentricity error is similar to that of the fractional eccentricity error shown in Fig. 212, except that it is multiplied
by a factor d and raised up by an amount d, according to the relation:
$d_{computed} = d + d\left(\dfrac{\Delta d}{d}\right)$. The same relation holds for the computed
spacing curve containing an absorption error, or both errors.

When one computes spacings from measurements of reflection positions, he obtains points along the total error curve of Fig. 213. If a
sufficiently representative number of points can be located, the form
of the curve can be plotted, extrapolated back to $\cos^2 \theta = 0$, and the
true value of the spacing, d, found. The form of this curve, as pointed
out before, is close to that of a straight line, except when absorption
is extreme and the crystal is large.

An example[38] of this extrapolation, computed for d values equal to
the hexagonal A and C axes of tourmaline, is shown in Fig. 214. The
absorption of this crystal was low, and errors in this example are due
almost entirely to eccentricity. The points used in Fig. 214 were computed as shown in Table 23. Such computations are based on the
relation

$$d = \frac{n\lambda}{2 \sin \theta}. \tag{66}$$

In this instance the data were derived from measuring the coordinates,
x_1 and x_2, of pairs of spots on each side of the center line of a Weissenberg photograph made on a camera of standard diameter ($D = 57.30$

mm.).   For this diameter,

$$1 \text{ mm. } x \text{ (film measure)} \approx 2° \text{ arc} \approx 2°(2\theta) \approx 1° \theta.$$

The distance from the spot to its mate is $x_2 - x_1 = 2x$.   Thus,

$$\theta_{\text{uncorrected}} = x = \frac{x_2 - x_1}{2}.$$

This allows for no shrinkage or radius correction.   In the camera in question, this correction amounted to 0.1 per cent; hence

$$\theta_{\text{corrected}} = x + 0.001x. \tag{67}$$

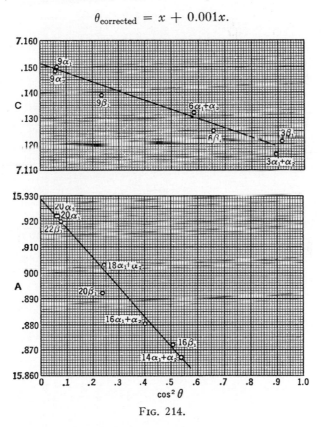

Fig. 214.

This supplies angular data for substitution into (66).   Note that, since x is measured by a millimeter scale, the resulting θ computed from (67) is in degrees and decimals, and that this ordinarily requires conversion into degrees and minutes.   The computation shown in Table 23 is the logarithmic computation of (66).

TABLE 23

COMPUTATION OF APPARENT LENGTH OF HEXAGONAL A-AXIS OF TOURMALINE

(Data for the plotting of Fig. 213)

	20	20	22	20	18	16	16	14
Order								
Radiation	Cu Kα₂	Cu Kα₁	Cu Kβ₁	Cu Kβ₁	Cu Kα₁+α₂	Cu Kα₁+α₂	Cu Kβ₁	Cu Kα₁+α₂
λ	1.54123Å	1.53739Å	1.38935Å	1.38935Å	1.53931Å	1.53931Å	1.38935Å	1.53931Å
Film reading, x_2	159.91 mm.	159.34 mm.	158.18 mm.	145.49 mm.	145.06 mm.	135.40 mm.	129.07 mm.	127.40 mm.
Film reading, x_1	9.17	9.65	10.86	23.72	24.01	33.81	40.26	41.95
$2x(=x_2-x_1)$	150.74	149.69	147.32	121.77	121.05	101.59	88.81	85.45
x	75.37	74.845	73.66	60.885	60.525	50.795	44.405	42.725
$\theta(=x+0.001x)$	75.445°	74.919°	73.733°	60.945°	60.585°	50.845°	44.449°	42.767°
θ	75°27′	74°55′	73°44′	60°57′	60°35′	50°51′	44°27′	42°46′
Log $(n/2)$	1.00000	1.00000	1.04139	1.00000	0.95424	0.90309	0.90309	0.84510
Log λ	0.18786	0.18678	0.14281	0.14281	0.18733	0.18733	0.14281	0.18733
Log $(n\lambda/2)$	1.18786	1.18678	1.18420	1.14281	1.14157	1.09042	1.04590	1.03243
Log sin θ	9.98584−10	9.98477−10	9.98226−10	9.94161−10	9.94005−10	9.88958−10	9.84528−10	9.83188−10
Log d	1.20202	1.20201	1.20194	1.20120	1.20152	1.20084	1.20062	1.20055
d	15.922Å	15.922Å	15.920Å	15.892Å	15.903Å	15.880Å	15.872Å	15.867Å
Cos² θ	0.0631	0.0677	0.0785	0.2355	0.2415	0.401	0.509	0.538

DISCUSSION OF BRADLEY AND JAY'S EXTRAPOLATION

Shrinkage errors. An important feature of Bradley and Jay's method is the use of a radius-calibrated film. The reason for this is that the fractional error due to shrinkage error varies in an unfortunate way with respect to $\cos^2 \theta$. According to (13) the *absolute* angular

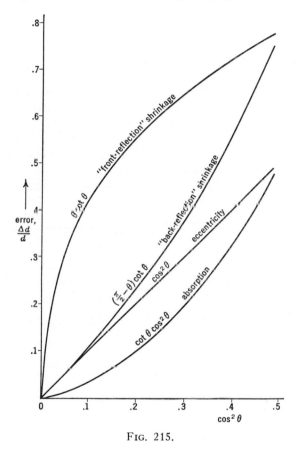

Fig. 215.

error due to shrinkage is proportional to the angle being measured, and it might be supposed from this that the error due to shrinkage error increases with θ. This, however, is not true because other factors are involved. The *fractional* spacing error involves both the angular error and the angle, and has the form, (63):

$$\frac{\Delta d}{d} = -\cot \theta \, \Delta \theta. \qquad (68)$$

Substituting for $\Delta\theta$ from (13) gives

$$\frac{\Delta d}{d} = (-\cot\theta)\left(\frac{\Delta R}{R}\theta\right)$$

$$= -\frac{\Delta R}{R}\theta\cot\theta. \tag{69}$$

The trigonometric part of (69), plotted against Bradley and Jay's extrapolation function, $\cos^2\theta$, is shown in Fig. 215. It is evident that while the spacing error due to shrinkage vanishes at $\theta = 90°$, $\cos^2\theta = 0$, it does so in a most unfortunate and non-linear manner. It is quite impossible, therefore, to extrapolate all the errors against an approximately straight line if any great amount of shrinkage is present. Note that, from Fig. 215, if shrinkage were the only error, an extrapolation against $\cos^2\theta$ would give such an acute intercept with the error axis that no accurate estimate of the extrapolated value of d could be made. These considerations doubtless decided Bradley and Jay to make a calibration allowance for shrinkage errors before applying their extrapolation method.

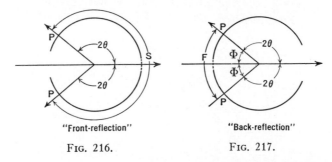

"Front-reflection"   "Back-reflection"

Fig. 216.   Fig. 217.

Unfortunately, this decision disregarded the earlier development of the " back-reflection " method by van Arkel,[25] von Göhler and Sachs,[26] and Braekken.[28] In the usual or " front-reflection " method, the x-ray beam enters the camera between the two loose ends of the film, Fig. 216. In the " back-reflection " method, Fig. 217, the x-ray beam *leaves* the camera between the two ends of the film, and, by necessity, enters the camera through a small hole in the center of the film. In this case, one measures reflection position, P, by measuring the film-arc F, not S. The relation between F and S is evidently

$$F + S = 2\pi R,$$

and so

$$F = 2\pi R - S. \tag{70}$$

Where the arc S corresponded with the angle 4θ in front reflection, the arc F corresponds with the angle 2Φ in back reflection. From Fig. 217,

$$\Phi = \pi - 2\theta, \tag{71}$$

and so

$$\theta = \frac{\pi}{2} - \frac{\Phi}{2}. \tag{72}$$

Now the major reason why the shrinkage function in (69) reaches zero at such a sharp angle and thus reduces its usefulness in extrapolation is that in its equivalent equation, (68), while one factor ($\cot \theta$) is decreasing, the other part ($\Delta\theta$, which is proportional to ΔS) is increasing, and the function becomes zero only because of the first factor. In back reflection, both factors simultaneously decrease toward zero, giving a more favorable condition. This may be quantitatively investigated by recasting the derivation of the shrinkage error in terms of Φ. (Both eccentricity and absorption errors are independent of whether d is measured through θ or Φ, i. e., computed from S or F.)

(8):

$$\Delta F = Q' - F. \tag{73}$$

(9):

$$Q' = 2\Phi (R + \Delta R). \tag{74}$$

(10):

$$F = 2\Phi \cdot R. \tag{75}$$

(11):

$$\Delta F = 2\Phi (R + \Delta R) - 2\Phi \cdot R \tag{76}$$

$$= 2\Phi\Delta R.$$

(12):

$$\Delta\Phi = \frac{\Delta(F/2)}{R} = \frac{\Delta F}{2R} \tag{77}$$

(13):

$$\Delta\Phi = \frac{2\Phi\Delta R}{2R} = \Phi \frac{\Delta R}{R}. \tag{78}$$

This can be recast in the form

$$\Delta\theta = - \left(\frac{\pi}{2} - \theta\right)\frac{\Delta R}{R}. \tag{78'}$$

The fundamental spacing-error equation, (63), may also be recast in terms of Φ by substitution from (72) as follows:

$$\frac{\Delta d}{d} = -\cot\theta\,\Delta\theta$$

$$= -\cot\left(\frac{\pi}{2} - \frac{\Phi}{2}\right)\Delta\left(\frac{\pi}{2} - \frac{\Phi}{2}\right)$$

$$= -\tan\frac{\Phi}{2}\Delta\left(-\frac{\Phi}{2}\right)$$

$$= \left(-\tan\frac{\Phi}{2}\right)\left(-\Delta\frac{\Phi}{2}\right)$$

$$= \tan\frac{\Phi}{2}\Delta\frac{\Phi}{2}. \tag{79}$$

Substituting from (78), this becomes

$$\frac{\Delta d}{d} = \left(\tan\frac{\Phi}{2}\right)\left(\frac{\Phi}{2}\frac{\Delta R}{R}\right)$$

$$= \left[\frac{\Delta R}{R}\right]\frac{\Phi}{2}\tan\frac{\Phi}{2}. \tag{80}$$

For purposes of comparing with earlier results, this is now recast into terms of θ, with the aid of (71):

$$\frac{\Delta d}{d} = \left[\frac{\Delta R}{R}\right]\left(\frac{\pi}{2} - \theta\right)\tan\left(\frac{\pi}{2} - \theta\right)$$

$$= \left[\frac{\Delta R}{R}\right]\left(\frac{\pi}{2} - \theta\right)\cot\theta. \tag{81}$$

A graph of the trigonometric part of this function against $\cos^2\theta$ is shown in Fig. 215. It will be observed that this function also vanishes at $\theta = 90°$, $\cos^2\theta = 0$, and that it cuts the vertical sharply, giving rise to a well-defined indication of the error zero. A most important property of this graph is that it is approximately linear, especially for small values of $\cos^2\theta$. This property was first pointed out by Cohen.[36] For this reason, both eccentricity and radius errors can be eliminated by plotting the total error against $\cos^2\theta$ and extrapolating back to $\cos^2\theta = 0$. The extrapolation is very approximately a straight line for small values of $\cos^2\theta$ provided that absorption is practically absent.

It is also possible to plot computed spacings against $\left(\dfrac{\pi}{2} - \theta\right) \cot \theta$, or against its equivalent in (80), namely, $\dfrac{\Phi}{2} \tan \dfrac{\Phi}{2}.$ The latter form is especially convenient because Φ is the immediate result obtained after measuring the film arc, F. If this function is used for extrapolation, the radius error is strictly linear, and the eccentricity error is

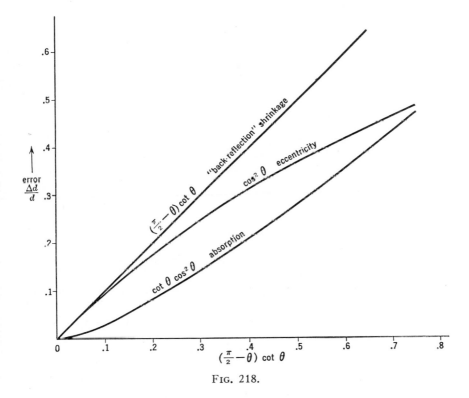

FIG. 218.

approximately linear, because it is proportional to $\cos^2 \theta$ and this is approximately linear with $\left(\dfrac{\pi}{2} - \theta\right) \cot \theta.$ A graph of the errors plotted against this function is shown in Fig. 218. Note that the absorption error is still non-linear, and tangent to the horizontal axis at $\left(\dfrac{\pi}{2} - \theta\right) \cot \theta = 0.$ If shrinkage errors are known to be great, it is better to plot against $\left(\dfrac{\pi}{2} - \theta\right) \cot \theta$ than against $\cos^2 \theta$, because then the shrinkage will be eliminated. This need not be done, however, because shrinkage calibration is extremely easy.

Absorption errors. It is unfortunate that the absorption error is not approximately linear with $\cos^2 \theta$. Suppose that the photograph made by a crystal of high absorption is being examined, and suppose, further,

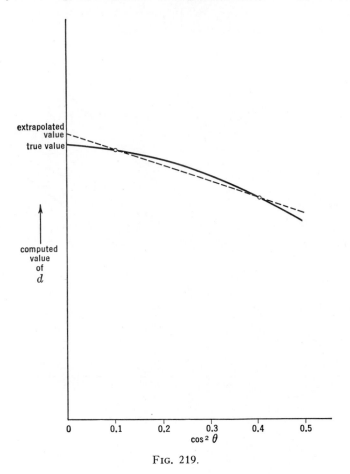

FIG. 219.

that, owing to good mechanical construction and calibration, errors due to eccentricity and shrinkage are absent. The entire error is then given by the absorption curve, Fig. 219. If two reflections give data within this range, represented by the two circles of Fig. 219, then the d determined by assuming that the absorption error is linear is always seriously too high by an unknown amount, as shown in Fig. 219. Even when other errors are present, and absorption is only a component of the total error, the extrapolated spacing is always too high.

There are two possible remedies to this condition.   The first is to fit the data points to an error curve of the form

$$d = K_1 \cot \theta \cos^2 \theta + K_2 \cos^2 \theta \tag{82}$$

plotted against θ. K_1 is an experimental constant which takes account of terms in brackets of (65) as well as the amount of absorption, while

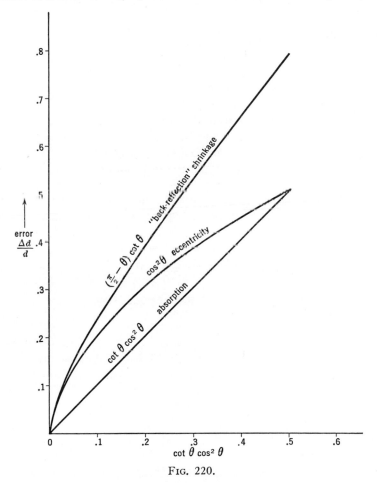

Fig. 220.

K_2 takes account of the experimental constants in the brackets of (64) and (81).   To graph (82) requires at least three well-separated data points.

The second method is to remove eccentricity errors mechanically by perfect instrument construction, to remove shrinkage errors experimentally by using a calibrated film, and thus leave only absorption error. The error may be plotted against the absorption error function, $\cot \theta \cos^2 \theta$, and may then be extrapolated away against a straight line. If the other errors have not been completely removed, they vary as shown in Fig. 220. Even if other errors are not first allowed for mechanically, it is better to plot against $\cot \theta \cos^2 \theta$ if a large crystal of high absorption is being investigated.

Miscellaneous remarks. Since

$$\sin^2 \theta = 1 - \cos^2 \theta, \tag{83}$$

$\sin^2 \theta$ varies linearly with $\cos^2 \theta$. It is therefore possible to plot the observed d's against $\sin^2 \theta$ instead of $\cos^2 \theta$, and extrapolate to $\cos^2 \theta = 0$, or to $\sin^2 \theta = 1$, according to (83). Plotting against $\sin^2 \theta$ involves a little less labor than plotting against $\cos^2 \theta$ if logarithms are not used, because $\sin \theta$ must be looked up anyway to compute (66) from θ. If logarithms are used, $\sin \theta$ is not looked up, but $\log \sin \theta$ is looked up directly.

An interesting feature of any plot of d against a function of θ is that the points corresponding with the α_1 and α_2 components of $K\alpha$ doublet are spread apart. This effect is seen in Fig. 231. It is evidently due to an optical illusion resulting in the apparent misplacing of the peaks of the α_1 and α_2 components with respect to their mean.

COHEN'S ANALYTICAL EXTRAPOLATION[31, 32, 36, 37]

Cohen's fundamental equation. Bradley and Jay's extrapolation method is a graphical one. Cohen has shown how the same extrapolation may be performed by analytical means. Cohen's fundamental equation takes account of the errors as they affect the ideal Bragg's law:

$$\frac{n\lambda}{2d} = \sin \theta. \tag{84}$$

If there are errors in θ, then, instead of the true θ, one measures $\theta + \Delta\theta$, and the values of d as computed from Bragg's law then become

$$\frac{n\lambda}{2d} = \sin (\theta + \Delta\theta). \tag{85}$$

In order to recast (85) in useful form, it is necessary to develop the right member in terms of sums of trigonometric forms of θ and $\Delta\theta$. It turns out that the only way of doing this leading to usable results is to square and expand by Taylor's theorem. Squaring (85) results in

$$\left(\frac{n\lambda}{2d}\right)^2 = \sin^2(\theta + \Delta\theta). \tag{86}$$

The form of Taylor's theorem desired in this case is

$$f(a + h) = f(a) + hf'(a) + \frac{h^2}{2}f''(a) \cdots \tag{87}$$

For this specific example,

$$\left.\begin{aligned}
f(a) &= \sin^2\theta. \\
f'(a) &= 2\sin\theta\cos\theta. \\
h &= \Delta\theta.
\end{aligned}\right\} \tag{88}$$

Expanding the right member of (86) according to this theorem, and neglecting terms involving powers of the error, h, (86) becomes

$$\left(\frac{n\lambda}{2d}\right)^2 = \sin^2\theta + \Delta\theta \cdot 2\sin\theta\cos\theta$$

$$= \sin^2\theta + \Delta\theta\sin 2\theta. \tag{89}$$

The term $\Delta\theta$ is made up of several systematic errors, due respectively to eccentricity, shrinkage, and absorption. Equation (89) may therefore be expanded into

$$\left(\frac{n\lambda}{2d}\right)^2 = \sin^2\theta + (\Delta\theta_{\text{ecc.}} + \Delta\theta_{\text{shr.}} + \Delta\theta_{\text{abs.}})\sin 2\theta$$

$$= \sin^2\theta + \Delta\theta_{\text{ecc.}}\sin 2\theta + \Delta\theta_{\text{shr.}}\sin 2\theta + \Delta\theta_{\text{abs.}}\sin 2\theta. \tag{90}$$

The values of the $\Delta\theta$'s have already been determined as the following equations, where the D's and E are experimental constants:

(7):
$$\Delta\theta_{\text{ecc.}} = -D_1\sin 2\theta.$$

Back-reflection shrinkage error,

(78′):
$$\Delta\theta_{\text{shr.}} = -D_2\left(\frac{\pi}{2} - \theta\right).$$

(37A):
$$\Delta\theta_{\text{abs.}} = -E\cos^2\theta.$$

Substituting these in (90), it becomes

$$\left(\frac{n\lambda}{2d}\right)^2 = \sin^2\theta - D_1 \sin^2 2\theta - D_2\left(\frac{\pi}{2} - \theta\right)\sin 2\theta$$
$$- E\cos^2\theta \sin 2\theta. \quad (90A)$$

It has already been shown that the shrinkage error is approximately linear with the eccentricity error, so that these errors may be bulked together as follows:

$$D_1 \sin^2 2\theta + D_2\left(\frac{\pi}{2} - \theta\right)\sin 2\theta \approx D \sin^2 2\theta. \quad (91)$$

The error equation then becomes

$$\left(\frac{n\lambda}{2d}\right)^2 = \sin^2\theta - D\sin^2 2\theta - E\cos^2\theta \sin 2\theta. \quad (92)$$

Cohen's basic method, neglecting absorption errors. In order to illustrate Cohen's extrapolation method, absorption is first neglected, and, with the last term thus omitted, (92) is rearranged as

$$\left(\frac{n\lambda}{2d}\right)^2 + D\sin^2 2\theta = \sin^2\theta. \quad (93)$$

For convenience of further manipulation, let

$$\left.\begin{array}{r}\left(\dfrac{\lambda}{2d}\right)^2 = A. \\[2mm] n^2 = \alpha. \\[2mm] \sin^2 2\theta = \delta.\end{array}\right\} \begin{array}{l}\text{(These are purposely different from} \\ \text{Cohen's substitution to take care} \\ \text{of variable wavelengths)}\end{array} \quad (94)$$

Then (93) may be rewritten

$$\alpha A + \delta D = \sin^2\theta. \quad (95)$$

For each reflection on a photograph, one knows the order of reflection and therefore α, and one measures θ, and therefore knows δ and $\sin^2\theta$. Therefore (95) is a homogeneous linear equation in two unknowns, namely A (function of the unknown spacing, d) and D (function of the unknown experimental error coefficient). If two different reflections on the same film are measured, there are two equations of the type (95) available, namely:

$$\begin{cases}\alpha_1 A + \delta_1 D = \sin^2\theta_1 \\ \alpha_2 A + \delta_2 D = \sin^2\theta_2.\end{cases} \quad (96)$$

This pair of equations in two unknowns can be solved for the unknowns A and D:

$$A = \frac{\begin{vmatrix} \sin^2 \theta_1 & \delta_1 \\ \sin^2 \theta_2 & \delta_2 \end{vmatrix}}{\begin{vmatrix} \alpha_1 & \delta_1 \\ \alpha_2 & \delta_2 \end{vmatrix}}$$

$$D = \frac{\begin{vmatrix} \alpha_1 & \sin^2 \theta_1 \\ \alpha_2 & \sin^2 \theta_2 \end{vmatrix}}{\begin{vmatrix} \alpha_1 & \delta_1 \\ \alpha_2 & \delta_2 \end{vmatrix}}$$

$$\left. \right\} \qquad (97)$$

With the aid of (94), A can be transformed into the desired spacing, d.

Cohen's basic method, including absorption errors. If absorption errors are taken into account, another correction term must be added to Cohen's equations. In order to simplify (92), let

$$\cos^2 \theta \sin 2\theta = \epsilon. \qquad (98)$$

Substituting from (98) and (94), equation (92) can be consolidated into the form

$$\alpha A + \delta D + \epsilon E = \sin^2 \theta. \qquad (99)$$

When θ is measured for a reflection, $\sin^2 \theta$, ϵ, and δ become known, and α is known when index and wavelength are known. Only A (function of the spacing), D (experimental constant of the eccentricity and shrinkage errors), and E (experimental constant of the absorption error) are unknown. If three reflections on the same film are measured, three equations of type (99) become available, namely:

$$\begin{cases} \alpha_1 A + \delta_1 D + \epsilon_1 E = \sin^2 \theta_1. \\ \alpha_2 A + \delta_2 D + \epsilon_2 E = \sin^2 \theta_2. \\ \alpha_3 A + \delta_3 D + \epsilon_3 E = \sin^2 \theta_3. \end{cases} \qquad (100)$$

These three equations may be solved for their three unknowns A, D, and E, of which only A is really wanted. The solution of A is

$$A = \frac{\begin{vmatrix} \sin^2 \theta_1 & \delta_1 & \epsilon_1 \\ \sin^2 \theta_2 & \delta_2 & \epsilon_2 \\ \sin^2 \theta_3 & \delta_3 & \epsilon_3 \end{vmatrix}}{\begin{vmatrix} \alpha_1 & \delta_1 & \epsilon_1 \\ \alpha_2 & \delta_2 & \epsilon_2 \\ \alpha_3 & \delta_3 & \epsilon_3 \end{vmatrix}} \qquad (101)$$

This value of A is easily transformed to d with the aid of (94).

Cohen's basic method extended to refinement of lattice constants.
Cohen's method is capable of a very important extension to eliminate
errors, not only from spacing measurements but also from the lattice
constants derived from them.   The method becomes too cumbersome,
however, with crystals less symmetrical than orthorhombic.   A tetragonal case will be used to illustrate the method.   For a tetragonal
crystal, the spacing of a plane is a function of a, c, and the indices h,
k, and l of the plane:

$$d = \frac{1}{\sqrt{\dfrac{h^2 + k^2}{a^2} + \dfrac{l^2}{c^2}}}. \tag{102}$$

For purposes of illustration, the absorption correction is neglected.
Equation (93) is now rewritten in such a form that h, k, and l absorb
the n; i.e., h, k, and l are no longer prime to one another:

$$\left(\frac{\lambda}{2d_{(hkl)}}\right)^2 + D \sin^2 2\theta = \sin^2 \theta. \tag{103}$$

Substituting for d from (102), this becomes

$$(h^2 + k^2)\left(\frac{\lambda}{2a}\right)^2 + l^2\left(\frac{\lambda}{2c}\right)^2 + D \sin^2 2\theta = \sin^2 \theta. \tag{104}$$

For compactness, let

$$\left.\begin{array}{cc} (h^2 + k^2) = \alpha, & \left(\dfrac{\lambda}{2a}\right)^2 = A, \\[2ex] l^2 = \gamma, & \left(\dfrac{\lambda}{2c}\right)^2 = C, \\[2ex] \multicolumn{2}{c}{\sin^2 2\theta = \delta.} \end{array}\right\} \tag{105}$$

Then (104) may be rewritten:

$$\alpha A + \gamma C + \delta D = \sin^2 \theta. \tag{106}$$

Here there are three unknowns, A, C, and D, while α and γ are known
from the indexing, and δ and $\sin^2 \theta$ are known from the measured position of the reflection.   When three reflections are measured on the
same film, three equations of type (106) become available, each having
the same A's, C's and D's, but different α's, γ's, δ's, and $\sin^2 \theta$'s.   These
are:

$$\begin{cases} \alpha_1 A + \gamma_1 C + \delta_1 D = \sin^2 \theta_1. \\ \alpha_2 A + \gamma_2 C + \delta_2 D = \sin^2 \theta_2. \\ \alpha_3 A + \gamma_3 C + \delta_3 D = \sin^2 \theta_3. \end{cases} \tag{107}$$

The unknown functions of the lattice constants, A and C, may be solved for:

$$A = \frac{\begin{vmatrix} \sin^2 \theta_1 & \gamma_1 & \delta_1 \\ \sin^2 \theta_2 & \gamma_2 & \delta_2 \\ \sin^2 \theta_3 & \gamma_3 & \delta_3 \end{vmatrix}}{\begin{vmatrix} \alpha_1 & \gamma_1 & \delta_1 \\ \alpha_2 & \gamma_2 & \delta_2 \\ \alpha_3 & \gamma_3 & \delta_3 \end{vmatrix}}$$

$$C = \frac{\begin{vmatrix} \alpha_1 & \sin^2 \theta_1 & \delta_1 \\ \alpha_2 & \sin^2 \theta_2 & \delta_2 \\ \alpha_3 & \sin^2 \theta_3 & \delta_3 \end{vmatrix}}{\begin{vmatrix} \alpha_1 & \gamma_1 & \delta_1 \\ \alpha_2 & \gamma_2 & \delta_2 \\ \alpha_3 & \gamma_3 & \delta_3 \end{vmatrix}} \qquad (108)$$

A and C can be easily transformed to a and c with the aid of (105).

This scheme can be extended to allow for absorption correction by adding a term ϵE as in (99) for each equation, and adding an additional equation. Solution of these four equations becomes a bit tedious, however.

Cohen's method including random errors. In the development just given, it was assumed that each measured θ was exact. The random errors of measurement are unavoidable, however, as evidenced by the scatter of points in Figs. 230 and 231, for example. In order to eliminate these, it is desirable to make θ measurements on *more* reflections than the number of equations required in (96), (100), or (107). The data are then handled by the method of least squares to obtain the best average representation of the data.

To apply the least-squares refinement to the first simple example discussed for Cohen's method, one would measure *more* than three reflections — the more the better. Each measurement would provide data for computing $\sin^2 \theta$ and δ of (95), α being known from the indexing. For the least-square solution of these data, there are needed the following additional computations based upon $\sin^2 \theta$, δ, and α:

$$\Sigma\alpha^2, \ \Sigma\alpha\delta, \ \Sigma\alpha \sin^2 \theta, \ \Sigma\delta^2, \ \Sigma\delta \sin^2 \theta.$$

From these values, there are set up two " normal equations ":

$$\begin{cases} A\Sigma\alpha^2 + D\Sigma\alpha\delta = \Sigma\alpha \sin^2 \theta. \\ A\Sigma\alpha\delta + D\Sigma\delta^2 = \Sigma\delta \sin^2 \theta. \end{cases} \qquad (109)$$

The normal equations are homogeneous linear equations in A and D

and can be solved for A:

$$A = \frac{\begin{vmatrix} \alpha \sin^2 \theta & \sum \alpha \delta \\ \delta \sin^2 \theta & \sum \delta^2 \end{vmatrix}}{\begin{vmatrix} \sum \alpha^2 & \sum \alpha \delta \\ \sum \alpha \delta & \sum \delta^2 \end{vmatrix}}. \tag{110}$$

The value of A, so found, can be transformed to d with the aid of (94). Some hints on the solution of (110) are:

1. The value of $\delta = \sin^2 2\theta$ is needed to only two significant figures.

2. To have the coefficients of the normal equations of the same magnitude, use $\delta = 10 \sin^2 2\theta$.

3. To decrease the number of significant figures needed in computations, adopt an approximate value of A, and find corrections to these values from the differences of $\sin^2 \theta_{(\text{exp.})}$ and $\sin^2 \theta$ from the adopted A. Letting $v = \sin^2 \theta_{(\text{adopted})} - \sin^2 \theta_{(\text{exp.})}$, the terms in the normal equations are $\sum \alpha v$, and $\sum \delta v$ instead of $\sum \alpha \sin^2 \theta$ and $\sum \delta \sin^2 \theta$, and ΔA instead of A.

Other forms of Cohen's equations. All the equations involving θ can be recast in terms of Φ. The important equation (92) then becomes

$$\left(\frac{n\lambda}{2d}\right)^2 + D \sin^2 \Phi + E \sin^2 \frac{\Phi}{2} \sin \Phi = \cos^2 \frac{\Phi}{2}. \tag{92'}$$

LITERATURE

Fundamental early publications

[1] ASSAR HADDING.   Über Störeinger der Linienabstände und der Linienbreite bei Debyediagrammen.   *Centr. Mineral. Geol.*, **1921,** 631–636.

[2] O. PAULI.   Die Debye-Scherrer-Methode zur Untersuchung von Kristallstrukturen.   *Z. Krist.* (*A*), **56** (1922), 604–607.

Error elimination by calibration

[3] H. OTT.   Präzisionsmessungen einiger Alkalihalogenide.   *Z. Krist.* (*A*), **63** (1926), 228–230.

Error elimination by the Straumanis method

[4] M. STRAUMANIS and O. MELLIS.   Präzisionsaufnahmen nach dem Verfahren von Debye und Scherrer.   *Z. Physik,* **94** (1935), 184–191.

[5] M. STRAUMANIS and A. IEVIŅŠ.   Präzisionsbestimmung von Glanzwinkeln und Gitterkonstanten nach der Methode von Debye und Scherrer.   *Naturwiss.,* **23** (1935), 833.

[6] M. STRAUMANIS and A. IEVIŅŠ.   Präzisionsaufnahmen nach dem Verfahren von Debye und Scherrer.   II.   *Z. Physik,* **98** (1936), 461–475.

[7] A. IEVIŅŠ and M. STRAUMANIS.   Die Gitterkonstante des reinsten Aluminiums.   *Z. physik. Chem.* (*B*), **33** (1936), 265–274; also **34** (1936), 402–403.

[8] M. STRAUMANIS and A. IEVIŅŠ.   Die Gitterkonstanten des NaCl und des Steinsalzes.   *Z. Physik,* **102** (1936), 353–359.

[9] M. Straumanis and E. Ence. Über das System Zn[Hg(CNS)₄]-Cu[Hg(CNS)₄].
Z. anorg. allegem. Chem., **228** (1936), 338–339.

[10] M. Straumanis and E. Mankovičs. Die Löslichkeit von Cu-Ionen im
Cd[Hg(CNS)₄]. *Z. anorg. allgem. Chem.*, **233** (1937), 201–208.

[11] A. Ieviņš and M. Straumanis. Experimentelle oder rechnerische Fehlerelimi-
nation bei Debye-Scherrer-Aufnahmen. *Z. Krist.* (A), **94** (1936) 40–52.

[12] Ieviņš and M. Straumanis. Bemerkung zur Arbeit von M. U. Cohen. "The
elimination of systematic errors in powder photographs." *Z. Krist.* (A), **95**
(1936), 451–454.

[13] A. Ieviņš, M. Straumanis, and K. Karlsons. Präzisionsbestimmung von Gitterkonstanten hygroskopischer Verbindungen (LiCl, NaBr). *Z. physik. Chem.*
(B), **40** (1938), 146–150.

[14] A. Ieviņš, M. Straumanis, and K. Karlsons. Die Präzisionsbestimmung von
Gitterkonstanten nichtkubischer Stoffe (Bi, Mg, Sn) nach der asymmetrischen
Methode. *Z. physik. Chem.* (B), **40** (1938), 347–356.

[15] M. Straumanis and A. Ieviņš. Die Bestimmung von Ausdehnungskoeffizien-
ten nach der Pulver- und der Drehkristall-Methode. *Z. anorg. allgem. Chem.*,
238 (1938), 175–188.

[16] M. Straumanis and A. Ieviņš. Die Drehkristallmethode als Präzisionsverfahren
und deren Vergleich mit der Pulvermethode. *Z. Physik*, **109** (1938), 728–743.

[17] M. Straumanis, A. Ieviņš, and K. Karlsons. Hängt die Gitterkonstante von
der Wellenlänge ab? *Z. physik. Chem.* (B), **42** (1939), 143–152.

[18] A. Ieviņš and K. Karlsons. Der Einfluss des Kameradurchmessers und der
Blendenform auf die Grösse der Gitterkonstante, bestimmt nach der asymmetrischen Methode. *Z. Physik*, **112** (1939), 350–351.

Error elimination by systematic allowance
[19] F. Regler. Neue Eichungsmethode zur Präzisionsbestimmung von Gitterkonstanten an polykristallinen Materialien. *Physik. Z.*, **32** (1931), 680–687.

[20] N. H. Kolkmeijer and A. L. Th. Moesveld. Präzisionsbestimmung der
Dimensionen von Kristallgittern. *Z. Krist.* (A), **80** (1931), 63–90.

[21] N. H. Kolkmeijer and A. L. Th. Moesveld. Über die Reglersche Eichungsmethode zur Präzisionsbestimmung von Gitterdimensionen und die unsrige.
Physik. Z., **33** (1932), 265–269.

[22] F. Lihl. Der Einfluss der Divergenz, der Präparatdicke und der Eindringtiefe
auf die Präzisionsbestimmung von Gitterdimensionen nach der Methode von
Debye und Scherrer. *Z. Krist.* (A), **83** (1932), 193–221.

[23] Franz Lihl. Der Einfluss der Divergenz der primären Röntgenstrahlung auf
der Auswertung von Diagrammen nach Rückstrahlverfahren. *Ann. Physik*,
(5) **19** (1934), 305–334.

Error elimination by utilizing high angle reflections
[24] Ulrich Dehlinger. Über die Verbreiterung der Debyelinien bei kaltbear-
beiteten Metallen. *Z. Krist.* (A), **65** (1927), 615–616.

[25] A. E. van Arkel. Eine einfache Methode zur Erhöhung der Genauigkeit bei
Debye-Scherrer-Aufnahmen. *Z. Krist.* (A), **67** (1928), 235–238.

[26] Frhr. von Göler and G. Sachs. Die Veredelung einer Aluminiumlegierung im
Röntgenbild. *Metallwirtschaft Wissenschaft und Technik*, **8** (July 12, 1929),
671–680.

[27] G. Sachs and J. Weerts. Die Gitterkonstanten der Gold-Silberlegierungen.
Z. Physik, **60** (1930), 481–490.

[28] H. Braekken. Präzisionsbestimmung von Gitterkonstanten nach der Drehkristallmethode. *Det Kongelige Norske Videnskabers Selskab. Forhandlinger*, I, Nr. 64, 1929.

Error elimination by graphical extrapolation
[29] Gustav Kettmann. Beiträge zur Auswertung von Debye-Scherrer-Aufnahmen. *Z. Physik*, 53 (1929), 198–209.
[30] A. J. Bradley and A. H. Jay. A method for deducing accurate values of the lattice spacing from x-ray powder photographs taken by the Debye-Scherrer method. *Proc. Phys. Soc.*, 44 (1932), 563–579.

Error elimination by analytical extrapolation
[31] M. U. Cohen. Precision lattice constants from x-ray powder photographs. *Rev. Sci. Instruments*, 6 (1935), 68–74.
[32] M. U. Cohen. Errata; Precision lattice constants from x-ray powder photographs. *Rev. Sci. Instruments*, 7 (1936), 155.
[33] Eric R. Jette and Frank Foote. Precision determination of lattice constants. *J. Chem. Phys.*, 3 (1935), 605–616.
[34] E. A. Owen and Llewelyn Pickup. The lattice constants of beryllium. *Phil. Mag.*, (7) 20 (1935), 1155–1158.
[35] George H. Walden and M. U. Cohen. An x-ray investigation of the solid solution nature of some nitrate-contaminated barium sulfate precipitates. *J. Am. Chem. Soc.*, 57 (1935), 2591–2597.
[36] M. U. Cohen. The elimination of systematic errors in powder photographs. *Z. Krist. (A)*, 94 (1936), 288–298.
[37] M. U. Cohen. The calculation of precise lattice constants from x-ray powder photographs. *Z. Krist. (A)*, 94 (1936), 306–310.

Error elimination by extrapolation, applied to single crystals
[38] M. J. Buerger and William Parrish. The unit cell and space group of tourmaline (an example of the inspective equi-inclination treatment of trigonal crystals). *Am. Mineral.*, 22 (1937), 1146–1148.

CHAPTER 21

THE PRECISION DETERMINATION OF THE LINEAR AND ANGULAR LATTICE CONSTANTS OF SINGLE CRYSTALS

EXTENSION OF PRECISION DETERMINATIONS TO NON-ISOMETRIC CRYSTALS

Crystallographic weakness of the powder method. The powder method is well adapted to precision determination of the lattice constants of isometric crystals for two reasons: each line can be unequivocally indexed when monochromatic radiation is used, and each line has a position which is fundamentally dependent only on its index and on the single unknown lattice parameter, a. When an attempt is made to extend the method to non-isometric crystals, grave difficulties are encountered. In the first place, unequivocal indexing is, in general, impossible even with monochromatic radiation. The indexing becomes increasingly uncertain as the very region is approached where the lines are of greatest value for precision work, namely in the high-θ region. Since the only reflections having their positions fixed solely by the spacing whose value is precisely sought, say the (100) spacing for sake of illustration, are the class $h00$, only the position of this class of reflections should be included as data in a straightforward precision determination of the (100) spacing. It is this very class of reflections which is least well resolved, for $(h0l)$, $(hl0)$, etc., reflect to the same immediate region when h is large, as it is in the region of $\theta \rightarrow 90°$. This lack of resolution is especially important in crystals with large axes.

In order to increase the resolution of reflections for indexing purposes, it is obvious that the data must be derived by some other method than the powder method, and this requires the use of single crystals. The rotation method offers some relief in this respect, but it is only partial, for lack of resolving power is inherent in this method also. What is obviously needed for the precision determination of single crystal lattice constants is one of the moving-film methods, in which all reflections are resolved.

It will be pointed out in the sequel that not only can great precision in interplanar spacing determinations be attained by comparatively simple straightforward means with a moving-film method, but also that great precision can be attained in the determination of the angles between crystallographic axes (or any other crystallographic angles) as well.   In other words, all the linear and angular lattice constants of a crystal may be precisely determined regardless of the complexity of the crystal.

Requirements of an apparatus for the precision determination of single-crystal lattice spacings.   From the discussion of precision methods in the last chapter, it is evident that any precision method must be based upon a symmetrical back-reflection type of recording, i.e., the x-ray beam must enter through the center of the film.   The symmetrical aspect does away with calibrated edges, and the back-reflection aspect permits taking advantage of the fact that the film-shrinkage errors are approximately linear with eccentricity errors.

The Weissenberg method is not obviously adaptable to the conditions set forth in the first paragraph of this section because of the physical difficulty of introducing the pinhole system through the center of the moving film.   If the ordinary long Weissenberg translation is retained, the slot in the film, necessary to permit entrance of the pin-hole system, would have to be so long that the film could not be expected to behave as a single integral sheet when removed from the camera.   Fortunately, however, as more fully demonstrated below by actual examples, only sufficient Weissenberg motion is needed to effect a detectable resolution of diffraction spots.   If the motion is thus restricted, only a short slot need be made in the film.   A great deal of additional film may be retained on each side of the slot to give strength to the film sheet and insure that it behaves as a unit.

In actual practice, as demonstrated below, this solution to the problem has been found to be entirely satisfactory.   The cylindrical Weissenberg camera with the crystal at its center permits not only sharp focusing but also a very accurate determination of the camera diameter.

THE INSTRUMENT

General features.   The instrument (Figs. 221 and 222) designed for single-crystal precision work is, briefly, a symmetrical, back-reflection, Weissenberg-type arrangement, utilizing a cylindrical camera.   The general features of the instrument are similar to those already described for the equi-inclination Weissenberg instrument, except that, since only equator photographs are of any value in the present problem, the inclination axis is omitted.   For the sake of eliminating all possible instrumental errors, care has been taken to attain ruggedness in the apparatus.   The entire aligning and centering of the mechanism and camera are built into the instrument by indicating and doweling. Only the adjustment of the pinhole system axis to intersection with the crystal rotation axis need be made after the instrument is built.

FIG. 221. Precision, back-reflection, Weissenberg instrument, in operating position.

FIG. 222. Precision, back-reflection, Weissenberg instrument, with camera and layer line screen removed.

Some features of interest in the instrument are the following: the entire electrical drive system, including motor, reversing switch, start-stop switch, and plug connection, are mounted on a small, easily detachable, individual insulating base. If any trouble develops in the electrical circuit, this unit may be instantly removed and all connections and motor repairs conveniently made without disturbing the adjustment of the main instrument to the x-ray beam or otherwise disturbing the exposure being made; the exposure may be completed when the electrical system is repaired and returned. This is important, because sometimes exposures are long and require several different radiations. The insulating base just mentioned, together with insulated mechanical coupling of the motor to the mechanism and to the switch trip which reverses the motion at the end of the translation, effectively insulates the electrical drive system from the metallic part of the instrument. This prevents motor burn-outs in the event of a static discharge arising from temporarily erratic x-ray-tube operation.

In order to keep the length of the slot in the film at a reasonably small size, the total Weissenberg translation motion is reduced to about an inch (with additional clearance for slight excess motion). The coupling constant, C_2, is 8° crystal rotation per millimeter film translation. This is four times the coupling constant of the equi-inclination instrument; it permits an exploration over a 180° rotation (the maximum ever necessary) with a translation of only $22\frac{1}{2}$ mm., or with a motion of less than an inch. Although this coupling constant would be much too large to be convenient for space-group determination, which is the chief purpose of the equi-inclination instrument, it gives ample resolution of diffraction spots for the present purpose and for the technique described below.

The upper base, bearing the Weissenberg motion, is capable of translation in and out of the x-ray beam so that the crystal may be centered by sighting directly through the pinhole system. The upper base may also be tilted to adjust the pinhole axis to the x-ray beam.

To facilitate accurate final centering of the crystal after the adjusting crystal holder has been transferred from the optical goniometer to its place on the rotation spindle, a lens is clipped over the x-ray-tube end of the pinhole system. This lens has a focal length equal to its distance from the crystal axis, and, with its aid, the eye can easily and conveniently frame the crystal in the center of the circular pinhole aperture. A 10-diopter lens is about right for this purpose. The pinholes are drilled in lead disks. The limiting aperture in the system is made with a 75 drill (0.021 inch diameter), and is located exactly at the intersection of the pinhole axis with the film surface, to satisfy the focusing condition. Small, well-defined diffraction spots result with this design (see Figs. 225, 226, and 227).

FIG. 223. Precision, back-reflection, Weissenberg camera.

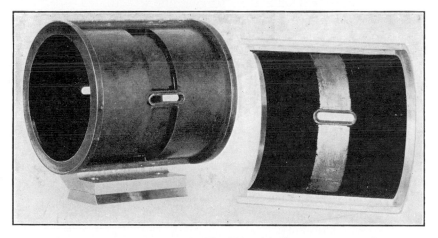

FIG. 224. Camera shown in Fig. 223, with film cover removed.

The camera. The success of the precision instrument herein described depends to a large extent upon the camera design (Figs. 223 and 224). This consists of a complete cylinder turned from a bronze casting; flanges are retained on the edges to increase resistance to distortion.

The cylinder seats on the translation carriage in such a way that it recovers the same setting each time. This is accomplished by means of a dovetail base seat, from which a central longitudinal clearance is cut, so that the camera actually rests on two long narrow surfaces. The dovetail is clamped home to the carriage by means of an eccentric dovetail clamp (Figs. 221 and 222).

The film is wrapped about the outside of the cylinder, and almost the entire surface, except the central section to be exposed, is held accurately against metal by a somewhat yielding velvet lining, shellacked to the inside of the hemicylindrical aluminum film cover (Fig. 224, right). The cover fits the camera cylinder very neatly and is easily manipulated in the darkroom. The fit is so good that no clamping device is necessary. By means of a special punch, an accurate, smooth slot is cut in the unexposed film before loading (Figs. 225, 226, and 227). This fits accurately about a metal rim surrounding the pinhole-system slot in the camera cylinder. The film is protected from direct light exposure in the slot region by means of an accurately fitting cover for this rim (the central dark rectangle in Fig. 223) which is fitted to the aluminum film-cover. The direct beam leaves the camera through a rear slot. An outer lead sheet over the aluminum film-cover protects the film from fogging due to scattered x-radiation. The film receives crystal reflections through a central zone milled out of the camera cylinder, Fig. 224, but is protected from light exposure by means of a black photographic paper shellacked to the inside of the opening. When making runs with highly scattered radiation (for example, Cu radiation with an iron-bearing crystal), a layer of 0.001 inch aluminum foil is also held across this opening with the aid of Scotch drafting tape.

The important dimensions of the camera are as follows: the diameter of the camera, which may be accurately determined, since the entire cylinder is retained, is 114.592 mm. film center to film center (1 mm. arc $= 1°$ for $2\theta = \frac{1}{2}°$ for θ) and about $4\frac{13}{16}$ inches long between flanges. This accommodates a standard 5-by-7-inch x-ray film. The 7-inch length would ordinarily cover 178° of arc, but, with the commercial film tolerance allowance, it actually covers only about 175°.

The central open exposed zone is $25\frac{1}{2}$ mm. wide. With a coupling constant $C_2 = 8$, this corresponds with a crystal rotation of 204°, which amply covers the necessary 180° range with an overlap and clearance area of 12° on each side. The circumferential extent of the exposed area of the film covers 161 mm. of arc, but has a blind area $13\frac{1}{2}$ mm. double width in the region of the pinhole slot. These dimensions give the reflection range the following characteristics:

	Lower limit	Upper limit
ϒ arc, measured from beam exit:	$99\frac{1}{2}°$	$173\frac{1}{4}°$
Reflection range $\begin{cases} \theta = \\ \sin\theta = \\ \xi = \\ \cos^2\theta = \\ \left(\dfrac{\pi}{2} - \theta\right)\cos\theta \end{cases}$	$49\frac{3}{4}°$	$86\frac{5}{8}°$
	0.763	0.99825
	1.526	1.9965
	0.418	0.00347
	0.595	0.00392

INTERPRETATION OF THE FILM

Three films taken with this instrument are shown in Figs. 225, 226, and 227. These photographs realize the portion of the Weissenberg diagram in which all values of ϒ are positive. The ovaloid forms of the reciprocal lattice line curves on such photographs have already been discussed in Chapter 14.

For the purpose of identifying reflections on these photographs, a back reflection chart of the general Wooster variety is very convenient. A chart of this type, accurately reproduced to correct scale for the instrument described, is given in Fig. 228. Data for labeling and locating its graduations are given in Table 24. The problem of interpreting photographs like Figs. 225, 226, and 227 is quite different from the interpretation of ordinary equi-inclination Weissenberg diagrams for space-group purposes, as will be demonstrated beyond.

The orders of reflection of a crystal plane appear on these photographs along straight lines slanting at an angle of

$$\text{Я} = \tan^{-1}\left(\frac{\Upsilon/C_1}{\omega/C_2}\right) = \tan^{-1}\left(2\frac{C_2}{C_1}\right)$$

from the horizontal, where C_1 is the camera constant and C_2 is the coupling constant of the instrument [see (15), Chapter 14]. For the constants given for this instrument, this slope is $\tan^{-1}16 = 86° \, 25'$. The orders of reflection occur parallel with the sloping coordinates of Fig. 228. These are so labeled that a crystal plane, originally parallel with the x-ray beam at the start of the motion, reflects along the sloping line labeled zero. For a photograph with a 180° crystal-rotation range, as in the instrument described, such a line of reflections crosses the center horizontal line of the film, $\xi = 2.00$, in the center of the film. Crystal planes which come to parallelism with the x-ray beam after a rotation of the crystal spindle of $+\bar{\varphi}°$ have a series of orders of reflections along the slanting line labeled $+\bar{\varphi}$ in Fig. 228. In this chart an

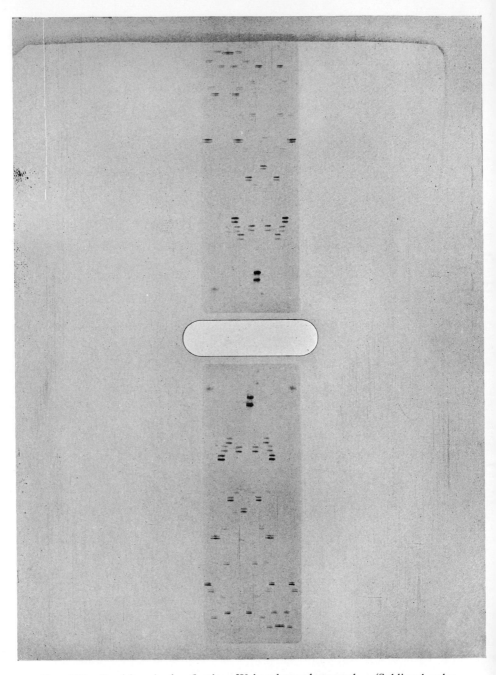

FIG. 225. Precision, back-reflection, Weissenberg photograph. (Sublimed valentinite, Sb_2O_3, orthorhombic; c-axis equator; unfiltered ZnK + CuK radiation from gas x-ray tube.)

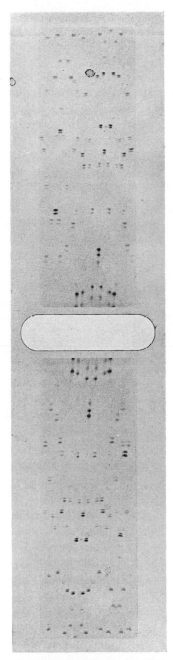

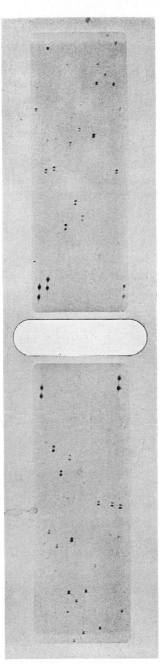

Fig. 226. Back-reflection Weissenberg photograph. (Gypsum, CaSO$_4$ · 2H$_2$O, monoclinic; c-axis equator; CuK + NiK radiation.)

Fig. 227. Back-reflection Weissenberg photograph. (Gypsum, b-axis equator; ZnK + CuK radiation.)

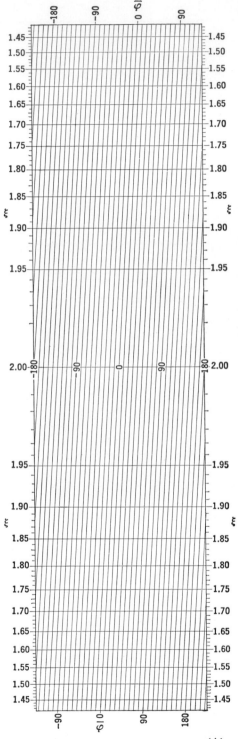

additional $\bar{\varphi}$ range of 90° is given on each side of the actual ideal photograph range of 180°.

Photographs taken by this back-reflection instrument display the plane point-group symmetry customarily shown by Weissenberg photographs (see next chapter). If a photograph displays two or more lines of symmetry (an equatorial photograph, such as taken with this instrument, cannot have only one line of symmetry) its interpretation is relatively simple, for the only reflections whose measurements will lead to lengths of crystal axes lie along the symmetry lines. Of these, the symmetry line intersecting the center of the film contains the reflections of the plane whose initial position was parallel with the x-ray beam. Examples of symmetrical photographs of this kind are Figs. 225 and 226. Figure 227 shows a photograph lacking symmetry lines. In such instances, the series of orders of the reflection desired must be identified from a knowledge of the location of the plane whose spacing interval is desired. An important case is that of a photograph taken for the b-axis rotation of a monoclinic crystal (or for the a-, b-, or c-axis rotation of a

Fig. 228. Wooster-type chart for interpretation of precision, back-reflection, Weissenberg photographs. (Camera diameter, 114.59 mm.)

444

TABLE 24

DATA FOR THE LOCATION AND LABELING OF THE ξ GRADUATIONS
OF BACK-REFLECTION WEISSENBERG CHART

The second column gives the distance in millimeters from the base line to the center for a chart to be used with a camera of diameter 114.592 mm. (1° arc = 1 mm. on film). For any other camera diameter, multiply the values in this column by the factor $\dfrac{\text{Camera diameter in mm.}}{114.592}$.

ξ label	$(\Upsilon - 90°)$	ξ label	$(\Upsilon - 90°)$	ξ label	$(\Upsilon - 90°)$
1.40	−1.146	1.60	16.260	1.80	38.316
1.41	−0.340	1.61	17.220	1.81	39.646
1.42	0.470	1.62	18.192	1.82	41.010
1.43	1.286	1.63	19.174	1.83	42.412
1.44	2.108	1.64	20.170	1.84	43.852
1.45	2.938	1.65	21.176	1.85	45.336
1.46	3.772	1.66	22.198	1.86	46.870
1.47	4.614	1.67	23.232	1.87	48.456
1.48	5.462	1.68	24.280	1.88	50.104
1.49	6.318	1.69	25.344	1.89	51.818
1.50	7.180	1.70	26.424	1.90	53.610
1.51	8.050	1.71	27.520	1.91	55.492
1.52	8.928	1.72	28.634	1.92	57.480
1.53	9.814	1.73	29.766	1.93	59.592
1.54	10.708	1.74	30.918	1.94	61.860
1.55	11.610	1.75	32.090	1.95	64.322
1.56	12.522	1.76	33.284	1.96	67.044
1.57	13.442	1.77	34.504	1.97	70.128
1.58	14.372	1.78	35.746	1.98	73.780
1.59	15.310	1.79	37.016	1.99	78.536
				2.00	90.000

triclinic crystal). The important series of reflections here (see below) are those from (001), (100), and ($\bar{1}$01). If the instrument is started with (001) initially parallel with the beam, and the rotation brings these three planes parallel with the beam in the order given, then the several series of orders of reflections from these planes will appear on the film with the following coordinates:

Series of reflections from	$\bar{\varphi}$ Coordinate
(001)	0°
(100)	(001) $\wedge$ (100) $= \beta*$
($\bar{1}$01)	(001) $\wedge$ ($\bar{1}$01)

CALCULATION OF SPACING VALUES FROM FILM MEASUREMENTS

Transforming measurement to spacings. The calculation of inter-planar spacings from film measurements is made by determining the equivalent of θ in the Bragg relation

$$n\lambda = 2d \sin \theta. \tag{1}$$

The actual measurement made is the vertical component of the distance between equivalent spots on each side of the center line, $\xi = 2.00$ (see

FIG. 229.   Device for the precision determination of distances, distance components, and spacings on x-ray photographs.   The figure shows the back-reflection Weissenberg film of Fig. 225 clipped in this measuring device.

Figs. 225, 226, 227, and 228).   This film distance, F, double the distance of a spot from the center line of the photograph, is measured with the device shown in Fig. 229.   In this arrangement, the film is clipped to the adjusted position on a glass surface, behind which is the ordinary illuminated opal glass background.   The double distance, F, is measured by means of a sliding frame, whose glass window bears a fine vertical scratch on its lower surface next to the film.   The position of a spot on the film may be determined directly to 0.05 mm. by means of a millimeter scale and vernier at the lower part of the measuring device.

The relation of the film measurement, F, to the Bragg glancing angle, θ, is as follows: Any angle, A, and its corresponding film arc, F, are proportioned, thus:

$$\frac{A}{360°} = \frac{F}{\pi D}, \tag{2}$$

where D is the camera diameter. For a measured F, then, the angle A is

$$A = \left[\frac{360°}{\pi D}\right] F, \tag{3}$$

where the term in brackets is the instrumental constant, C_1 (Chapter 12). The particular angle, in this case, is 2Φ. Consequently,

$$\Phi = \frac{A}{2} = \left[\frac{360°}{\pi D}\right]\frac{F}{2}. \tag{4}$$

The total deviation angle, 2θ, and the angle Φ are supplementary, hence

$$180° - 2\theta = \Phi$$

and

$$90° - \theta = \frac{\Phi}{2}. \tag{5}$$

In terms of film measurements, according to (4), this is equal to

$$90° - \theta = \left[\frac{360°}{\pi D}\right]\frac{F}{4}. \tag{6}$$

This can now be substituted into (1), rearranged to give the spacing value

$$d = \frac{\frac{n}{2}\lambda}{\sin \theta}$$

$$= \frac{\frac{n}{2}\lambda}{\cos (90° - \theta)} \tag{7}$$

$$= \frac{\frac{n}{2}\lambda}{\cos \left(\left[\frac{360°}{\pi D}\right]\frac{F}{4}\right)}. \tag{8}$$

TABLE 25

Radiation	n						
	1	2	3	4	5	6	7
Ti α_2	0.1377986	0.4388286	0.6149199	0.7398586	0.8367686	0.9159499	0.9828966
Ti α_1	0.1372227	0.4382527	0.6143440	0.7392827	0.8361927	0.9153740	0.9823207
Ti β_1	0.0984707	0.3995007	0.5755920	0.7005307	0.7974407	0.8766220	0.9435687
V α_2	0.0972799	0.3983099	0.5744012	0.6993399	0.7962499	0.8754312	0.9423779
V α_1	0.0966233	0.3976533	0.5737446	0.6986833	0.7955933	0.8747746	0.9417213
V β_1	0.0568477	0.3578777	0.5339690	0.6589077	0.7558177	0.8349990	0.9019457
Cr α_2	0.0585987	0.3596287	0.5357200	0.6606587	0.7575687	0.8367500	0.9036967
Cr α_1	0.0578619	0.3588919	0.5349832	0.6599219	0.7568319	0.8360132	0.9029599
Cr β_1	0.0171586	0.3181886	0.4942799	0.6192186	0.7161286	0.7953099	0.8622566
Mn α_2	0.0214973	0.3225273	0.4986186	0.6235573	0.7204673	0.7996486	0.8665953
Mn α_1	0.0206741	0.3217041	0.4977954	0.6227341	0.7196441	0.7988254	0.8657721
Mn β_1	9.9791385	0.2801685	0.4562598	0.5811985	0.6781085	0.7572898	0.8242365
Fe α_2	9.9858781	0.2869081	0.4629994	0.5879381	0.6848481	0.7640294	0.8309761
Fe α_1	9.9849942	0.2860242	0.4621155	0.5870542	0.6839642	0.7631455	0.8300922
Fe β_1	9.9427551	0.2437851	0.4198764	0.5448151	0.6417251	0.7209064	0.7878531
Co α_2	9.9516265	0.2526565	0.4287478	0.5536865	0.6505965	0.7297778	0.7967245
Co α_1	9.9506788	0.2517088	0.4278001	0.5527388	0.6496488	0.7288301	0.7957768
Co β_1	9.9077982	0.2088282	0.3849195	0.5098582	0.6067682	0.6859495	0.7528962
Ni α_2	9.9186462	0.2196762	0.3957675	0.5207062	0.6176162	0.6967975	0.7637442
Ni α_1	9.9176368	0.2186668	0.3947581	0.5196968	0.6166068	0.6957881	0.7627348
Ni β_1	9.8742063	0.1752363	0.3513276	0.4762663	0.5731763	0.6523576	0.7193043
Cu α_2	9.8868380	0.1878680	0.3639593	0.4888980	0.5858080	0.6649893	0.7319360
Cu α_1	9.8857555	0.1867855	0.3628768	0.4878155	0.5847255	0.6639068	0.7308535
Cu β_1	9.8417817	0.1428117	0.3189030	0.4438417	0.5407517	0.6199330	0.6868797
Zn α_2	9.8561335	0.1571635	0.3332548	0.4581935	0.5551035	0.6342848	0.7012315
Zn α_1	9.8549646	0.1559946	0.3320859	0.4570246	0.5539346	0.6331159	0.7000626
Zn β_1	9.8104174	0.1114474	0.2875387	0.4124774	0.5093874	0.5885687	0.6555154
Zr α_2	9.5957772	9.8968072	0.0728985	0.1978372	0.2947472	0.3739285	0.4408752
Zr α_1	9.5934522	9.8944822	0.0705735	0.1955122	0.2924222	0.3716035	0.4385502
Zr β_1	9.5442417	9.8452717	0.0213630	0.1463017	0.2432117	0.3223930	0.3893397
Nb α_2	9.5733880	9.8744180	0.0505093	0.1754480	0.2723580	0.3515393	0.4184860
Nb α_1	9.5709222	9.8719522	0.0480435	0.1729822	0.2698922	0.3490735	0.4160202
Nb β_1	9.5213866	9.8224166	9.9985079	0.1234466	0.2203566	0.2995379	0.3664846
Mo α_2	9.5519408	9.8529708	0.0290621	0.1540008	0.2509108	0.3300921	0.3970388
Mo α_1	9.5488996	9.8499296	0.0260209	0.1509596	0.2478696	0.3270509	0.3939976
Mo β_1	9.4989842	9.8000142	9.9761055	0.1010442	0.1979542	0.2771355	0.3440822
	1	2	3	4	5	6	7
Radiation	n						

$$\text{Log}\left(\frac{n}{2}\lambda\right).$$

			n				
8	9	10	11	12	13	14	15
1.0408886	1.0920411	1.1377986	1.1791913	1.2169799	1.2517420	1.2839266	1.3138449
1.0403127	1.0914652	1.1372227	1.1786154	1.2164040	1.2511661	1.2833507	1.3132690
1.0015607	1.0527132	1.0984707	1.1398634	1.1776520	1.2124141	1.2445987	1.2745170
1.0003699	1.0515224	1.0972799	1.1386726	1.1764612	1.2112233	1.2434079	1.2733262
0.9997133	1.0508658	1.0966233	1.1380160	1.1758046	1.2105667	1.2427513	1.2726696
0.9599377	1.0110902	1.0568477	1.0982404	1.1360290	1.1707911	1.2029757	1.2328940
0.9616887	1.0128412	1.0585987	1.0999914	1.1377800	1.1725421	1.2047267	1.2346450
0.9609519	1.0121044	1.0578619	1.0992546	1.1370432	1.1718053	1.2039899	1.2339082
0.9202486	0.9714011	1.0171586	1.0585513	1.0963399	1.1311020	1.1632866	1.1932049
0.9245873	0.9757398	1.0214973	1.0628900	1.1006786	1.1354407	1.1676253	1.1975436
0.9237641	0.9749166	1.0206741	1.0620668	1.0998554	1.1346175	1.1668021	1.1967204
0.8822285	0.9333810	0.9791385	1.0205312	1.0583198	1.0930819	1.1252665	1.1551848
0.8889681	0.9401206	0.9858781	1.0272708	1.0650594	1.0998215	1.1320061	1.1619244
0.8880842	0.9392367	0.9849942	1.0263869	1.0641755	1.0989376	1.1311222	1.1610405
0.8458451	0.8969976	0.9427551	0.9841478	1.0219364	1.0566985	1.0888831	1.1188014
0.8547165	0.9058690	0.9516265	0.9930192	1.0308078	1.0655699	1.0977545	1.1276728
0.8537688	0.9049213	0.9506788	0.9920715	1.0298601	1.0646222	1.0968068	1.1267251
0.8108882	0.8620407	0.9077982	0.9491909	0.9869795	1.0217416	1.0539262	1.0838445
0.8217362	0.8728887	0.9186162	0.9600389	0.9978275	1.0325896	1.0647742	1.0946925
0.8207268	0.8718793	0.9176368	0.9590295	0.9968181	1.0315802	1.0637648	1.0936831
0.7772963	0.8284488	0.8742063	0.9155990	0.9533876	0.9881497	1.0203343	1.0502526
0.7899280	0.8410805	0.8868380	0.9282307	0.9660193	1.0007814	1.0329660	1.0628843
0.7888455	0.8399980	0.8857555	0.9271482	0.9649368	0.9996989	1.0318835	1.0618018
0.7448717	0.7960242	0.8417817	0.8831744	0.9209630	0.9557251	0.9879097	1.0178280
0.7592235	0.8103760	0.8561335	0.8975262	0.9353148	0.9700769	1.0022615	1.0321798
0.7580546	0.8092071	0.8549646	0.8963573	0.9341459	0.9689080	1.0010926	1.0310109
0.7135074	0.7646599	0.8104174	0.8518101	0.8895987	0.9243608	0.9565454	0.9864637
0.4988672	0.5500197	0.5957772	0.6371699	0.6749585	0.7097206	0.7419052	0.7718235
0.4965422	0.5476947	0.5934522	0.6348449	0.6726335	0.7073956	0.7395802	0.7694985
0.4473317	0.4984842	0.5442417	0.5856344	0.6234230	0.6581851	0.6903697	0.7202880
0.4764780	0.5276305	0.5733880	0.6147807	0.6525693	0.6873314	0.7195160	0.7494343
0.4740122	0.5251647	0.5709222	0.6123149	0.6501035	0.6848656	0.7170502	0.7469685
0.4244766	0.4756291	0.5213866	0.5627793	0.6005679	0.6353300	0.6675146	0.6974329
0.4550308	0.5061833	0.5519408	0.5933335	0.6311221	0.6658842	0.6980688	0.7279871
0.4519896	0.5031421	0.5488996	0.5902923	0.6280809	0.6628430	0.6950276	0.7249459
0.4020742	0.4532267	0.4989842	0.5403769	0.5781655	0.6129276	0.6451122	0.6750305
8	9	10	11	12	13	14	15

n

TABLE 25 — *Continued*

Radiation	n						
	16	17	18	19	20	21	22
Ti α_2	1.3419186	1.3682475	1.3930711	1.4165522	1.4388286	1.4600179	1.4802213
Ti α_1	1.3413427	1.3676716	1.3924952	1.4159763	1.4382527	1.4594420	1.4796454
Ti β_1	1.3025907	1.3289196	1.3537432	1.3772243	1.3995007	1.4206900	1.4408934
V α_2	1.3013999	1.3277288	1.3525524	1.3760335	1.3983099	1.4194992	1.4397026
V α_1	1.3007433	1.3270722	1.3518958	1.3753769	1.3976533	1.4188426	1.4390460
V β_1	1.2609677	1.2872966	1.3121202	1.3356013	1.3578777	1.3790670	1.3992704
Cr α_2	1.2627187	1.2890476	1.3138712	1.3373523	1.3596287	1.3808180	1.4010214
Cr α_1	1.2619819	1.2883108	1.3131344	1.3366155	1.3588919	1.3800812	1.4002846
Cr β_1	1.2212786	1.2476075	1.2724311	1.2959122	1.3181886	1.3393779	1.3595813
Mn α_2	1.2256173	1.2519462	1.2767698	1.3002509	1.3225273	1.3437166	1.3639200
Mn α_1	1.2247941	1.2511230	1.2759466	1.2994277	1.3217041	1.3428934	1.3630968
Mn β_1	1.1832585	1.2095874	1.2344110	1.2578921	1.2801685	1.3013578	1.3215612
Fe α_2	1.1899981	1.2163270	1.2411506	1.2646317	1.2869081	1.3080974	1.3283008
Fe α_1	1.1891142	1.2154431	1.2402667	1.2637478	1.2860242	1.3072135	1.3274169
Fe β_1	1.1468751	1.1732040	1.1980276	1.2215087	1.2437851	1.2649744	1.2851778
Co α_2	1.1557465	1.1820754	1.2068990	1.2303801	1.2526565	1.2738458	1.2940492
Co α_1	1.1547988	1.1811277	1.2059513	1.2294324	1.2517088	1.2728981	1.2931015
Co β_1	1.1119182	1.1382471	1.1630707	1.1865518	1.2088282	1.2300175	1.2502209
Ni α_2	1.1227662	1.1490951	1.1739187	1.1973998	1.2196762	1.2408655	1.2610689
Ni α_1	1.1217568	1.1480857	1.1729093	1.1963904	1.2186668	1.2398561	1.2600595
Ni β_1	1.0783263	1.1046552	1.1294788	1.1529599	1.1752363	1.1964256	1.2166290
Cu α_2	1.0909580	1.1172869	1.1421105	1.1655916	1.1878680	1.2090573	1.2292607
Cu α_1	1.0898755	1.1162044	1.1410280	1.1645091	1.1867855	1.2079748	1.2281782
Cu β_1	1.0459017	1.0722306	1.0970542	1.1205353	1.1428117	1.1640010	1.1842044
Zn α_2	1.0602535	1.0865824	1.1114060	1.1348871	1.1571635	1.1783528	1.1985562
Zn α_1	1.0590846	1.0854135	1.1102371	1.1337182	1.1559946	1.1771839	1.1973873
Zn β_1	1.0145374	1.0408663	1.0656899	1.0891710	1.1114474	1.1326367	1.1528401
Zr α_2	0.7998972	0.8262261	0.8510497	0.8745308	0.8968072	0.9179965	0.9381999
Zr α_1	0.7975722	0.8239011	0.8487247	0.8722058	0.8944822	0.9156715	0.9358749
Zr β_1	0.7483617	0.7746906	0.7995142	0.8229953	0.8452717	0.8664610	0.8866644
Nb α_2	0.7775080	0.8038369	0.8286605	0.8521416	0.8744180	0.8956073	0.9158107
Nb α_1	0.7750422	0.8013711	0.8261947	0.8496758	0.8719522	0.8931415	0.9133449
Nb β_1	0.7255066	0.7518355	0.7766591	0.8001402	0.8224166	0.8436059	0.8638093
Mo α_2	0.7560608	0.7823897	0.8072133	0.8306944	0.8529708	0.8741601	0.8943635
Mo α_1	0.7530196	0.7793485	0.8041721	0.8276532	0.8499296	0.8711189	0.8913223
Mo β_1	0.7031042	0.7294331	0.7542567	0.7777378	0.8000142	0.8212035	0.8414069
	16	17	18	19	20	21	22
Radiation	n						

$$\text{Log}\left(\frac{n}{2}\lambda\right).$$

				n			
23	24	25	26	27	28	29	30
1.4995264	1.5180098	1.5357386	1.5527720	1.5691624	1.5849566	1.6001966	1.6149199
1.4989505	1.5174339	1.5351627	1.5521961	1.5685865	1.5843807	1.5996207	1.6143440
1.4601985	1.4786819	1.4964107	1.5134441	1.5298345	1.5456287	1.5608687	1.5755920
1.4590077	1.4774911	1.4952199	1.5122533	1.5286437	1.5444379	1.5596779	1.5744012
1.4583511	1.4768345	1.4945633	1.5115967	1.5279871	1.5437813	1.5590213	1.5737446
1.4185755	1.4370589	1.4547877	1.4718211	1.4882115	1.5040057	1.5192457	1.5339690
1.4203265	1.4388099	1.4565387	1.4735721	1.4899625	1.5057567	1.5209967	1.5357200
1.4195897	1.4380731	1.4558019	1.4728353	1.4892257	1.5050199	1.5202599	1.5349832
1.3788864	1.3973698	1.4150986	1.4321320	1.4485224	1.4643166	1.4795566	1.4942799
1.3832251	1.4017085	1.4194373	1.4364707	1.4528611	1.4686553	1.4838953	1.4986187
1.3824019	1.4008853	1.4186141	1.4356475	1.4520379	1.4678321	1.4830721	1.4977954
1.3408663	1.3593497	1.3770785	1.3941119	1.4105023	1.4262965	1.4415365	1.4562598
1.3476059	1.3660893	1.3838181	1.4008515	1.4172419	1.4330361	1.4482761	1.4629994
1.3467220	1.3652054	1.3829342	1.3999676	1.4163580	1.4321522	1.4473922	1.4621155
1.3044829	1.3229663	1.3406951	1.3577285	1.3741189	1.3899131	1.4051531	1.4198764
1.3133543	1.3318377	1.3495665	1.3665999	1.3829903	1.3987845	1.4140245	1.4287478
1.3124066	1.3308900	1.3486188	1.3656522	1.3820426	1.3978368	1.4130768	1.4278001
1.2695260	1.2880094	1.3057382	1.3227716	1.3391620	1.3549562	1.3701962	1.3849195
1.2803740	1.2988574	1.3165862	1.3336196	1.3500100	1.3658042	1.3810442	1.3957675
1.2793646	1.2978480	1.3155768	1.3326102	1.3490006	1.3647948	1.3800348	1.3947581
1.2359341	1.2544175	1.2721463	1.2891797	1.3055701	1.3213643	1.3366043	1.3513276
1.2485658	1.2670492	1.2847780	1.3018114	1.3182018	1.3339960	1.3492360	1.3639593
1.2474833	1.2659667	1.2836955	1.3007289	1.3171193	1.3329135	1.3481535	1.3628768
1.2035095	1.2219929	1.2397217	1.2567551	1.2731455	1.2889397	1.3041797	1.3189030
1.2178613	1.2363447	1.2540735	1.2711069	1.2874973	1.3032915	1.3185315	1.3332548
1.2166924	1.2351758	1.2529046	1.2699380	1.2863284	1.3021226	1.3173626	1.3320859
1.1721452	1.1906286	1.2083574	1.2253908	1.2417812	1.2575754	1.2728154	1.2875387
0.9575050	0.9759884	0.9937172	1.0107506	1.0271410	1.0429352	1.0581752	1.0728985
0.9551800	0.9736634	0.9913922	1.0084256	1.0248160	1.0406102	1.0558502	1.0705735
0.9059695	0.9244529	0.9421817	0.9592151	0.9756055	0.9913997	1.0066397	1.0213630
0.9351158	0.9535992	0.9713280	0.9883614	1.0047518	1.0205460	1.0357860	1.0505093
0.9326500	0.9511334	0.9688622	0.9858956	1.0022860	1.0180802	1.0333202	1.0480435
0.8831144	0.9015978	0.9193266	0.9363600	0.9527504	0.9685446	0.9837846	0.9985079
0.9136686	0.9321520	0.9498808	0.9669142	0.9833046	0.9990988	1.0143388	1.0290621
0.9106274	0.9291108	0.9468396	0.9638730	0.9802634	0.9960576	1.0112976	1.0260209
0.8607120	0.8791954	0.8969242	0.9139576	0.9303480	0.9461422	0.9613822	0.9761055
23	24	25	26	27	28	29	30

n

The term in brackets is an instrumental constant. For the dimensions of the instrument described, this is unity, and (8) reduces to

$$d = \frac{\dfrac{n}{2}\lambda}{\cos\left(\dfrac{F}{4}\right)}. \tag{8'}$$

The data obtained from film measurements may be extrapolated by means of Bradley and Jay's graphical method against $\cos^2\theta$ or by Cohen's analytical method.

The values of d for various parts of the film may be calculated either by means of a computing machine or with the aid of logarithms. In the latter case, seven-place logarithms are necessary.† The linear measurements of F in equation (8) or (8') is made in millimeter units and decimals. The angle $(90° - \theta)$ of equation (7) thus automatically appears in the form of degrees and decimals. Direct seven-place trigonometric functions of angles in this form may be found in Peters' tables.‡ This form of trigonometric table is necessary for computations made by computing machine. For computations with the aid of logarithms, the lack of seven-place logarithm tables of angles expressed in degrees and decimals necessitates a conversion of the decimal form of $(90° - \theta)$ to degrees, minutes, and seconds. The logarithm may then be looked up in von Vega's tables. It is about as easy to leave the angle in its original decimal form, however, look up its cosine in Peters' decimal tables, and then find the logarithm of this number in von Vega's tables, and there is a somewhat smaller chance of making an error in this sequence.

It should be observed that, in (8) and (8'), n can have only the few particular values of certain small integers (the integers from 1 to about 30 for inorganic crystals having spacings of the magnitudes already recorded), and λ can have only those few values of wavelength which are of practical use with present-day x-ray diffraction technique (the K radiations of the elements from about Mo to Ti). It follows that the numerator of (8) and (8') can have only a limited number of definite values, and that therefore all the partial solutions of this

† Von Vega's tables are recommended: Logarithmic tables of numbers and trigonometrical functions (translated by W. L. Fischer), 85th stereotyped edition (D. Van Nostrand Company, New York).

‡ J. Peters, Siebenstellige Werte der trigonometrischen Funktionen von Tausendstel zu Tausendstel des Grades. (Verlag der Optischen Anstalt C. P. Goerz Akt.-Ges., Berlin-Friedenau, 1918.) (Obtainable in the United States from G. B. Stechert Co., 33 East 10 St., New York, N. Y.)

part of the spacing calculations can be made once and for all, thus saving needless labor.   Table 25 gives the logarithms of all possible values of the numerator, $\frac{n}{2} \lambda$, for all values of n and λ ever likely to be used.   The wavelengths of the K radiations of the elements between Zr and Zn (namely, Y, Sr, Rb, Kr, Br, Se, As, Ge, and Ga) have been omitted from this table because, although the wavelengths are of appropriate magnitude, the elements themselves form poor x-ray-target material for physical reasons.   In the preparation of Table 25, Siegbahn's wavelength values have been used, all initial logarithm

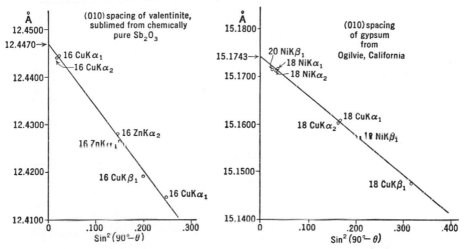

FIG. 230.   Interpretation of x-ray photograph shown in Fig. 225.

FIG. 231.   Interpretation of x-ray photograph shown in Fig. 226.

interpolations have been checked by different men, and all additions have been performed by means of a recording adding machine, whose record has been checked.   The values given in the table are therefore believed to be reasonably accurate.

Table 25 has been found to be a very important aid in decreasing the labor of computation.   It eliminates two of the five searches in a book of seven-place tables and eliminates one of the four interpolations, per spacing computation.

In Figs. 230 and 231, the graphical determinations of the precise lattice constants of two crystals are illustrated.   It will be observed that a precision of about six significant figures is attainable.   The precision is limited by the precision of x-ray wavelengths, and by the fact that greater precision requires accurate temperature control, since the lattice spacings vary some $0.0000x$ Å per °C.

Remarks on the application of Cohen's method. No special remarks regarding the application of Cohen's method to single crystal investigations are necessary provided that: (1) one uses only one wavelength for a single film, and (2) one uses only the orders of reflection from a single plane. If one oversteps either of these conditions, however, then the following comments apply:

1. It is quite advantageous, as discussed below, to use several wavelengths for the same film, in order to locate reflections strategically (see Figs. 230 and 231). When more than one wavelength is used, it is desirable to redistribute the terms in the definitions of A and α given by (94), Chapter 20. They were purposely distributed as given, in order that the α's would be simple integers and thus facilitate the computations in the determinants of (97) Chapter 20. This throws the highly precise λ value into the A, which need be computed only once. When the wavelength is variable, however, it cannot be bulked with the constant d, whose solution is desired. In this case, λ must be placed with the variable n, and A and α redefined:

$$\left(\frac{1}{d}\right)^2 = A',$$

$$\left(\frac{n\lambda}{2}\right)^2 = \alpha'.$$

This makes the computation of both (97) and (110) Chapter 20, more difficult.

2. In accordance with the scheme outline in the last chapter (page 430), it is usual to compute lattice constants from powder photographs by using reflections involving indices h, k, and l. With a single crystal, reflections involving all indices lying in an equatorial plane are available for measurement. The commonest case of this is to rotate the crystal about a crystallographic axis, say c, and thus have in the zero layer the indices corresponding with the remaining axes, in this example, $hk0$. This permits the use of all indices in the simultaneous determination of the a- and b-axes. The danger in using *all* reflections, however, is that, because of the crystal habit, the effective diameter of the crystal may be greater between a pair of planes (h_1k_10) and $(\bar{h}_1\bar{k}_10)$ than between another pair of planes (h_2k_20) and $(\bar{h}_2\bar{k}_20)$. To take an extreme example, suppose that the crystal has extreme absorption and gives rise to only surface reflections; then the radius, r, to the reflecting plane surface for substitution in the absorption error coefficient, (65), Chapter 20, is different

for the order of reflection from $(h_1 k_1 0)$ and for the orders of reflection from $(h_2 k_2 0)$. For this reason, it is safe to use Cohen's method for orders of reflection from the same plane, but dangerous to use it for reflections from different planes if the absorption is high, and if the crystal shape departs appreciably from cylindrical.

PLANNING A RUN

In order to distribute points to the best advantage on a graphical solution like Fig. 230 or 231, it is necessary to plan the run carefully in advance. Indeed, failure to take this precaution may result in the appearance of no points at all within the recording range of the apparatus.

In order to plan a run, the value of the spacing to be determined precisely must be approximately known in advance. Since, for the zero layer line,

$$\xi = 2 \sin \theta, \tag{9}$$

Bragg's law,

$$n\lambda = 2d \sin \theta, \tag{1}$$

can be conveniently expressed

$$n\lambda = d\xi. \tag{10}$$

The approximate value of d is known, and the ξ range which can be recorded by the apparatus is also known. The possible range of $n\lambda$ values which will give reflections within the range of the apparatus may therefore be determined. (In this connection account must be taken of the absence of certain orders due to space-group extinctions. In many cases the odd orders of reflection are not available for measurement purposes because of the presence of lattice centering, a glide plane, or even-fold screw axis.) The appropriate wavelengths which will give rise to recordable reflections are accordingly determined. Since, as already explained, only a limited number of definite $n\lambda$ values is possible, an $n\lambda$ table may be prepared which will show all available values at a glance. Such a table is of the greatest value in planning the run. Table 26 gives the required data for the precision required in the present connection. It is compiled for use with a 20-inch slide rule, so that values less than 5 are given to four significant figures, and values greater than 5 are given to three significant figures. All the values listed are adding-machine sums and are believed to be accurate.

In selecting wavelengths, precaution should be taken not to include various wavelengths for which the atoms of the crystal have highly

TABLE 26

$n\lambda$

Radiation	1	2	3	4	5	6	7	8	9	10	11	12	13	14	15
								n							
Ti α_2	2.747	5.49	8.24	10.99	13.73	16.48	19.23	21.97	24.72	27.47	30.21	32.96	35.71	38.46	41.20
Ti α_1	2.743	5.49	8.23	10.97	13.72	16.46	19.20	21.95	24.69	27.43	30.18	32.92	35.66	38.40	41.15
Ti β_1	2.509	5.02	7.53	10.04	12.55	15.05	17.56	20.07	22.58	25.09	27.60	30.11	32.62	35.13	37.64
V α_2	2.502	5.00	7.51	10.01	12.51	15.01	17.51	20.02	22.52	25.02	27.52	30.03	32.53	35.03	37.53
V α_1	2.498	4.997	7.50	9.99	12.49	14.99	17.49	19.99	22.49	24.98	27.48	29.98	32.48	34.98	37.48
Cr α_2	2.289	4.578	6.87	9.16	11.44	13.73	16.02	18.31	20.60	22.89	25.18	27.47	29.76	32.04	34.33
Cr α_1	2.285	4.570	6.86	9.14	11.43	13.71	16.00	18.28	20.57	22.85	25.14	27.42	29.71	31.99	34.28
V β_1	2.280	4.559	6.84	9.12	11.40	13.68	15.96	18.24	20.52	22.80	25.08	27.36	29.64	31.92	34.20
Mn α_2	2.102	4.203	6.30	8.41	10.51	12.61	14.71	16.81	18.91	21.02	23.12	25.22	27.32	29.42	31.52
Mn α_1	2.098	4.195	6.29	8.39	10.49	12.59	14.68	16.78	18.88	20.98	23.07	25.17	27.27	29.37	31.46
Cr β_1	2.081	4.161	6.24	8.32	10.40	12.48	14.56	16.64	18.73	20.81	22.89	24.97	27.05	29.13	31.21
Fe α_2	1.936	3.872	5.81	7.74	9.68	11.62	13.55	15.49	17.42	19.36	21.30	23.23	25.17	27.10	29.04
Fe α_1	1.932	3.864	5.80	7.73	9.66	11.59	13.52	15.46	17.39	19.32	21.25	23.19	25.12	27.05	28.98
Mn β_1	1.906	3.812	5.72	7.62	9.53	11.44	13.34	15.25	17.16	19.06	20.97	22.87	24.78	26.69	28.59
Co α_2	1.789	3.578	5.37	7.16	8.95	10.74	12.52	14.31	16.10	17.89	19.68	21.47	23.26	25.05	26.84
Co α_1	1.785	3.571	5.36	7.14	8.93	10.71	12.50	14.28	16.07	17.85	19.64	21.42	23.21	24.99	26.78
Fe β_1	1.753	3.506	5.26	7.01	8.77	10.52	12.27	14.02	15.78	17.53	19.28	21.04	22.79	24.54	26.30
Ni α_2	1.658	3.317	4.975	6.63	8.29	9.95	11.61	13.27	14.93	16.58	18.24	19.90	21.56	23.22	24.88
Ni α_1	1.655	3.309	4.964	6.62	8.27	9.93	11.58	13.24	14.89	16.55	18.20	19.85	21.51	23.16	24.82
Co β_1	1.617	3.235	4.852	6.47	8.09	9.70	11.32	12.94	14.56	16.17	17.79	19.41	21.03	22.64	24.26
Cu α_2	1.541	3.082	4.624	6.16	7.71	9.25	10.79	12.33	13.87	15.41	16.95	18.49	20.04	21.58	23.12
Cu α_1	1.537	3.075	4.612	6.15	7.69	9.22	10.76	12.30	13.84	15.37	16.91	18.45	19.99	21.52	23.06
Ni β_1	1.497	2.994	4.491	5.99	7.49	8.98	10.48	11.98	13.47	14.97	16.47	17.97	19.46	20.96	22.46
Zn α_2	1.436	2.872	4.308	5.74	7.18	8.62	10.05	11.49	12.92	14.36	15.80	17.23	18.67	20.10	21.54
Zn α_1	1.432	2.864	4.297	5.73	7.16	8.59	10.03	11.46	12.89	14.32	15.75	17.19	18.62	20.05	21.48
Cu β_1	1.389	2.779	4.168	5.56	6.95	8.34	9.73	11.12	12.50	13.89	15.28	16.67	18.06	19.45	20.84
Zn β_1	1.293	2.585	3.878	5.17	6.46	7.76	9.05	10.34	11.63	12.93	14.22	15.51	16.80	18.10	19.39
Zr α_2	.789	1.577	2.366	3.154	3.943	4.731	5.52	6.31	7.10	7.89	8.67	9.46	10.25	11.04	11.83
Zr α_1	.784	1.569	2.353	3.137	3.922	4.706	5.49	6.27	7.06	7.84	8.63	9.41	10.20	10.98	11.76
Nb α_2	.749	1.498	2.247	2.996	3.745	4.493	5.24	5.99	6.74	7.49	8.24	8.99	9.74	10.48	11.23
Nb α_1	.745	1.489	2.234	2.979	3.724	4.468	5.21	5.96	6.70	7.45	8.19	8.94	9.68	10.43	11.17
Mo α_2	.713	1.426	2.138	2.851	3.564	4.277	4.990	5.70	6.42	7.13	7.84	8.55	9.27	9.98	10.69
Mo α_1	.708	1.416	2.123	2.831	3.539	4.247	4.955	5.66	6.37	7.08	7.79	8.49	9.20	9.91	10.62
Zr β_1	.700	1.401	2.101	2.801	3.502	4.202	4.902	5.60	6.30	7.00	7.70	8.40	9.10	9.80	10.50
Nb β_1	.664	1.329	1.993	2.658	3.322	3.986	4.651	5.32	5.98	6.64	7.31	7.97	8.64	9.30	9.97
Mo β_1	.631	1.262	1.893	2.524	3.155	3.786	4.417	5.05	5.68	6.31	6.94	7.57	8.20	8.83	9.47
Radiation	1	2	3	4	5	6	7	8	9	10	11	12	13	14	15
								n							

TABLE 26 — *Continued*

n

Radiation	16	17	18	19	20	21	22	23	24	25	26	27	28	29	30
Ti α_2	43.95	46.70	49.44	52.2	54.9	57.7	60.4	63.2	65.9	68.7	71.4	74.2	76.9	79.7	82.4
Ti α_1	43.89	46.63	49.38	52.1	54.9	57.6	60.4	63.1	65.8	68.6	71.3	74.1	76.8	79.6	82.3
Ti β_1	40.14	42.65	45.16	47.67	50.2	52.7	55.2	57.7	60.2	62.7	65.2	67.7	70.3	72.8	75.3
V α_2	40.03	42.54	45.04	47.54	50.0	52.5	55.0	57.5	60.1	62.6	65.1	67.6	70.1	72.6	75.1
V α_1	39.97	42.47	44.97	47.47	49.97	52.5	55.0	57.5	60.0	62.5	65.0	67.5	70.0	72.5	75.0
Cr α_2	36.62	38.91	41.20	43.49	45.78	48.07	50.4	52.6	54.9	57.2	59.5	61.8	64.1	66.4	68.7
Cr α_1	36.56	38.85	41.13	43.42	45.70	47.99	50.3	52.6	54.8	57.1	59.4	61.7	64.0	66.3	68.6
V β_1	36.48	38.75	41.03	43.31	45.59	47.87	50.2	52.4	54.7	57.0	59.3	61.6	63.8	66.1	68.4
Mn α_2	33.62	35.73	37.83	39.93	42.03	44.13	46.23	48.33	50.4	52.5	54.6	56.7	58.8	60.9	63.0
Mn α_1	33.56	35.66	37.76	39.85	41.95	44.05	46.15	48.24	50.3	52.4	54.5	56.6	58.7	60.8	62.9
Cr β_1	33.29	35.37	37.45	39.53	41.61	43.69	45.77	47.85	49.93	52.0	54.1	56.2	58.3	60.3	62.4
Fe α_2	30.98	32.91	34.85	36.78	38.72	40.66	42.59	44.53	46.46	48.40	50.3	52.3	54.2	56.1	58.1
Fe α_1	30.91	32.85	34.78	36.71	38.64	40.57	42.51	44.44	46.37	48.30	50.2	52.2	54.1	56.0	58.0
Mn β_1	30.50	32.41	34.31	36.22	38.12	40.03	41.94	43.84	45.75	47.66	49.56	51.5	53.4	55.3	57.2
Co α_2	28.63	30.42	32.21	33.99	35.71	37.57	39.36	41.15	42.94	44.73	46.52	48.31	50.1	51.9	53.7
Co α_1	28.56	30.35	32.14	33.92	35.64	37.49	39.28	41.06	42.85	44.63	46.42	48.20	49.99	51.8	53.6
Fe β_1	28.05	29.80	31.55	33.31	35.05	36.81	38.57	40.32	42.07	43.83	45.58	47.33	49.08	50.8	52.6
Ni α_2	26.53	28.19	29.85	31.51	33.17	34.83	36.48	38.14	39.80	41.46	43.12	44.78	46.44	48.09	49.75
Ni α_1	26.47	28.13	29.78	31.44	33.09	34.74	36.40	38.05	39.71	41.36	43.02	44.67	46.33	47.98	49.64
Co β_1	25.88	27.50	29.11	30.73	32.35	33.97	35.58	37.20	38.82	40.44	42.05	43.67	45.29	46.90	48.52
Cu α_2	24.66	26.20	27.74	29.28	30.82	32.37	33.91	35.45	36.99	38.53	40.07	41.61	43.15	44.69	46.24
Cu α_1	24.60	26.14	27.67	29.21	30.75	32.29	33.82	35.36	36.90	38.44	39.97	41.51	43.05	44.58	46.12
Ni β_1	23.95	25.45	26.95	28.44	29.94	31.44	32.94	34.43	35.93	37.43	38.92	40.42	41.92	43.42	44.91
Zn α_2	22.98	24.41	25.85	27.28	28.72	30.16	31.59	33.03	34.46	35.90	37.34	38.77	40.21	41.64	43.08
Zn α_1	22.92	24.35	25.78	27.21	28.64	30.08	31.51	32.94	34.37	35.81	37.24	38.67	40.10	41.53	42.97
Cu β_1	22.23	23.62	25.01	26.40	27.79	29.18	30.57	31.96	33.35	34.74	36.12	37.51	38.90	40.29	41.68
Zn β_1	20.68	21.97	23.27	24.56	25.85	27.14	28.44	29.73	31.02	32.32	33.61	34.90	36.19	37.49	38.78
Zr α_2	12.62	13.40	14.19	14.98	15.77	16.56	17.35	18.14	18.92	19.71	20.50	21.29	22.08	22.87	23.66
Zr α_1	12.55	13.33	14.12	14.90	15.69	16.47	17.25	18.04	18.82	19.61	20.39	21.18	21.96	22.74	23.53
Nb α_2	11.98	12.73	13.48	14.23	14.98	15.73	16.48	17.22	17.97	18.72	19.47	20.22	20.97	21.72	22.47
Nb α_1	11.92	12.66	13.40	14.15	14.89	15.64	16.38	17.13	17.87	18.62	19.36	20.11	20.85	21.60	22.34
Mo α_2	11.40	12.12	12.83	13.54	14.26	14.97	15.68	16.39	17.11	17.82	18.53	19.25	19.96	20.67	21.38
Mo α_1	11.32	12.03	12.74	13.45	14.16	14.86	15.57	16.28	16.99	17.70	18.40	19.11	19.82	20.53	21.23
Zr β_1	11.20	11.91	12.61	13.31	14.01	14.71	15.41	16.11	16.81	17.51	18.21	18.91	19.61	20.31	21.01
Nb β_1	10.63	11.29	11.96	12.52	13.29	13.95	14.62	15.28	15.95	16.61	17.27	17.94	18.60	19.27	19.93
Mo β_1	10.10	10.73	11.36	11.99	12.62	13.25	13.88	14.51	15.14	15.78	16.41	17.04	17.67	18.30	18.93
Radiation	16	17	18	19	20	21	22	23	24	25	26	27	28	29	30

n

different absorptions.   If such radiations are included, the absorption error will change erratically, instead of gradually with θ.

When the possible orders and wavelengths have been selected, it is necessary to calculate the ξ value of each combination to identify it later on the film.   This is done by rearranging (10):

$$\xi = \frac{nl}{d}. \tag{11}$$

This entire computation is best arranged in a table.   An example is given in Table 27, in which radiations of wavelength shorter than Zn $K\alpha_1$ have been neglected.   The right-hand column of Table 27 gives the ξ coordinates of all possible reflections which can arise with the crystal spacing in question.   The middle column shows the corre-

TABLE 27

DETERMINATION OF APPROPRIATE RADIATIONS AND LOCATION OF
REFLECTIONS IN PRECISION DETERMINATION OF b-AXIS OF
LÖLLINGITE (NEW ORIENTATION)

$d \approx 2.85\text{Å}$

$$n\lambda \approx d\xi \begin{bmatrix} 1.9965 \\ \\ 1.526 \end{bmatrix} \approx 4.35 \text{ to } 5.69$$

only even values of n available

$n\lambda$ values (derived from Table 26) between recordable limits and for even values of n	Order of reflection and radiation (derived from Table 26)	Expected approximate value of ξ $\left(= \dfrac{n\lambda}{d} \right)$
5.56	4Cu β_1	1.951
5.49	2Ti α_2	1.926
5.49	2Ti α_1	1.926
5.17	4Zn β_1	1.814
5.02	2Ti β_1	1.762
5.00	2V α_2	1.755
4.997	2V α_1	1.753
4.578	2Cr α_2	1.606
4.570	2Cr α_1	1.603
4.559	2V β_1	1.599

sponding radiation giving rise to each possible reflection. After the appropriate radiations for the run have been selected, and the run made, the radiation and order of each reflection can be identified by superposing the film on the chart of Fig. 228 and comparing the ξ coordinate of each reflection of the series with the expected approximate coordinates of possible reflections listed in Table 27.

It may be pointed out that the number of radiations which can give rise to appropriate reflections for small interplanar spacings is quite limited; for large spacings, the choice is less restricted. This is partly because the $n\lambda$ range itself is directly proportional to d, as indicated by (10), and partly because, with larger values of $n\lambda$, new, greater values of n become available, each with an appropriate value of λ.

The production of x-radiation of appropriate wavelength to give a sufficient number of reflections for a desired precision may be accomplished in several ways:

1. By the use of short wavelengths such as Mo K, Nb K and Zr K. Each of these gives a relatively large number of orders because of its short wavelength. The disadvantages of using these radiations are that:

(a) The scattering power falls off directly as $\dfrac{\sin \theta}{\lambda}$, so that these short wavelengths require very long exposures. Certain crystals with complicated structures give no reflections in the correct range with these radiations no matter how long the exposure.

(b) These radiations are relatively penetrating and hard to shield, especially with long exposures.

(c) The radiations are not very clean.

(d) Nb and Zr are at present difficult to obtain in appropriate form for target metals.

2. By the use of alloy targets. This method has the disadvantage that the appropriate alloy to use changes with each new interatomic spacing, according as Table 27 indicates different strategic wavelength combinations. About the only way out of this difficulty is to use one alloy containing all the nine metals from Zn through Ti.

3. By changing targets. Since the instrument described gives rise to focused reflections whose positions are determined by the limiting aperture of the pinhole system and are independent of the position of the focal spot on the x-ray target, this procedure is a permissible one. The number of changes per run can be somewhat reduced by the use of alloy targets. The changes, however, are

annoying, since they require letting down the vacuum, cleaning and regreasing the target seal, re-evacuating, etc. The changing of targets is so important in precision technique that a special target design is desirable for x-ray tubes (Fig. 232) used for precision determinations such that rotation of the target in its vacuum seal brings different target materials to face the cathode-ray beam, thus giving rise to different radiations. This permits the change of radiations during the progress of the run without any trouble whatever.

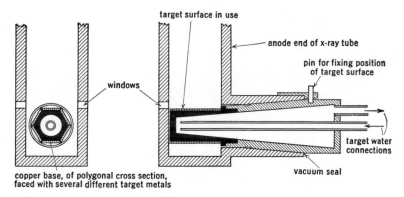

FIG. 232

THE PRECISION DETERMINATION OF INTERPLANAR ANGLES AND INTERLINEAR ANGLES

General theory and method.   It was pointed out in Chapter 19 that a method of triangulation may be used for the determination of interfacial angles. This method is ideally adapted to the determination of all crystallographic angles with a precision equal to the precision of determining interplanar spacings, already discussed.

Given any two planes, $(h_1k_1l_1)$ and $(h_2k_2l_2)$, of any crystal; these planes determine the zone $[(h_1k_1l_1)(h_2k_2l_2)]$. The plane $(h_2-h_1; k_2-k_1; l_2-l_1)$ is also in this zone.† The spacings of these three planes may be precisely determined with the aid of the instrument and methods discussed, from measurements taken from an equatorial photograph with the zone $[(h_1k_1l_1)(h_2k_2l_2)]$ as a rotation axis. A reciprocal lattice with proportionality constant unity may now be imagined, Fig. 233. In this reciprocal lattice, the reciprocals of the spacings $d_{(h_1k_1l_1)}$, $d_{(h_2k_2l_2)}$, and $d_{(h_2-h_1; k_2-k_1; l_2-l_1)}$ are vectors from the origin to the points $[[h_1k_1l_1]]$, $[[h_2k_2l_2]]$, and $[[h_2-h_1; k_2-k_1; l_2-l_1]]$.

† Austin F. Rogers, The addition and subtraction rule in geometrical crystallography. *Am. Mineral.*, **11** (1926), 303–315.

The latter vector is also the vector connecting the points $[[h_1k_1l_1]]$ and $[[h_2k_2l_2]]$. The reciprocals of the three spacing measurements, $d_{(h_1k_1l_1)}$, $d_{(h_2k_2l_2)}$, and $d_{(h_2-h_1;\, k_2-k_1;\, l_2-l_1)}$, thus form a triangle whose angle at the origin is the interfacial angle $(h_1k_1l_1) \wedge (h_2k_2l_2)$. The three sides of this triangle have been precisely determined; therefore the

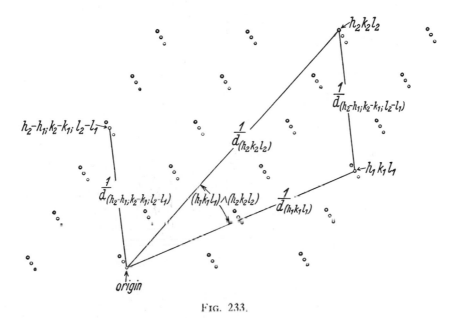

Fig. 233.

interfacial angle may be precisely computed. The following relations are appropriate for precision computations:

$$(h_1k_1l_1) \wedge (h_2k_2l_2) =$$
$$2\cos^{-1}\sqrt{\frac{\Sigma\frac{1}{d}}{2}\left(\frac{\Sigma\frac{1}{d}}{2} - \frac{1}{d_{(h_2-h_1;\, k_2-k_1;\, l_2-l_1)}}\right)d_{(h_1k_1l_1)} \cdot d_{(h_2k_2l_2)}} \quad (12)$$

$$(h_1k_1l_1) \wedge (h_2k_2l_2) =$$
$$2\sin^{-1}\sqrt{\left(\frac{\Sigma\frac{1}{d}}{2} - \frac{1}{d_{(h_1k_1l_1)}}\right)\left(\frac{\Sigma\frac{1}{d}}{2} - \frac{1}{d_{(h_2k_2l_2)}}\right)d_{(h_1k_1l_1)} \cdot d_{(h_2k_2l_2)}}. \quad (12')$$

Application to monoclinic cells. The determination of the axial ratio and the crystallographic angle, β, of monoclinic crystals provides a very important application of the general case discussed above.

Figure 234A shows a cell of a monoclinic reciprocal lattice, having proportionality constant unity, projected on (010)*. The three vectors, $\dfrac{1}{d_{(100)}}$, $\dfrac{1}{d_{(\bar{1}01)}}$, and $\dfrac{1}{d_{(001)}}$ form a triangle having the supplement of the crystallographic angle β at the origin. This may be computed from precision spacing determination of $d_{(100)}$, $d_{(\bar{1}01)}$, and

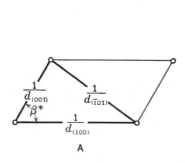

FIG. 234A. Cell of monoclinic reciprocal lattice.

FIG. 234B. Cell of corresponding monoclinic direct lattice.

$d_{(001)}$, derived from the b-axis zero-layer photograph, with the aid of the relation:

$$\beta^* = 2\cos^{-1}\sqrt{\frac{\Sigma\dfrac{1}{d}}{2}\left(\frac{\Sigma\dfrac{1}{d}}{2} - \frac{1}{d_{(\bar{1}01)}}\right)}\,d_{(100)} \cdot d_{(001)}. \qquad (13)$$

When β^* has been precisely determined in this fashion, the direct axes of the monoclinic crystal may be computed with the aid of the relations (see Fig. 234B):

$$a = \frac{d_{(100)}}{\sin \beta^*}. \qquad (14)$$

$$b = d_{(010)}. \qquad (15)$$

$$c = \frac{d_{(001)}}{\sin \beta^*}. \qquad (16)$$

$$\beta = 180° - \beta^*. \qquad (17)$$

Application to triclinic cells.   For triclinic crystals, the three reciprocal cell angles, α^*, β^*, and γ^*, are computable from spacing measurements with the aid of the following relations analogous to (13):

$$\alpha^* = 2\cos^{-1}\sqrt{\dfrac{\Sigma\dfrac{1}{d}}{2}\left(\dfrac{\Sigma\dfrac{1}{d}}{2} - \dfrac{1}{d_{(01\bar{1})}}\right)}\, d_{(001)} \cdot d_{(010)} \qquad (18)$$

$$\beta^* = 2\cos^{-1}\sqrt{\dfrac{\Sigma\dfrac{1}{d}}{2}\left(\dfrac{\Sigma\dfrac{1}{d}}{2} - \dfrac{1}{d_{(\bar{1}01)}}\right)}\, d_{(100)} \cdot d_{(001)} \qquad (19)$$

$$\gamma^* = 2\cos^{-1}\sqrt{\dfrac{\Sigma\dfrac{1}{d}}{2}\left(\dfrac{\Sigma\dfrac{1}{d}}{2} - \dfrac{1}{d_{(1\bar{1}0)}}\right)}\, d_{(010)} \cdot d_{(100)}. \qquad (20)$$

The standard crystallographic interaxial angles, α, β, and γ, may be calculated from the reciprocal interaxial angles, α^*, β^*, and γ^*, (18), (19), and (20), with the aid of the relations

$$\alpha = 2\sin^{-1}\sqrt{\dfrac{\sin\left(\dfrac{\alpha^* + \beta^* + \gamma^*}{2}\right)\sin\left(\dfrac{\alpha^* + \beta^* + \gamma^*}{2} - \alpha^*\right)}{\sin\beta^*\,\sin\gamma^*}} \qquad (21)$$

$$\beta = 2\sin^{-1}\sqrt{\dfrac{\sin\left(\dfrac{\alpha^* + \beta^* + \gamma^*}{2}\right)\sin\left(\dfrac{\alpha^* + \beta^* + \gamma^*}{2} - \beta^*\right)}{\sin\alpha^*\,\sin\gamma^*}} \qquad (22)$$

$$\gamma = 2\sin^{-1}\sqrt{\dfrac{\sin\left(\dfrac{\alpha^* + \beta^* + \gamma^*}{2}\right)\sin\left(\dfrac{\alpha^* + \beta^* + \gamma^*}{2} - \gamma^*\right)}{\sin\beta^*\,\sin\alpha^*}}. \qquad (23)$$

The direct cell axes are related to the reciprocal cell axes and angles by relations of the type

$$a = \dfrac{b^*c^*\sin\alpha^*}{V^*}. \qquad (24)$$

For the present purposes, V^* may be expanded as follows:

$$V^* = a^*b^*c^*\sin\alpha^*\sin\beta^*\sin\gamma \qquad (25)$$

$$= a^*b^*c^*\sin\alpha^*\sin\beta\sin\gamma^* \qquad (26)$$

$$= a^*b^*c^*\sin\alpha\sin\beta^*\sin\gamma^*. \qquad (27)$$

By combining (24) with (25), (26), and (27), the several linear axes of the triclinic unit may therefore be calculated from spacing measurements as follows:

$$a = \frac{d_{(100)}}{\sin \beta^* \sin \gamma} = \frac{d_{(100)}}{\sin \beta \sin \gamma^*}. \tag{28}$$

$$b = \frac{d_{(010)}}{\sin \alpha^* \sin \gamma} = \frac{d_{(010)}}{\sin \alpha \sin \gamma^*}. \tag{29}$$

$$c = \frac{d_{(001)}}{\sin \alpha^* \sin \beta} = \frac{d_{(001)}}{\sin \alpha \sin \beta^*}. \tag{30}$$

DISCUSSION

In this chapter an instrument has been described and an account given of a thoroughly general technique for the direct precision determination of all the geometrical features of crystal lattices.   These include the general geometrical features:

1. The interplanar spacing, $d_{(hkl)}$, of any index, (hkl).
2. The interfacial angle, $(h_1k_1l_1) \wedge (h_2k_2l_2)$, for any indices $(h_1k_1l_1)$ and $(h_2k_2l_2)$.

The technique also leads to the determination of the following special geometrical quantities, important in systematic crystallography and crystallographic description:

3. The linear and angular lattice constants a^*, b^*, c^*, α^*, β^*, γ^* of the reciprocal lattice.
4. The linear and angular lattice constants, a, b, c, α, β, γ of the direct crystal lattice.

A precision of about six significant figures is possible and may be assured by the appropriate selection of wavelengths to give many, well-located, reflection data.   This permits computation of crystallographic angles with an accuracy of a few seconds of arc.   This accuracy is independent of minor crystal imperfections of the type which prevent the attainment of greater accuracy than a few minutes of arc in the ordinary optical goniometry.   Not only is the accuracy obtainable by the x-ray methods spectacular compared with that ordinarily achieved in optical goniometry of surface morphology, but also the results are much more reliable than those obtained by the older methods.   This is because accuracy in surface morphological studies depends on identity of surface angles with true lattice angles.   Crystal imperfection disturbs this identity and renders the meaning of surface measurements uncertain.   Accuracy in the x-ray method, on the other hand, depends on the measurements of interplanar spacings, and these are independent of crystal imperfections.

The instrument described is ideally suited to further development in the direction of crystal temperature control. The Weissenberg layer line screen affords an excellent opportunity for arranging abnormal temperatures near the crystal while maintaining normal temperatures in the region of the film.

LITERATURE

M. J. Buerger. The precision determination of the linear and angular lattice constants of single crystals. *Z. Krist. (A)*, **97** (1937), 433–468.

CHAPTER 22

THE THEORY AND INTERPRETATION OF RECIPROCAL LATTICE PROJECTIONS

INTRODUCTION

The reciprocal lattice embodies two important properties: it partakes of the point-group symmetry of the crystal, and it is intimately related to the translation-group of the crystal lattice. The reciprocal space lattice may be considered as composed of stacks of plane lattices. In several of the rotation methods employing monochromatic x-rays, these reciprocal lattice sheets can be studied only in a collapsed state, and therefore by rather indirect inference, but the moving-film methods have the enormous advantage of giving sets of direct projections of these reciprocal lattice levels. Each photograph is the projection of one level of the reciprocal lattice lying normal to the axis of rotation. The moving-film methods are, so to speak, devices for resolving the reciprocal lattice into stacks of plane lattices. It follows that a consideration of plane groups is important to the study of moving-film photographs, for each reciprocal lattice level may have one of the symmetries of a possible plane point-group and a possible plane translation-group.

In Weissenberg photographs taken with the equi-inclination method, for example, the projection pattern of each of the groups is invariant with change in layer line. This follows from the fact that central lines invariably project as straight lines on the photograph. Central lattice lines and symmetry lines (which are all central, and correspond with planes and axes in space symmetry) in the plane of the lattice level, therefore, project the same regardless of level. The projection of each of the groups is unique, and, furthermore, the appearance of each group pattern is distinctive enough to allow of recognition by inspection. Since space point-groups and space translation-groups may be regarded as appropriately combined plane point-groups and plane translation-groups, it is therefore possible to determine the (centrosymmetrical) point-group and reciprocal lattice type of the crystal by inspection of several Weissenberg photographs.

In this chapter, the use of plane point-groups and plane translation-groups for this purpose is discussed. These results are directly applicable to the interpretation of de Jong and Bouman photographs, which are undistorted pictures of the reciprocal lattice levels. Criteria are also given for the recognition of these groups in Weissenberg photographs.

THE POINT GROUP

The plane point-group. Any figure in a plane may be brought into coincidence with any equivalent figure by a combination of a rotation about a point (rotor) in the plane, a translation in the plane, and, in the case of an enantiomorphous equivalent figure, by reflection across a line in the plane. Consequently the crystallographic point-groups in a plane (here designated " plane point-groups "†) may be developed from appropriately combining 1-, 2-, 3-, 4-, and 6-fold rotors and symmetry lines. There are ten such combinations: five cyclical groups, C_1, C_2, C_3, C_4, and C_6, characterized by an n-fold rotor in the plane; and five more groups, C_l, C_{2l}, C_{3l}, C_{4l}, and C_{6l}, obtained by adding symmetry lines in the plane to each of the first groups. These groups are illustrated in Fig. 235.

These are not the only ways of bringing a figure into coincidence with an equivalent figure in the same plane; there are other operations which accomplish the same purpose. Any such operation must, of course, leave the plane itself unmoved. Such permissible operations include:

Function as n-fold rotor:

> n-fold axis normal to plane.

Function as 2-fold rotor:

> inversion center in plane.

Function as symmetry line:

> reflection plane normal to plane,
> 2-fold rotation axis in plane.

---

† In the present application of geometrical groups, two geometrical aspects are of considerable importance, namely: (1) the invariants of the transformation, and (2) the dimensional aspect of the space in which the geometry of the group is confined. The invariants of the transformation are indicated in the present discussion by a hyphen, thus:

> point-group (point unmoved),
> line-group (line unmoved),
> plane-group (plane unmoved),
> space-group (space unmoved), etc.

This nomenclature does not depart from customary usage. In order to specify the dimensions of the space to which the geometry of the group is confined, a modifying adjective is prefixed to the above designations, thus:

> plane point-group (point-group in a plane),
> space point-group (point-group in space), etc.

Point-groups can be developed by appropriately combining the above operations. Thus Polya has developed extended plane groups isomorphous with one of the alternative combinations of operations,

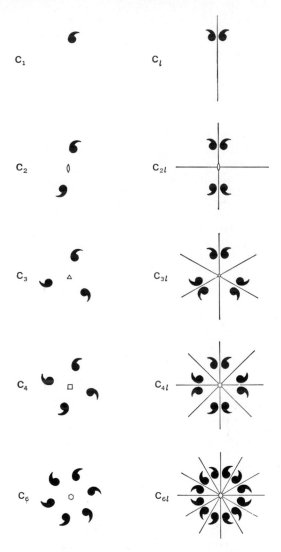

FIG. 235.   The 10 point-groups in a plane.

namely, n-fold axes normal to the plane, combined with 2-fold axes in the plane. These are cyclical groups, C_n, and dihedron groups, D_n. Niggli has developed extended surface groups isomorphous with another alternative combination, namely, n-fold axes normal to the plane, combined with " vertical " symmetry planes normal to the plane. These are cyclical groups, C_n, and the groups C_{nv}. Such

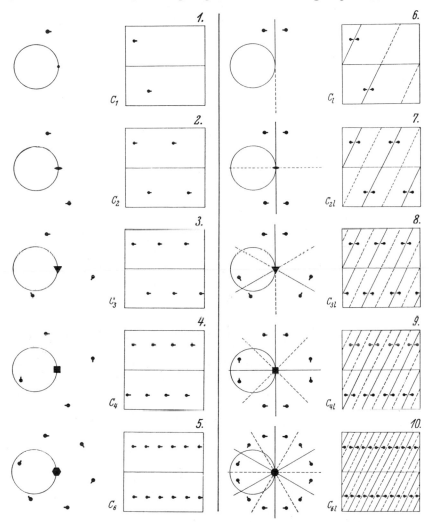

FIG. 236.   The appearances of the 10 plane point-groups in equi-inclination Weissenberg photographs.

groups rank higher than plane point-groups, however, for the influence of the symmetry elements extends outside the realm of the plane; i.e., they are groups in space.   They may be regarded as either point-groups in space, subject to certain restrictions (restricted space point-groups), or as line-groups in space (space line-groups) according to their utilization. In the present connection, it is desirable to regard them as space line-groups.

For the purpose of classifying reciprocal lattice level photographs, it is desirable to have the activity of the coincidence operation confined entirely to the plane and not to operate outside of it. For this purpose, groups strictly confined to a plane and derived from rotors and symmetry lines are most appropriate.

Recognition of the plane point-groups in equi-inclinationWeissenberg photographs. The ten plane point-groups and their appearances on the Weissenberg equi-inclination projection are illustrated in Fig. 236. The tails on the spots represent the directions of similar sequences of reflection intensities on the films.

The Weissenberg projections can be easily distinguished by noting the angular (ω) repetition period of equivalent spots on the same half of the film. Thus, C_1 and C_l repeat in 360°, C_2 and C_{2l} repeat in 180°, C_3 and C_{3l} repeat in 120°, C_4 and C_{4l} repeat in 90°, and C_6 and C_{6l} repeat in 60° ω. The groups C_n and C_{nl} are distinguished by the presence in the latter of $2n$ symmetry lines per 360° ω, each sheared Weissenberg-fashion to make an angle Я with the center line of the film, i.e., there is a symmetry line every $\dfrac{360°}{2n}$ of ω. This is because two equivalent symmetry lines always involve another symmetry line bisecting the angle between them. When n is odd, the bisecting line is the negative end of one of the first equivalent lines. It follows from the above discussion that all plane point-groups may be distinguished on Weissenberg films taken for a 360° rotation and also on films taken for a 180° rotation, provided that in the latter case there is a slight overlap. On films taken with rotations of less than 180°, however, it is impossible to distinguish C_1 from C_2, which have repetition periods of 180° or greater.

It should be pointed out that, in determining the plane point-group represented by a photograph, it is important to consider not only the positions but also the intensities of the spots representing the reciprocal lattice points. This is because the position of a spot is controlled by the translation symmetry pattern of the lattice level (see below), which may be higher than its point-group symmetry.

Attention should also be called to the possibility of false intensity distribution which may result from several causes and give rise to erroneous determinations of plane point-group symmetry. Photographs showing degraded symmetry commonly result if the shape of the crystal is less symmetrical, owing to distorted habit or to irregularity

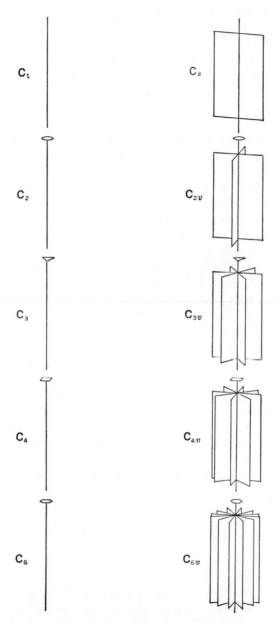

F̲ɪɢ. 237.   The 10 Niggli space line-groups.

imposed by the conditions of preparing the specimen, than permitted by the internal structural symmetry. This is especially important in crystals having considerable absorption. The presence of lineage structure in the crystal specimen frequently gives rise to unequal focusing effects in the reflections of structurally equivalent planes, which thus degrade the apparent symmetry of the photograph. On the other hand, the presence of twinning in the specimen may give rise to a false increase in symmetry.

Distribution of the plane point-groups among the space point-groups.
The 10 Niggli space line-groups mentioned above, page 468, consist of 5 cyclical groups, C_1, C_2, C_3, C_4, and C_6, each containing an n-fold axis, and the 5 groups, C_s, C_{2v}, C_{3v}, C_{4v}, and C_{6v}, derived from these by adding the operations of n " vertical " reflection planes intersecting in the n-fold axis. These are illustrated in Fig. 237. It is evident that the normal cross section of each space line-group, C_n, is the plane point-group, C_n, and that the normal cross section of each space line-group, C_{nv}, is the plane point-group, C_{nl}.

Now, the crystal classes are space point-groups and may be thought of as bundles of space line-groups intersecting at a common point. The space line-group symmetry of any rational axis, $[uvw]$, of the space point-group, therefore, is the symmetry to which the non-zero levels of the reciprocal lattice must conform. In the case of zero reciprocal lattice levels, the special symmetry operations of the intersection at the origin may become, if properly located, operations in the zero level of the space line-group in question, thus specializing the symmetry of the zero levels. The possible additional operations must be among those listed on page 467. A reflection plane normal to the lattice level, or an n-fold axis normal to the lattice level, if present in the space point-group, is already a member of the space line-group and has consequently received consideration in the treatment just given for the symmetry of the n-layers. In addition, an even-fold axis in the space point-group normal to the line-group axis introduces the operations of the group C_l into the zero level, and a symmetry center in the space point-group introduces the operations of the group C_2 into the zero level.

If the symmetry of the n-level of the reciprocal lattice normal to the $[uvw]$ axis is represented by the symbol C_x, then the zero level normal to the same axis contains the subgroups:

(1) C_x if the space point-group has neither an even-fold rotation axis normal to $[uvw]$ nor an inversion center.

(2) C_x, C_l if the space point-group has an even-fold rotation axis normal to $[uvw]$ but no inversion center.

(3) C_x, C_2 if the space point-group has an inversion center but no even-fold rotation axis normal to [uvw].

(4) C_x, C_l, C_2 if the space point-group has both an even-fold rotation axis normal to [uvw] and an inversion center.

This last case may be simplified: If a centrosymmetrical space point-group contains an even-fold rotation axis normal to [uvw], it also includes a reflection plane containing [uvw], by virtue of the relation $C_2 \cdot C_i = C_2 \cdot C_v = C_i \cdot C_v = C_{2h}$. Since this plane contains the line-group direction, [uvw], the space line-group must already contain this subgroup, and hence C_x must already include the subgroup C_l. Only the first three cases are therefore distinct for crystals of centrosymmetrical space point-groups.

Since the x-ray diffraction effects are themselves centrosymmetrical, except in certain instances employing wavelengths in the immediate vicinity of an absorption edge, only centrosymmetrical space point-groups need be considered for practical purposes. From the above it follows, then, that the symmetry of the zero level of the reciprocal lattice is enhanced over that of the n-level by the subgroup C_2 if not already present, i. e., if the rotor of the group is not even-fold. The symmetries of reciprocal lattice levels, therefore, occur in the pairs given in the first two columns of Table 28.

The possible plane point-group symmetries which may be displayed in reciprocal lattice level photographs by crystals of the various crystal classes may be determined by allowing the axis of rotation to occupy the various possible special symmetry locations of the corresponding centrosymmetrical symmetry classes. The symmetry of the n-level is then the corresponding plane point-group symmetry of the space line-group of the axis. The symmetry of the zero level is determined as indicated in the preceding paragraph.

Table 28 indicates the distribution, so obtained, of reciprocal lattice level symmetry among the various centrosymmetrical crystal classes. This table is useful in surveying the possible interpretations of symmetry data for a crystal whose crystal class or orientation, or both, are imperfectly known or are under discussion. Another arrangement of the same material is given in Table 29. This is of service in quickly checking the crystal class of a crystal whose symmetry is presumably correctly known from other crystallographic investigations. It is apparent from Table 29 that the centrosymmetrical crystal class to which the crystal may be referred may be determined from, at most, about three appropriately chosen level photographs.

TABLE 28

CENTROSYMMETRICAL CRYSTAL CLASSES CONSISTENT WITH LEVEL
SYMMETRY COMBINATIONS

(Distribution of the plane point-groups among space point-groups)

Level symmetry		Centro-symmetrical crystal class	Crystal-rotation axis (reciprocal lattice level normal to:)
n-level	0-level		
C_1	C_2	All classes	$[uvw]$
C_2	C_2	C_{2h}	$[010]$
		D_{3d}	$[10.0]$, $[01.0]$, $[11.0]$ or $[10\bar{1}]$, $[\bar{1}10]$, $[0\bar{1}1]$
C_3	C_6	C_{3i}	$[00.1]$ or $[111]$
		T_h	$[111]$
C_4	C_4	C_{4h}	$[001]$
C_6	C_6	C_{6h}	$[00.1]$
C_l	C_{2l}	C_{2h}	$[u0w]$
		V_h	$[uv0]$, $[u0w]$, $[0vw]$
		C_{4h}	$[uv0]$
		D_{4h}	$[uv0]$, $[u0w]$, $[0vw]$, $[uuw]$
		D_{3d}	$[21.w]$, $[\bar{1}1.w]$, $[12.w]$, or $[uuw]$, $[uvv]$, $[uvu]$
		C_{6h}	$[uv.0]$
		D_{6h}	$[10.w]$, $[01.w]$, $[11.w]$ $[21.w]$, $[\bar{1}1.w]$, $[12.w]$
		T_h	$[uv0]$, $[u0w]$, $[0vw]$
		O_h	$[uv0]$, $[u0w]$, $[0vw]$, $[uuw]$, $[uvv]$, $[uvu]$.
C_{2l}	C_{2l}	V_h	$[100]$, $[010]$, $[001]$
		D_{4h}	$[100]$, $[010]$, $[110]$
		D_{6h}	$[10.0]$, $[01.0]$, $[11.0]$
		T_h	$[100]$, $[010]$, $[001]$
		O_h	$[110]$, $[101]$, $[011]$
C_{3l}	C_{6l}	D_{3d}	$[00.1]$ or $[111]$
		O_h	$[111]$
C_{4l}	C_{4l}	D_{4h}	$[001]$
		O_h	$[100]$, $[010]$, $[001]$
C_{6l}	C_{6l}	D_{6h}	$[00.1]$

TABLE 29

LEVEL SYMMETRIES DISPLAYED FOR VARIOUS ROTATION AXES OF
CENTROSYMMETRICAL CRYSTAL CLASSES

(Symmetries of various levels of the reciprocal lattice normal to
direct lattice axes)

Crystal system	Centro-symmetri-cal crystal class	Crystal-rotation axis (reciprocal lattice level normal to:)	Lattice level symmetry for layer	
			n	0
Triclinic	C_i	$[uvw]$	C_1	C_2
Mono-clinic	C_{2h}	$[uvw]$	C_1	C_2
		$[010]$	C_2	C_2
		$[u0w]$	C_l	C_{2l}
Ortho-rhombic	V_h	$[uvw]$	C_1	C_2
		$[uv0]$, $[u0w]$, $[0vw]$	C_l	C_{2l}
		$[100]$, $[010]$, $[001]$	C_{2l}	C_{2l}
Tetrag-onal	C_{4h}	$[uvw]$	C_1	C_2
		$[uv0]$	C_l	C_{2l}
		$[001]$	C_4	C_4
	D_{4h}	$[uvw]$	C_1	C_2
		$[uv0]$, $[u0w]$, $[0vw]$, $[uuw]$	C_l	C_{2l}
		$[100]$, $[010]$, $[110]$	C_{2l}	C_{2l}
		$[001]$	C_{4l}	C_{4l}
Hexag-onal	C_{3i}	$[uv.w]$ or $[uvw]$	C_1	C_2
		$[00.1]$ or $[111]$	C_3	C_6
	D_{3d}	$[uv.w]$ or $[uvw]$	C_1	C_2
		$[10.0]$, $[01.0]$, $[11.0]$ or $[10\bar{1}]$, $[\bar{1}10]$, $[0\bar{1}1]$	C_2	C_2
		$[21.w]$, $[\bar{1}1.w]$, $[12.w]$ or $[uuw]$, $[uvv]$, $[uvu]$	C_l	C_{2l}
		$[00.1]$ or $[111]$	C_{3l}	C_{6l}
	C_{6h}	$[uv.w]$	C_1	C_2
		$[uv.0]$	C_l	C_{2l}
		$[00.1]$	C_6	C_6
	D_{6h}	$[uv.w]$	C_1	C_2
		$[uv.0]$ $[10.w]$, $[01.w]$, $[11.w]$ $[21.w]$, $[11.w]$, $[12.w]$	C_l	C_{2l}
		$[10.0]$, $[01.0]$, $[11.0]$ $[21.0]$, $[\bar{1}1.0]$, $[12.0]$	C_{2l}	C_{2l}
		$[00.1]$	C_{6l}	C_{6l}
Iso-metric	T_h	$[uvw]$	C_1	C_2
		$[111]$	C_3	C_6
		$[uv0]$, $[u0w]$, $[0vw]$	C_l	C_{2l}
		$[100]$, $[010]$, $[001]$	C_{2l}	C_{2l}
	O_h	$[uvw]$	C_1	C_2
		$[uv0]$, $[u0w]$, $[0vw]$ $[uuw]$, $[uvv]$, $[uvu]$	C_l	C_{2l}
		$[110]$, $[101]$, $[011]$	C_{2l}	C_{2l}
		$[111]$	C_{3l}	C_{6l}
		$[100]$, $[010]$, $[001]$	C_{4l}	C_{4l}

THE SPACE LATTICE TYPE

Analytical and geometrical methods. The determination of the crystal lattice type by the rotation or oscillation methods requires a blank reciprocal framework already available from other data (layer line spacing measurements). The reciprocal lattice is then reconstructed by the process of trial fitting of the radial coordinates of each lattice point, as obtained from the photograph with the aid of a suitable scale, to integral points of the blank framework. The direct crystal lattice is then determined by the following rather roundabout procedure: the reciprocal lattice points present on the framework are indexed in terms of the framework coordinate system, and some rule is sought which expresses analytically the systematic absence of framework points. Then the reciprocal lattice concept is discarded, the absent framework points being identified with, and regarded as, absent reflections from the *direct* lattice, from which the crystal lattice type may be identified by reference to suitable tables. This is an analytical identification of the crystal lattice.

It is possible to determine the crystal lattice type by direct geometrical means, once the reciprocal lattice has been reconstructed from data obtained by any method whatever: rotation, oscillation, or moving film. The time taken in indexing and classifying absent framework points is thereby saved. The full benefit of the geometrical method, however, is realized only in the moving-film methods, where it is unnecessary to reconstruct the reciprocal lattice. Moving-film photographs give a level-by-level projection picture of the reciprocal lattice. The reciprocal lattice may therefore be identified by inspection of the photographs. To each reciprocal lattice there is a unique direct lattice. It thus becomes possible to *see* a representation of the crystal lattice in the photographs and thus to determine it by inspection.

The plane net patterns. The possible moving-film photograph patterns are the corresponding projections of the possible plane net patterns. The derivation of the possible patterns which can be formed by plane nets resolves itself into a problem of investigating the possible combinations of each of the 10 plane point-groups with two translations in the plane. There are several ways of doing this. A simple one is as follows (Fig. 238):

In plane point-group C_1, place a random point; no further points are generated by symmetry, and so no translation is implied. A second random point provides a total of only two points, and therefore only one translation is implied; a third random point is necessary to

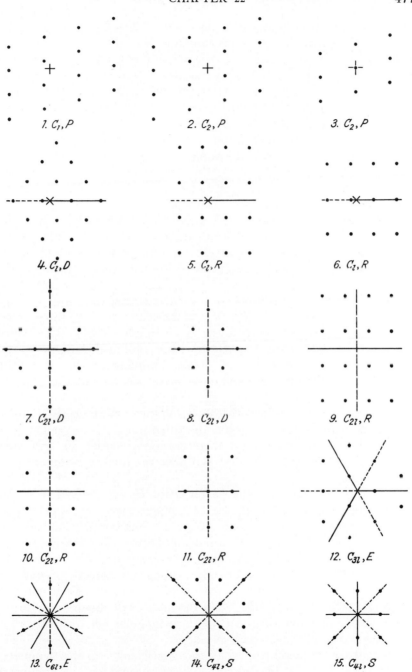

Fig. 238.   The 15 possible level patterns.

supply the two translations of a plane net.   Thus a parallelogram net in any position and orientation is consistent with C_1.

One point placed in C_2 generates another point, implying one translation.   It is necessary to supply another random point to imply the second necessary translation.   Carrying out the indicated translations gives a parallelogram net with the origin at the center of a parallelogram.   In this and the two following C_n groups, the special nature of the origin permits the location of an additional point at the origin without violating the original symmetry.   There are no other special positions in C_n groups.   In C_2, this new point implies translations giving rise to a parallelogram net with a point at the origin, and therefore a second C_2 net pattern.

One point in C_3 generates three points, implying a trigonal or equilateral triangular net with the origin at the center of a triangle.   Placing an additional point at the origin implies translations which generate a second equilateral triangular net pattern with a point at the origin.

One point in C_4 generates four points, implying translations which generate a square net with the origin at the center of one of the squares. Placing an additional point at the origin implies translations which generate a second square net pattern with a point at the origin.

One point in C_6 generates six points, implying translations which generate an equilateral triangular net with a point automatically at the origin.

Net patterns consistent with the C_{nl} groups may be easily derived by noting that the group C_{nl} has the subgroup C_n.   The net unit of C_{nl} containing the origin must therefore have at least the symmetry of the corresponding C_n net unit and have the further requirement of n symmetry lines.

Following this method, the patterns for C_l are derived from the pattern for C_1 by requiring the latter to have a symmetry line through its unit containing the origin.   Now, a symmetry line must pass through a line of lattice points or halfway between two lines of points. The parallelogram unit of C_1 can therefore acquire a symmetry line in only three ways: along an edge, along a diagonal, or halfway between edges.   This results in patterns of (1) a rectangular net with the origin on a rectangle side, (2) a rhombus or diamond net with the origin on the diamond diagonal, (3) a rectangular net with the origin on a line halfway between parallel sides.

The only other possible net patterns consistent with this symmetry would be derived by placing an additional point on the symmetry line, which is the only special position.   Inconsistent translations fail to appear only when the new point occupies the cell center or the rectangle edge-center.   No new net patterns arise in this way.

The plane point-group C_{2l} includes the subgroups C_2 and C_l.   The patterns for C_{2l} may therefore be derived by specializing the central unit of either of these groups by requiring it to have also the symmetry of the other.   Starting with C_2, there are three possible cells, one with a point at the origin, one with the origin at the center of a parallelogram, and a third cell, chosen in a different way from the second net, with the origin at the center of a parallelogram side.   Applying the symmetry line as above to these cells results in five net patterns: (1) a rectangular net with the origin at the center of a rectangle; (2) a rectangular net with the origin at the center of one of the sides of a rectangle; (3) a rectangular net with the origin at one of the lattice points; (4) a diamond net with the origin at the center of one of the diamonds; (5) a diamond net with the origin at a lattice point.   New net patterns do not arise by placing additional points on permissible points of the special positions.

The plane point-group C_{3l} includes subgroups C_3 and C_l.   The patterns of the group C_3 must therefore be derivable from the two equilateral net patterns of C_3, specialized to include the three symmetry lines.   These net patterns, however, already contain the symmetry lines, and the symmetry lines of the group C_{3l} simply fix the orientation of the nets in one of two possible positions differing by a $60°$ (or $180°$) rotation.

In the same manner, the plane point-group C_{4l} may have the two net patterns of C_4, but each with two possible fixed orientations differing from one another by $45°$.

C_{6l} similarly may have the equilateral net of C_6, with the orientation fixed to two possible positions differing by $30°$ from each other.

The last five patterns are geometrical duplicates of patterns generated in less symmetrical plane point-groups.   There are therefore 15 possible distinct plane net patterns and accordingly 15 possible moving-film, point-position patterns.   It will be observed that each of the patterns is based on one of the five possible two-dimensional translation groups with the origin at a possible special symmetry position.   Location of the origin in all possible symmetry positions of each plane lattice is covered in the above derivation.   The 15 possible level patterns are listed in Table 30, illustrated in Fig. 238, and their distribution among the plane point-groups and plane lattices indicated in Table 31.

TABLE 30

THE 15 POSSIBLE LEVEL PATTERNS

Plane point-group of lattice level	Plane lattice type	Position of origin in central unit	Pattern number	Position of center line of film in lattice line sequences
C_1	P	Random	1	At random in all central lattice line sequences
C_2	P	Halfway between lattice points	2	At $\frac{1}{2}$ spacing in all central lattice line sequences
	P	At lattice point	3	Included in all central lattice line sequences
$C_{\bar{1}}$	D	Along diagonal	4	At random along symmetry line sequence
	R	Along line halfway between rectangle sides	5	Symmetry line not a lattice line
	R	Along side	6	At random along symmetry line sequence
C_{2I}	D	At diamond center	7	At $\frac{1}{2}$ spacing in both symmetry line sequences (and all central lattice line sequences).
	D	At lattice point	8	Included in both symmetry line sequences (and all central lattice line sequences).
	R	At rectangle center	9	Neither symmetry line a lattice line, at $\frac{1}{2}$ spacing in all central lattice line sequences.
	R	Halfway along rectangle sides	10	At $\frac{1}{2}$ spacing in one symmetry line sequence, other symmetry line not a lattice line.
	R	At lattice point	11	Included in both symmetry line sequences (and all central lattice line sequences).
C_{3I}, C_3	E	At triangle center	12	At $\frac{1}{3}$ in one symmetry line sequence, at $\frac{2}{3}$ in other symmetry line sequences.
C_{6I}, C_6 C_{3I}, C_3	E	At lattice point	13	Included in both symmetry line sequences (and all central lattice line sequences).
C_{4I}, C_4	S	At center of square	14	At $\frac{1}{2}$ spacing in one symmetry line, other symmetry line not a lattice line.
	S	At lattice point	15	Included in both symmetry line sequences (and all central lattice line sequences).

Heavy type indicates position symmetry of level pattern

P = parallelogram net
D = diamond net
R = rectangular net

E = equilateral triangular net
S = square net

TABLE 31

DISTRIBUTION OF POSSIBLE LEVEL PATTERNS

Plane point-group of photograph	Plane lattice basis of pattern					
	P	D	R	S	E	Totals
C_1	1					1
C_2	2					2
C_l		1	2			3
C_{2l}		2	3			5
C_{3l}, C_3					2	2
C_{4l}, C_4				2		2
C_{6l}, C_6					(1)	(1)
Totals	3	3	5	2	2	15

Heavy type indicates position symmetry of photograph.

The equi-inclination Weissenberg projections of the plane net patterns. The equi-inclination Weissenberg projections of the plane net patterns are illustrated in Fig. 239. They can easily be distinguished from one another by inspection on the basis of (1) the symmetry of the spot positions, (2) the plane lattice type, and (3) the position of the plane point-group origin (the rotation axis) in the unit containing the origin.

The first step in the identification of the plane net pattern on the Weissenberg photograph is to determine the plane point-group of the pattern as already discussed on pages 470–472. For each plane point-group there are a very limited number of plane lattice types, as indicated in Table 30. Thus, for C_1, C_2, C_3, C_{3l}, C_4, C_{4l}, C_6 and C_{6l}, there is only one possible plane lattice each; C_l and C_{2l} may both be referred to either of two lattices, diamond or rectangular. The following discussion will indicate incidentally the method of easily distinguishing between these.

It has been brought out in the foregoing section that the spot position patterns required for C_3, C_4, and C_6 cannot be further specialized for C_{3l}, C_{4l}, and C_{6l}, respectively. So far as spot positions are concerned, these patterns have the symmetries of the latter groups containing lines of symmetry. It is convenient to discuss the 15 possible patterns in terms of this position symmetry alone, disregarding, for this purpose, the possible quality (intensity) non-equivalence of the spots. For this reason, the symmetry lines have been drawn in the figures, but it should be understood that the pure rotor groups are referable to the same position patterns.

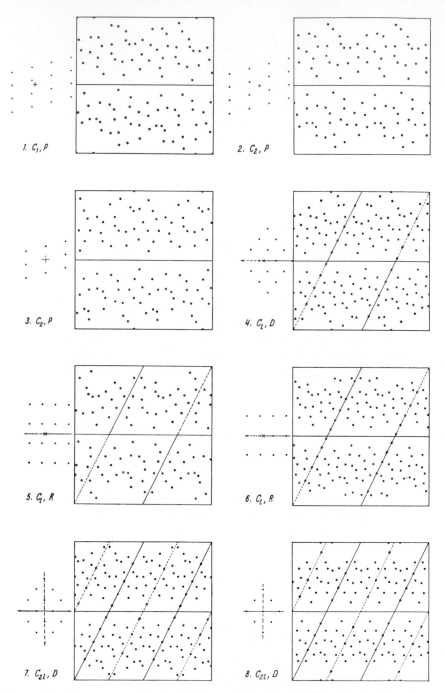

FIG. 239. The appearances of the 15 possible level patterns in equi-inclination Weissenberg photographs.

482

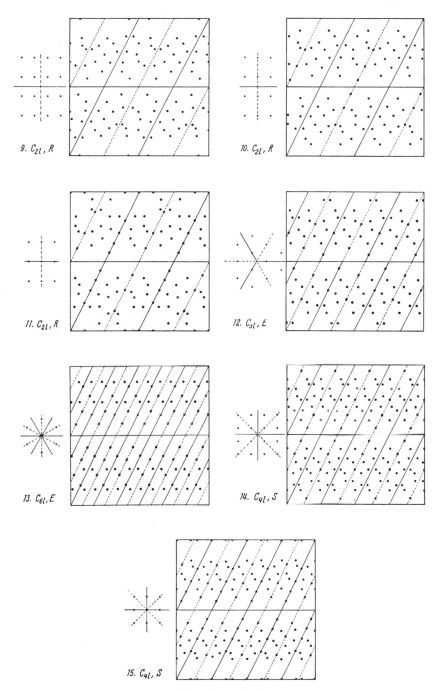

9. C_{2l}, R

10. C_{2l}, R

11. C_{2l}, R

12. C_{3l}, E

13. C_{6l}, E

14. C_{4l}, S

15. C_{4l}, S

Fig. 239, *continued.*

It follows from the Weissenberg projection transformation that an *n*-sided polygon on a reciprocal lattice level projects as an *n*-sided polygon in the Weissenberg projection, Fig. 240. The transformation includes a distortion such that a pair of lines normal to one another, one a central line (including the origin or rotation axis), becomes a pair of lines inclined an angle Я to one another. A polygon containing a symmetry line, Fig. 240, therefore appears on the Weissenberg pro-

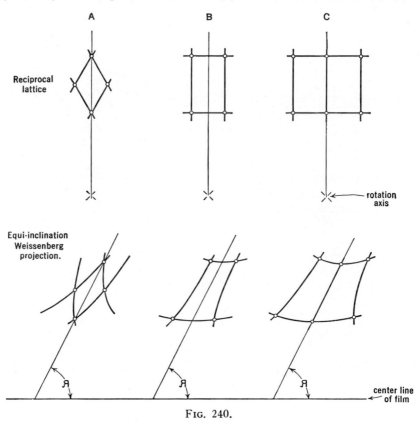

Fɪɢ. 240.

jection as a polygon containing a symmetry line and subsequently sheared until a line joining the reflected equivalent points makes an angle Я with the projected symmetry line. The line joining reflected equivalent points is therefore parallel to the center line of the film. The transformation also includes a distortion which spreads points near the origin owing to the larger angle they subtend, and stretches the radial distance between points near the edges of the film owing to the rapid variation of Υ with ξ in this region. In the region roughly $\frac{1}{3}$ or $\frac{1}{2}$

way from the center to the edge of the film, these distortions are not serious and a reasonably undistorted projection of the polygon may be recognized, sheared Weissenberg-fashion as described above. It is a relatively easy matter to pick out four neighboring spots in this region and recognize the plane lattice unit.

The relation of the symmetry line, if present, to the lattice unit is important. If the lattice is diamond, Fig. 240A (or triangular), a

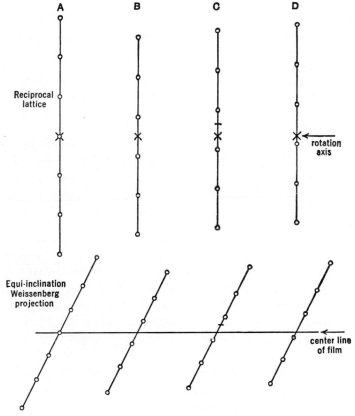

Fig. 241.

sheared diamond unit is plainly discernible in the region of the symmetry line, with the latter along a diagonal. If the lattice is rectangular, Figs. 240B and C (or square), the sheared rectangle with the symmetry line between sides, or two rectangles symmetrically linked on each side of the common symmetry and lattice line, are discernible.

The location of the position of the origin in the net, listed in the third column of Table 30, is the final step in determining the pattern type. The origin, which corresponds with the rotation axis, appears as the center line in the projection. It may be purposely recorded by a short exposure of the direct beam on the actual film. The position of the origin in the net is easily determined for the nets containing lattice point symmetry lines. The origin can have only four special positions on a lattice line, Fig. 241: A, at a lattice point; B, halfway between lattice points; C, $\frac{1}{3}$ or $\frac{2}{3}$ along the space between lattice points in the equilateral triangular net; or D, completely at random. In the first case, the normal ξ spacing sequence of spots along the symmetry line includes the center line; in the second, the center line halves the normal spacing sequence between the two central spots; in the third, the center line thirds the normal spacing sequence between the two central spots; and in the fourth, the two centermost spots on either side of the center are a different ξ distance from the center line. These criteria for the various patterns are included in the last column of Table 30.

It should be observed that the recognition of the Weissenberg patterns by the strict application of the criteria just given presupposes the non-absence of spots on the film due to fortuitous combinations of atom parameters or scattering powers or both. It is well to check the possibility of such absent spots in the general regions of important central lattice lines (for example, symmetry lines) before applying the criteria. This can usually be done easily by following the important non-central lattice line curves on the film in the region of important central lattice lines. Missing spots can ordinarily be detected by this simple inspection, and their places indicated on the film by rings.

The plane net stacks. If, to the two translations, $t_1 + t_2$, of any net pattern there be added a third non-coplanar translation, t_3, the pattern will be indefinitely repeated in space, i. e., a space lattice will result. In other words, plane lattice patterns may be stacked in space to form space lattices. This derivation of space lattices indicates that plane lattice patterns may be stacked to form space lattices provided that:

1. Identical plane nets are stacked in parallel position.
2. All the stacking intervals are equal.
3. The shear displacements between corresponding points of neighboring levels are identical.

The possible kinds of symmetrical lattice stacks may be easily investigated by making use of the possible crystallographic symmetries of space about a line, which in this case is the common normal to the plane lattice sheets. These are the 10 space line-groups, Fig. 237. Each space line-group may have lattice stacks composed of the permissible stacking combinations of the plane net patterns of corresponding plane symmetry. These are illustrated in Fig. 242 and derived as follows.

There is only one type of pattern corresponding with plane point-groups C_1. Hence this pattern may be stacked only with itself to produce a staggered stack of parallelograms.

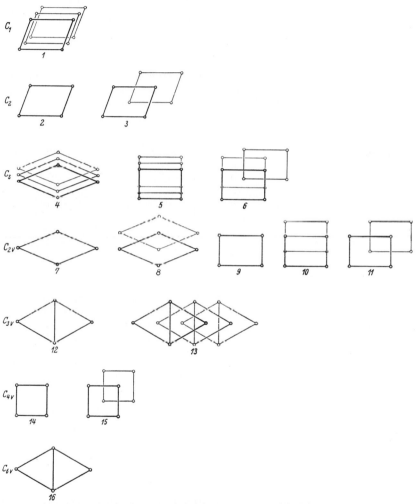

FIG. 242. The 15 possible plane-net stacks. (Note that numbers 12 and 16 are duplicates.)

Two patterns are possible in C_2, both parallelogram nets, one with a lattice point at the origin and another with a parallelogram center at the origin. In space line-group C_2, one may start with the first pattern for the zero level and take the same pattern again for the first level. The translation t_3 being thus established, all succeeding levels are then the same, and there results a stack of coincident parallelogram nets as seen along the line-group line. Or the first level may be taken as the second possible pattern. The projected pattern displacement per level is $\frac{1}{2}t_1 + \frac{1}{2}t_2$, which brings the second level into a position corresponding with the zero level. This results in a parallelogram-

TABLE 32

PLANE LATTICE STACKS

Space line-group symmetry of stack	Stack designation	Stack description	
		Net	Displacement projected on net plane
C_1	1	P	$mt_1 + nt_2$
C_2	2	P	$0 + 0$
	3	P	$\frac{1}{2}t_1 + \frac{1}{2}t_2$
C_s	4	D	$mt_1 + mt_1$
	5	R	$mt_1 + 0$
	6	R	$mt_1 + \frac{1}{2}t_2$
C_{2v}	7	D	$0 + 0$
	8	D	$\frac{1}{2}t_1 + \frac{1}{2}t_1$
	9	R	$0 + 0$
	10	R	$\frac{1}{2}t_1 + 0$
	11	R	$\frac{1}{2}t_1 + \frac{1}{2}t_2$
C_{3v}, C_3	12	E	$0 + 0$
	13	E	$\frac{1}{3}t_1 + \frac{1}{3}t_1$
C_{4v}, C_4	14	S	$0 + 0$
	15	S	$\frac{1}{2}t_1 + \frac{1}{2}t_1$
C_{6v}, C_6	12	E	$0 + 0$

centered alternating stack. Combining the second plane pattern with itself gives a stack geometrically identical with combining the first pattern with itself. No further stacking arrangements are therefore possible in this space line-group (except by the alternative choice of the parallelogram cell, which results in no new point-distributions).

In the line-group C_s, there may be stacks resulting from permissible combinations of C_l patterns. There are two types of units in C_l:

diamond and rectangle, each of which may be combined only with itself. Since there is only one kind of diamond pattern, it can be combined only with itself. Given a general projected displacement, consistent with the symmetry of C_l, namely, $mt_1 + mt_1$, this generates a staggered diamond stack. C_l has two kinds of rectangle patterns, one with the symmetry line along a side and the other with the symmetry line along the middle between sides. Two combinations of these are possible: the staggered rectangle, and the center alternating rectangle.

The other possible combinations for the other space line-groups may be developed following the lines indicated above. There result, from these considerations, 15 possible distinct plane-net stacks. These are tabulated in Table 32 and illustrated in Fig. 242.

Space lattices. The plane-net stacks may be distributed among the crystal classes on the following basis: since translations are centrosymmetrical, the stacks are centrosymmetrical; hence they can have point-position symmetries no lower than those of one of the 11 centrosymmetrical crystal classes (Table 1): C_i, C_{2h}, D_{2h}, C_{4h}, D_{4h}, T_h, O_h, C_{3i}, D_{3d}, C_{6h}, D_{6h}. The space nets of each of these crystal classes are stacks of plane lattices specialized by the several space requirements of symmetry of the class.

Thus the space point-group C_{2h} has the symmetry of the line-group C_2 along the symmetry axis. Hence it can be based on only one of two lattice stacks as seen from this aspect: the coincident parallelogram stack, 2, Fig. 242, or the parallelogram-centered alternating stack, 3. Seen along the symmetry plane, C_{2h} can be based only on the C_s stacks. Numbers 4 and 6 of these are two aspects or projections of the same space nets. From this it follows that only simple and body-centered ($=$ end-centered, $=$ face-centered) lattices are possible for this class.

In a similar manner, the distribution of plane net stacks may be made among all the centrosymmetrical classes. Table 33 indicates the distribution of plane net stacks along axial directions of the centrosymmetrical classes.

Reciprocity relations. Several important reciprocity relations now require brief mention because they have very practical application in the geometrical interpretation of level photographs. These concern the relationship between plane reciprocal lattices and space reciprocal lattices. Fundamental to these relationships are several simple projection theorems:

1. The projection of a line lattice on a plane normal to its translation vector is a point.

TABLE 33

Distribution of Reciprocal Plane Lattice Stacks among
Axial Directions of Centrosymmetrical Crystal Classes

Centrosymmetrical crystal class	Space line-group along				Possible reciprocal plane-lattice stacking combinations				Reciprocal lattice type
	[100]	[010]	[001]	[111]	[100]	[010]	[001]	[111]	
C_i	C_1	C_1	$\underline{C_1}$		1	1	$\underline{1}$		Simple triclinic
C_{2h}	C_s	$\underline{C_2}$	C_s		5	$\underline{2}$	5		Simple monoclinic
					6	$\underline{3}$	6		Body-centered monoclinic
V_h	C_{2v}	C_{2v}	$\underline{C_{2v}}$		10	10	$\underline{7}$		Base-centered orthorhombic
					8	8	$\underline{8}$		Face-centered orthorhombic
					9	9	$\underline{9}$		Simple orthorhombic
					(10	7	$\underline{10}$		End-centered orthorhombic)
					11	11	$\underline{11}$		Body-centered orthorhombic
C_{4h}	C_v	C_v	$\underline{C_4}$		(5	5	$\underline{14}$)		
	C_{2v}	C_{2v}	$\underline{C_4}$		9	9	$\underline{14}$		Simple tetragonal
					(6	6	$\underline{15}$)		
					11	11	$\underline{15}$		Body-centered tetragonal
D_{4h}	C_{2v}	C_{2v}	$\underline{C_{4v}}$		9	9	$\underline{14}$		Simple tetragonal
					11	11	$\underline{15}$		Body-centered tetragonal
C_{3i}	C_2	C_2		$\underline{C_3}$	(1	1		$\underline{13}$)	
	C_s	C_s		$\overline{C_{3v}}$	4	4		$\underline{13}$	Rhombohedral
	C_{2v}	C_{2v}	C_{6v}		10	10	12		Hexagonal
D_{3d}	C_s	C_s		$\underline{C_{3v}}$	4	4		$\underline{13}$	Rhombohedral
	C_s	C_s	C_{3v}						
	C_{2v}	C_{2v}	C_{6v}		10	10	12		Hexagonal
C_{6h}	C_s	C_s	$\underline{C_6}$		(5	5	$\underline{12}$)		
	C_{2v}	C_{2v}	$\underline{C_{6h}}$		10	10	$\underline{12}$		Hexagonal
D_{6h}	C_{2v}	C_{2v}	$\underline{C_{6h}}$		10	10	$\underline{12}$		Hexagonal
T_h	C_{2v}	C_{2v}	C_{2v}	C_3					
	C_{4v}	C_{4v}	C_{4v}	$\overline{C_{3v}}$	14	14	14	$\underline{13}$	Simple cubic
					15	15	15	$\underline{13}$	Face-centered cubic
					15	15	15	$\underline{13}$	Body-centered cubic
O_h	C_{4v}	C_{4v}	C_{4v}	C_{3v}	14	14	14	$\underline{13}$	Simple cubic
					15	15	15	$\underline{13}$	Face-centered cubic
					15	15	15	$\underline{13}$	Body-centered cubic

Principal axis underlined.   Heavy type indicates position symmetry.

2. The projection of a plane lattice on a plane normal to a rational lattice line is a line lattice.   This follows easily by regarding a plane lattice as composed of "stacks" of coplanar, equally spaced line lattices, all parallel to the rational direction (in the same way that space lattices were treated as stacks of plane lattices, pages 486–489).

3. The projection of a space lattice on a plane normal to a rational lattice line is a plane lattice.   This follows by regarding a space lattice as a stack of plane lattices as previously discussed (pages 486–489), the rational lattice direction in this case being the common rational direction in each of the plane lattices.   Each of the plane lattices projects according to theorem 2 as a linear lattice, and since the displacement of the plane lattices from neighbors is a constant, the projection of the displacements, and therefore the displacements of the linear lattices, are constants.   Consequently the set of linear lattices in the projection constitutes a plane lattice.

4. From the above theorems, it follows that the reciprocal of the planes in a zone, [uvw], of a space lattice is the same as the reciprocal of the projection of the space lattice on a plane normal to the zone [uvw].   This relationship affords a simple method of dealing with space lattices by means of the reciprocal plane lattices available in the moving-film projections.

5. To each zone, [uvw], in the direct lattice, there corresponds a zero-level plane, $(uvw)^*$, in the reciprocal lattice.   From the stack concept developed on pages 486–489, it follows that all the other parallel levels in the reciprocal lattice are of the same net character (neglecting extinctions; see below).

It was shown in Chapter 6 that the reciprocal of a plane net, whose unit is a parallelogram with edges in the ratio $c : a$ and making an angle β with one another, is a parallelogram net with corresponding sides in the ratio $a : c$ and making an angle $180° - \beta$ with one another.   From symmetry considerations, it follows that: (1) the reciprocal of a diamond net whose sides make an angle β with one another is another diamond net whose corresponding sides make an angle $180° - \beta$ with one another; (2) the reciprocal of an equilateral triangular net is another equilateral triangular net rotated 30° from the original; (3) the reciprocal of a rectangular net of side ratio $c : a$ is a second parallel, rectangular net of side ratio $a : c$; and (4) the reciprocal of a square net is a second, parallel, square net.

It is important to emphasize the fact that the reciprocal relations of lattices are based on reciprocity of sheeting period, not identity period.   This means that, regardless of the way the cell is chosen, the identity period of the reciprocal cell is the reciprocal of the projection of *one sheeting period*. In the determination of the unit of the symmetrical cells, of which the rectangle and diamond may be taken as general examples, it is especially important to avoid confusion on this point. Thus, for the sake of orthogonality, the diagonals of the diamond unit, rather than its unit translations, are customarily used as standard identity periods. The diagonals embrace *two* sheeting periods, as indicated in Fig. 243. In the case of the rectangle, the sides of the unit are the standard identity periods. This identity period is a single sheeting period (Fig. 243).

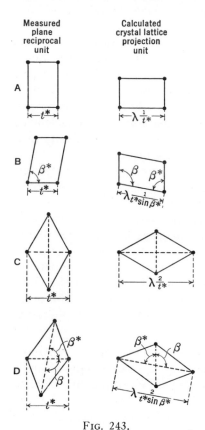

FIG. 243.

Reciprocals of space lattice types. From the principles discussed in the previous section, the dimensional relations between reciprocal space lattices may easily be found. These are indicated in Fig. 244 for the reciprocals of a simple lattice, an end-centered lattice, a face-centered lattice, and a body-centered lattice. The illustrations are drawn for orthorhombic lattices; they may be immediately specialized for more symmetrical lattices and generalized for monoclinic lattices. The four types have no significance for the triclinic system, where only a simple cell is necessary.

Figure 244 indicates that the reciprocal of a simple lattice is a simple lattice; that the reciprocal of a base-centered lattice is a base-centered lattice; that the reciprocal of a face-centered lattice is a body-centered lattice; and that the reciprocal of a body-centered lattice is a face-centered lattice. These reciprocity relationships are of highest importance in the present discussion.

Determination of unit cell dimensions. The level photographs made by any of the moving-film methods provide projections of the points of the reciprocal lattice, and by following the rules of the particular projection, the translations within a level of the reciprocal

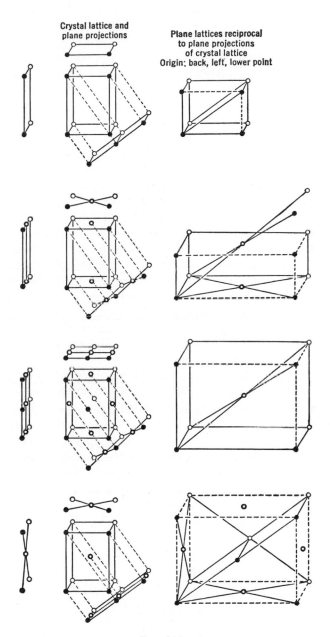

Fɪɢ. 244.

lattice may be measured.   This can be most easily done for the de Jong and Bouman photograph, where the desired translation may be directly measured and need only be multiplied by an instrumental constant to transform it into the reciprocal lattice translation.

With the reciprocal lattice translation measured, Fig. 243 shows quantitatively how this reciprocal translation is related to the *projection unit* of the direct cell (note that this is the projection unit of the direct cell, not the base of the direct cell).   Then Fig. 244 shows how the projection unit for the various kinds of cells are related to the actual cell translations.   To bring out the significance of this by a particular example, note the body-centered reciprocal cell of Fig. 244. The translations in the (001)* base of this cell can be measured from the second level photograph (for simplicity, suppose that they are directly measured in a de Jong and Bouman photograph).   The reciprocals of these reciprocal cell translations are the spacings of the projection unit shown above the direct cell.   The translations of this projection unit are related quantitatively to the measured reciprocal translations as indicated in Fig. 243 (the correct case being one of the upper two, depending on the crystal system involved).   The translations of the projection unit are now known, but note that, according to the third left illustration of Fig. 244, these translations are only half the true cell translations.   In order to pass to the latter, therefore, both projection unit translations must be multiplied by 2.

If one is working with Weissenberg photographs, then the measurement of the original reciprocal level unit is made on the second level (to avoid possible extinctions), with the aid of a ξ scale.   This measurement depends not only on the scale, but also on the accuracy with which the layer inclination, μ, has been determined, and therefore the measurement is not very accurate.   It is well to regard this as a preliminary measurement of the reciprocal lattice translation and refine it somewhat as follows: Select the zero-level translation corresponding with that approximately determined, and measure the film distance, x, to the most remote reflection in the line.   The ξ value of this reflection is, then,

$$\xi = 2 \sin \frac{\Upsilon}{2}$$

$$= 2 \sin \left(\left[\frac{360°}{\pi D} \right] \frac{x}{2} \right),$$

where the term in brackets is an instrumental constant.   The ξ value computed in this way is an integral number of times the desired reciprocal cell translation, already found approximately from the second-level photograph. This method leads to a moderately refined value of the reciprocal cell constants and eventually to the direct cell constants. In order to derive really refined values, the methods of eliminating systematic errors discussed in Chapter 20 should be employed.

INSPECTIVE INTERPRETATION OF THE CELLS OF THE SEVERAL SYSTEMS

Introduction. Since two adjacent levels and the spacing between them fix the character of the lattice, it is evident that the space lattice of a crystal can be completely determined by taking a rotation photograph and two adjacent level photographs. Because of the possibility of extinctions in the zero level, it is desirable not to include the zero level in these two, but to take, rather, the first- and second-level photographs. The following directions are suggested for the selecting and interpreting of these photographs for the several crystal systems. It should be understood that numerous alternative procedures are possible.

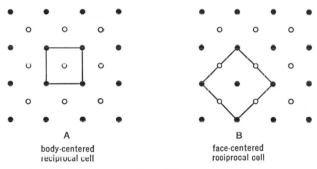

A
body-centered
reciprocal cell

B
face-centered
reciprocal cell

Fig. 245.

Isometric crystals. The crystal should preferably be rotated about a crystallographic axis. If the first- and second-level photographs are of the same pattern, the stacking sequence is 14, Fig. 242, and the cell is primitive. The direct cell edge, a, is then most easily determined approximately from the rotation photograph. The reciprocal cell edge is the shortest translation in the level. In Weissenberg photographs, it is along the central line of least reflection interval.

If the first- and second-level photographs are different, the stacking sequence is 15, Fig. 242, and the reciprocal cell is either body centered or face centered, Fig. 245. If the orientation of the level with respect to the direction of the crystallographic axes has been marked, then the correct alternative can be picked out at once.   If the level orientation is unknown, then the direction of the crystallographic axis must be picked out from its length.   This can be obtained by placing the rotation photograph on the Bernal chart and taking the reciprocal lattice level spacing, which is the same as the reciprocal cell edge.   This cell edge will be found to correspond with one of two alternatives, Fig. 245, i. e., either with the shortest or the second shortest translation in the level.   In the first case the reciprocal cell is body centered and the direct cell *is face centered;* in the second, the reciprocal cell is face centered and *the direct cell is body centered.*

Tetragonal crystals.   It is desirable but not necessary to rotate a tetragonal crystal about the c-axis.   If the adjacent level photographs are alike, the stacking sequence is 14, Fig. 242, and lattice is primitive. The reciprocal cell edge, a^*, is then the shortest translation in the level.

If adjacent level photographs are different, then the stacking sequence is 15, Fig. 242, and the reciprocal cell is either body centered or face centered, depending on how the edge of the square in the level unit is chosen, Fig. 245.   In this connection it should be recalled that body-centered cells are doubly primitive and face-centered cells are quadruply primitive.   In order to select the smallest direct cell, therefore, the body-centered direct cell should be aimed at.   Since body-centered and face-centered cells are reciprocals, this desired result is obtained if a face-centered reciprocal unit is selected.   In order to pick this out from the level photograph, one should select as the reciprocal cell axis, Fig. 245B, the second shortest translation in the level.

Orthorhombic crystals.   Select an axis for rotation, say the c-axis for sake of definiteness.   If the adjacent levels have similar patterns, then the level sequence is either 7 or 9, Fig. 242, corresponding with end-centered and primitive cells, respectively, both reciprocal and direct.   The direct unit cell translations can be computed from the reciprocal cell translations with the aid of the third and first cases of Fig. 243, respectively.

If the adjacent levels have different patterns, then the level sequence is either 8, 10, or 11, Fig. 242.   In the first case the reciprocal cell is face centered, in the second side centered, and in the third body centered.   These correspond with body-centered, side-centered, and face-centered direct cells, respectively.   With the aid of Fig. 243, direct cell projection units can be computed from the reciprocal cell translations, and with the aid of Fig. 244, these can be transformed to direct cell axes.

In the above cases, the recognition of the plane lattice type is essential.   This causes no difficulty with de Jong and Bouman photographs, where the plane lattice may be seen in undistorted form. With Weissenberg photographs, the plane lattice type can be recognized as suggested in Fig. 240. The diagonals of the diamond, if the pattern is diamond, or the sides of the rectangle, if the pattern is rectangular, appear along the symmetry lines. These point intervals are measured by means of the ξ scale.

Monoclinic crystals. For a monoclinic crystal, the preferable procedure is to investigate first the (010)* levels of the reciprocal lattice. Accordingly the crystal is set up for b-axis rotations. The b-axis of either the direct or reciprocal cell is then given by the layer line spacing of the b-axis rotation.

If adjacent levels have similar patterns, then the stacking sequence is 2, Fig. 242, and the lattice is primitive. In this case, the a^*- and c^*-axes are selected by choosing the shortest translations in the level net; they are separated by an angle β^*. From measurements of these, the a- and c-axes are easily computed with the aid of Fig. 243B.

If adjacent levels have different patterns, the stacking sequence is 3, Fig. 242, and the cell is either end centered, body centered, or face centered, depending on how the level translations are chosen. The way to be chosen depends upon the fact that one wishes to have the direct lattice of as low a multiplicity as possible. End-centered and body-centered cells are doubly primitive, while face-centered cells are quadruply primitive. One aims, therefore, to choose an end-centered or body-centered direct cell. The reciprocals of these are end centered and face centered, respectively. Body-centered reciprocal cells are to be avoided because they lead to the quadruply primitive direct cell.

Two cases now arise, owing to the necessity for selecting the reduced cell (see page 364). Since reduced cells are reciprocals in the monoclinic system (see page 365), the reduced direct cell will result from choosing the reduced reciprocal cell. Accordingly, one chooses from the level photographs the two shortest translations for a^* and c^*. If one of these turns out to be the centered translation, Fig. 246A, then the cell as selected is end centered. If the diagonal of the cell chosen turns out to be the centered translation, Fig. 246B, then the reciprocal cell is body centered, and this is to be avoided, as pointed out above. *In this case, a face-centered reciprocal cell should be chosen.* The axes of this cell are the diagonals of the body-centered cell. The original, primitive reciprocal unit may be retained, however, by thinking of it, not as rectangularlike, Fig. 243B, but as diamondlike, Fig. 243D. The direct lattice projection unit may be computed with the aid of Fig. 243, and the direct lattice cell edges computed from this with the aid of Fig. 244.

Note that, if Weissenberg photographs are being used, the edges of the primitive plane unit are chosen by selecting, from the second-layer photograph, the two central lattice lines of shortest interval. The central lattice lines of the next two shortest intervals are the two diagonals. If the cell proves to be non-primitive, one or two of these lines will have its interval halved by the corresponding line on the first-layer photograph. This identifies the centered translation.

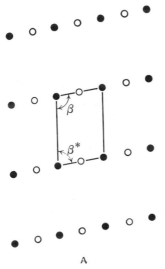

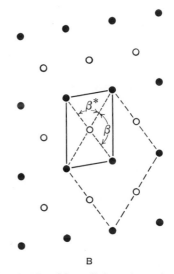

FIG. 246A. Monoclinic reciprocal cell which proves to be end-centered. Solid dots are points on even levels; open dots are points in odd levels.

FIG. 246B. Monoclinic reciprocal cell which proves to be body-centered (solid lines), also alternative face-centered cell (dashed lines). Solid dots are points on even levels; open dots are points on odd levels.

Furthermore, a study of Fig. 246A will show that, if the reciprocal cell is end centered, one of the axial lines of the second layer photograph has its interval halved by the corresponding line of the first-layer photograph, and the second axial line is missing on the first-layer photograph. Similarly, a study of Fig. 246B will show that, if both axial central lattice lines of the second-layer photograph correspond with absent lines on the first-layer photograph, the reciprocal cell is body centered with the cell chosen. The face-centered cell can then be picked out by choosing, as axial lines, those on the second-layer photograph of third and fourth shortest interval.

Hexagonal crystals. Crystals belonging to the hexagonal system (including rhombohedral crystals) are best investigated by studying the (001) levels of the reciprocal lattice. Accordingly, the crystal is set up for rotations about the hexagonal c-axis.   The length of the direct or reciprocal c-axis can be determined directly from the rotation photograph.

Crystals belonging to the hexagonal system are based upon either a hexagonal or a rhombohedral lattice.   The reciprocal of a hexagonal lattice is another hexagonal lattice, and the reciprocal of a rhombohedral lattice is another rhombohedral lattice.   These relationships may be arrived at in the following way: The hexagonal cell is primitive; hence its reciprocal is primitive.   According to the rules developed in Chapter 6, page 116, the bases of these cells are reciprocal plane cells of 60° diamond shape, but have orientations rotated 90° from one another.   The reciprocal of a hexagonal cell is therefore a hexagonal cell rotated 90° from the first.   The rhombohedral cell may be regarded as a distorted cube.   The reciprocal of a cube is a parallel cube.   The reciprocal of a cube distorted by stretching along the [111] axis is another cube distorted by compressing along the [111] axis.   Therefore the reciprocal of an acute rhombohedral cell is an obtuse rhombohedral cell in the same orientation.

A series of level photographs gives an easy criterion for hexagonal *versus* rhombohedral lattice.   Both the hexagonal and rhombohedral lattices consist of stacks of equilateral triangular nets normal to the c-axis, Fig. 242, numbers 12, 13, and 16.   In the case of the hexagonal lattice, the nets project on a plane parallel with the net plane in exactly superposed position, Fig. 242, numbers 12 and 16.   In the case of the rhombohedral lattice, they project in such a way as to be displaced by $\frac{1}{3}$ of a long mesh diagonal for each succeeding layer, Fig. 242, number 13.   In the interpretation of de Jong and Bouman photographs, it is an easy matter to distinguish between hexagonal and rhombohedral crystals by superposing the photographs of adjacent levels.   If the crystal has a hexagonal lattice, the triangle points register exactly in succeeding layers, whereas if the crystal is rhombohedral, the displacement of triangles is according to number 13, Fig. 242, and registry occurs only in layers separated by two other layers.

The hexagonal and rhombohedral lattice can be easily distinguished on sight also in a pair of adjacent equi-inclination Weissenberg photographs.   The displacement of the plane cells in the rhombohedral case is along one of the two non-equivalent position-symmetry lines of the photograph for the zero layer, or along the one set of unique position-symmetry lines of the n-layer.   Figure 247A shows the hexagonal and rhombohedral net stacks of diamonds, and Fig. 247B shows the Weissenberg projections of chains of these diamond-shaped cells which occur along the position-symmetry lines of the equi-inclination Weissenberg photograph.   The main thing to note in

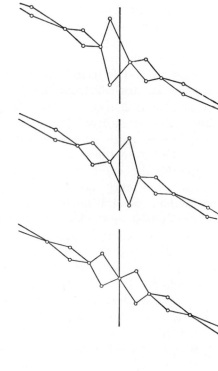

3 level   2 level   1 level   0 level

Fig. 247B.

Plane net stack

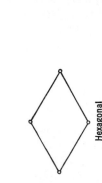

Hexagonal

Rhombohedral

Fig. 247A

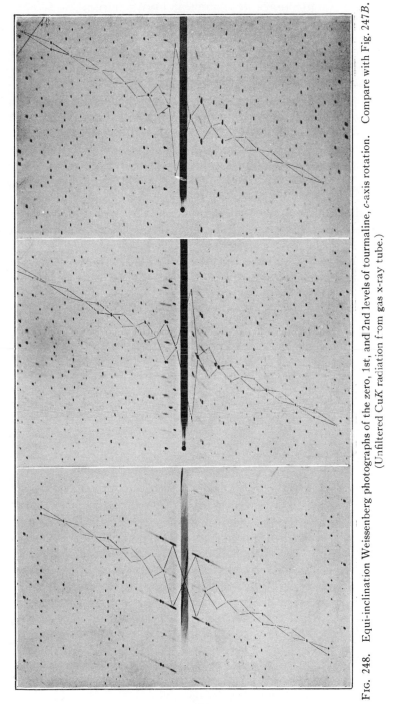

FIG. 248. Equi-inclination Weissenberg photographs of the zero, 1st, and 2nd levels of tourmaline, *c*-axis rotation. Compare with Fig. 247*B*. (Unfiltered Cu*K* radiation from gas x-ray tube.)

Fig. 247*B* is the displacement of the point of the diamond from the origin (the center line of the photograph) with succeeding layers. In the hexagonal lattice, Fig. 247 above, there is no such displacement. Figure 248 shows the appearance of a set of three succeeding levels of the hexagonal mineral tourmaline, in equi-inclination Weissenberg projection. It is evident, from comparison with Fig. 247*B*, that this crystal has a rhombohedral lattice.

The cell dimensions of crystals with a hexagonal lattice can be easily determined by measuring the hexagonal reciprocal translations and applying the relations indicated in Fig. 243*B*, taking the sides of the cell equal and $\beta^* = 60°$.

Cell dimensions of rhombohedral crystals are conveniently derived by first referring the crystal to hexagonal coordinates and then transforming to rhombohedral coordinates. A knowledge of the relations between reciprocal and direct lattices is of fundamental importance in the determination of the dimensions of the direct lattice. These

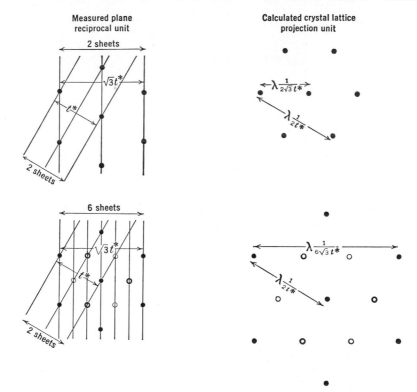

FIG. 249.   Reciprocity relations in the hexagonal lattice (above) and in the rhombohedral lattice (below).

relations have been discussed from this viewpoint for all important cases except the hexagonal ones. The reciprocal relations for the hexagonal system are illustrated in Fig. 249. It will be observed that the reciprocal of a hexagonal lattice is another hexagonal lattice whose orientation is rotated 90° from that of the first. The reciprocal of a hexagonal coordinate system to which a rhombohedral crystal has been referred, however, is another hexagonal coordinate system of identical orientation. This is to be expected because the reciprocal of a rhombohedral lattice is another rhombohedral lattice of different dimensions but identical orientation, as discussed before.

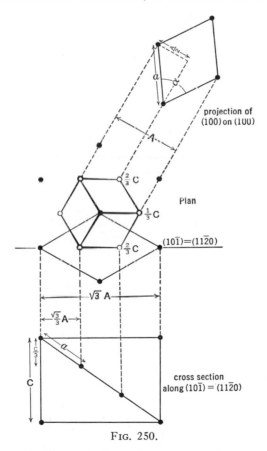

projection of
(100) on (1U0)

Plan

$(10\bar{1}) = (11\bar{2}0)$

cross section
along $(10\bar{1}) = (11\bar{2}0)$

FIG. 250.

In Fig. 250, the projection of part of a rhombohedral lattice upon $(111) = (0001)$ and a section along $(10\bar{1}) = (11\bar{2}0)$ are shown. From these it is apparent that a crystal based upon a rhombohedral lattice can be referred to a cell having hexagonal identity periods. If the hexagonal identity periods are designated A and C, it is evident from Fig. 250 that the rhombohedral identity period is related to it as follows:

$$a^2 = \left(\frac{\sqrt{3}A}{3}\right)^2 + \left(\frac{C}{3}\right)^2, \tag{1}$$

from which

$$a = \tfrac{1}{3}\sqrt{3A^2 + C^2}. \tag{2}$$

Figure 250 also shows one of the rhombohedral faces of the rhombohedral unit cell projected on a plane parallel with itself. From this projection it can be seen that the rhombohedral interaxial angle, α, may be derived from the following relation:

$$\sin\frac{\alpha}{2} = \frac{A/2}{a}. \tag{3}$$

Substituting, in (3), the value of a given by (2) yields the useful relation:

$$\sin\frac{\alpha}{2} = \frac{3A}{2\sqrt{3A^2 + C^2}}. \tag{4}$$

Another form of this relation may be obtained by substituting the value of $\sin\frac{\alpha}{2}$ given by (4) in the trigonometric identity

$$2\sin^2\frac{\alpha}{2} = 1 - \cos\alpha. \tag{5}$$

This gives the rhombohedral interaxial angle in a form dependent only on the hexagonal axial ratio, $\dfrac{C}{A}$, thus:

$$\alpha = \cos^{-1}\left[\frac{2\left(\dfrac{C}{A}\right)^2 - 3}{2\left(\dfrac{C}{A}\right)^2 + 6}\right]. \tag{6}$$

Triclinic crystals. The centered lattices have no significance in triclinic crystals, and its two space groups cannot be distinguished by diffraction methods. The interpretation of photographs of triclinic crystals is essentially a matter of selecting the simplest cell. This is done according to the discussion in Chapter 19.

THE SPACE GROUP

The determination of the space group directly from moving-film photographs by the geometrical method is relatively easy. It also has the advantage that the method of derivation automatically leads to the space-group symbol in the orientation notation.

The space group may have translations characteristic of the group in addition to the translations implied by the coincidence operations of the space lattice itself. These are included in characteristic glide planes and screw axes.

Glide planes. The effect of a glide plane is to produce an enantiomorphous point halfway between translation equivalents of the same point, on the other side of the glide plane. From the geometry of reflection it follows that these three non-collinear points project on the glide plane as three identical, non-enantiomorphous, collinear points separated by two equal spaces. The projection on the glide plane of a glide in the direction of a translation is therefore a second translation in the same direction and of half the magnitude of the first. It follows that the projection on the glide plane of a lattice plus a glide is a plane lattice, which is a duplicate of the plane projection of the space lattice alone, except that the translation in the glide direction is halved. Because of the general nature of the distance between the material components of the physical lattice and the glide plane, and the general non-equivalence of the glide equivalents as projected on any plane except the glide plane, the projection of a lattice containing a glide plane, on any plane normal to a rational direction, except the glide plane, consists of a pattern whose identity periods are the normal lattice projection for that direction.

All levels of the lattice reciprocal to one lacking characteristic glide planes are identical plane lattices, as discussed on pages 486–487. If the direct lattice contains characteristic glide planes, it follows, from what has just been said, that the zero level, reciprocal to the projection on the glide plane, contains certain reciprocal halved translations. Every non-zero level (n-level) parallel to this zero level, however, consists of points also contained in zero levels of other index $(uvw)^*$. These zero levels of other index $(uvw)^*$ are normal to other rational directions $[uvw]$, which, according to the above discussion, contain normal translations. It follows that the n-levels of the reciprocal lattice contain the reciprocals of the normal lattice translations, while the zero level of the reciprocal lattice parallel to the glide plane contains the reciprocals of the glide translations. The existence of a glide plane is therefore revealed as doubled translations in the zero level of the reciprocal lattice parallel to the glide plane, as compared with the parallel n-levels. Even-layers form the most convenient comparison standard because they are duplicates of the zero level (except in rhombohedral crystals) minus a possible glide. The direction of the glide is indicated on the same film by the direction of the doubled translation.

A plane projection cell having been selected, there are only two kinds of glide directions, i. e., in the direction of a cell side or a cell diagonal. The distribution of possible glide directions among the possible lattice types is given in Fig. 251. The illustration is given for plane lattice patterns of cell symmetry C_{2l}, which includes only rectangular and diamond cell shapes. It is convenient to regard all

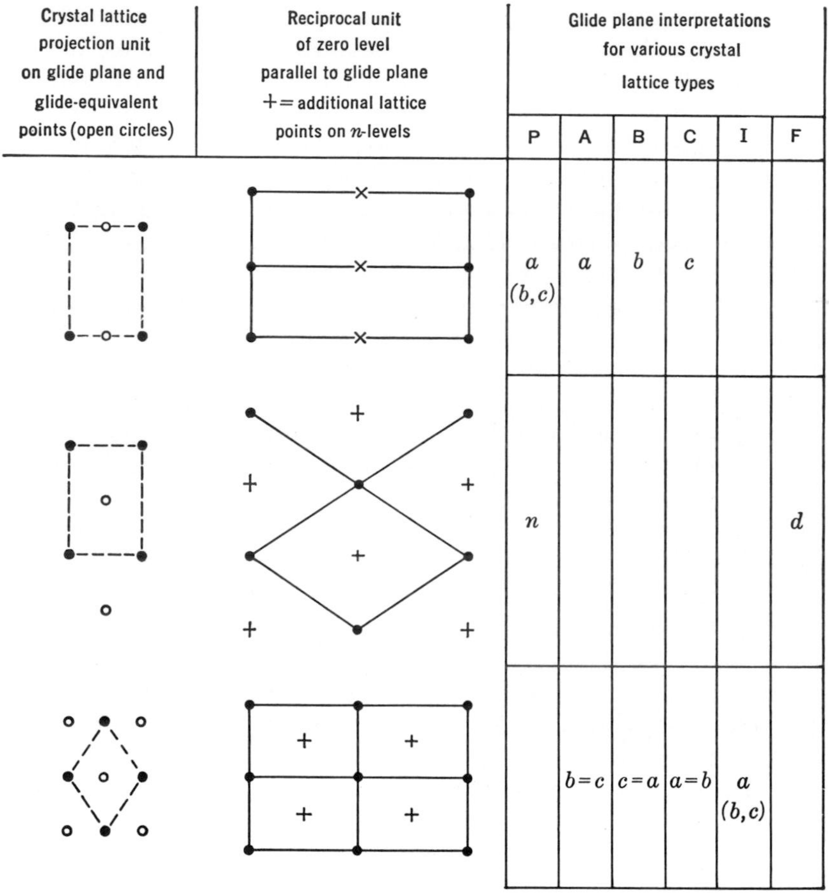

Crystal lattice projection unit on glide plane and glide-equivalent points (open circles)	Reciprocal unit of zero level parallel to glide plane += additional lattice points on n-levels	Glide plane interpretations for various crystal lattice types					
		P	A	B	C	I	F
		a (b,c)	a	b	c		
		n					d
		$b=c$	$c=a$	$a=b$	a (b,c)		

FIG. 251.

plane lattice cells as rectanglelike or diamondlike. Thus, one may think of these cells slightly distorted or generalized to become rectanglelike and diamondlike parallelograms, or specialized to become squares of 0° orientation and 45° orientation, respectively. Figure 251 may then be easily extended to these other cell types. The symbols of

Fig. 251 are in the orientation notation:

$(m =$ "glide" plane with glide component a, or b, or c).

$a =$ glide plane with glide component $\dfrac{a}{2}$.

$b =$ glide plane with glide component $\dfrac{b}{2}$.

$c =$ glide plane with glide component $\dfrac{c}{2}$.

$n =$ glide plane with glide component $\dfrac{a}{2} + \dfrac{b}{2}$ or $\dfrac{a}{2} + \dfrac{c}{2}$

$$\text{or} \quad \dfrac{b}{2} + \dfrac{c}{2}.$$

$d =$ glide plane with glide component $\dfrac{a}{4} + \dfrac{b}{4}$ or $\dfrac{a}{4} + \dfrac{c}{4}$

$$\text{or} \quad \dfrac{b}{4} + \dfrac{c}{4}.$$

Screw axes. Translations other than lattice operations may also be introduced into the space lattice complex by means of screw axes. The possible screw axes have the following characteristics:

	Symbol	Translation in terms of unit cell length
2-fold	$(2$	$1)$
	2_1	$\frac{1}{2}$
3-fold	$(3$	$1)$
	3_1	$\frac{1}{3}$
	3_2	$\frac{2}{3} = \left\lvert\frac{1}{3}\right\rvert$
4-fold	$(4$	$1)$
	4_1	$\frac{1}{4}$
	4_2	$\frac{2}{4} = \frac{1}{2}$
	4_3	$\frac{3}{4} = \left\lvert\frac{1}{4}\right\rvert$
6-fold	$(6$	$1)$
	6_1	$\frac{1}{6}$
	6_2	$\frac{2}{6} = \frac{1}{3}$
	6_3	$\frac{3}{6} = \frac{1}{2}$
	6_4	$\frac{4}{6} = \frac{2}{3} = \left\lvert\frac{1}{3}\right\rvert$
	6_5	$\frac{5}{6} = \left\lvert\frac{1}{6}\right\rvert$

The combination of a screw axis and a set of lattice translations generates a collection of points which project on the screw axis as a linear lattice of translation equal to the pitch of the screw. Including

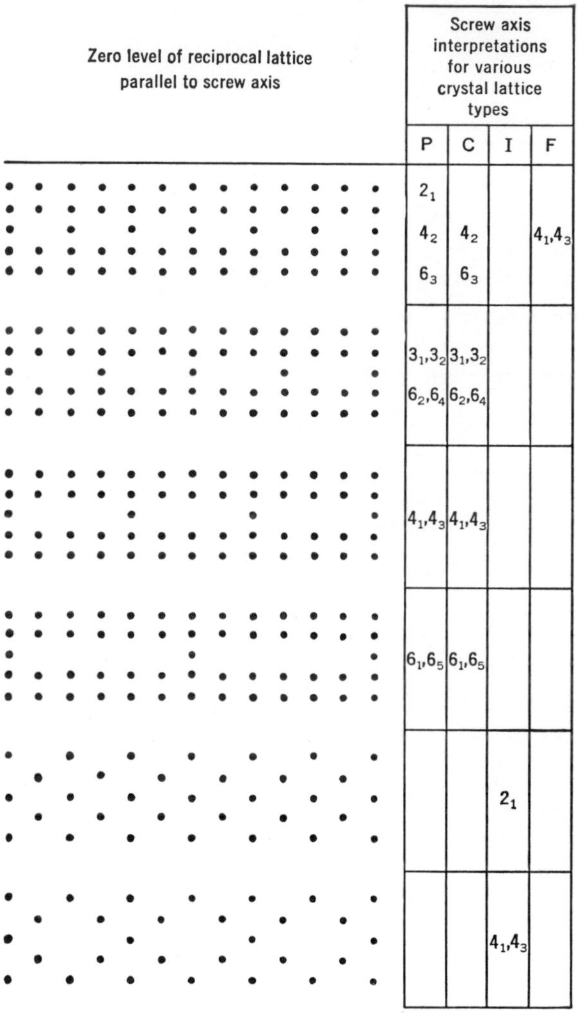

Zero level of reciprocal lattice parallel to screw axis	Screw axis interpretations for various crystal lattice types			
	P	C	I	F
(dot pattern)	2_1			
	4_2	4_2		$4_1,4_3$
	6_3	6_3		
(dot pattern)	$3_1,3_2$	$3_1,3_2$		
	$6_2,6_4$	$6_2,6_4$		
(dot pattern)	$4_1,4_3$	$4_1,4_3$		
(dot pattern)	$6_1,6_5$	$6_1,6_5$		
(dot pattern)			2_1	
(dot pattern)			$4_1,4_3$	

Fig. 252.

the special case of a rotation axis, this translation is 1, $\frac{1}{2}$, $\frac{1}{3}$, $\frac{1}{4}$, or $\frac{1}{6}$ times that of the axial identity period of the unit cell. Points generated by lattice translations of the centered lattices also project on axial directions to form a linear lattice whose translation is 1 or $\frac{1}{2}$ (or $\frac{1}{3}$ in the case of the rhombohedral lattice) times that of the axial identity period of the unit cell. The projection of the screw complex on any other rational direction generates a pattern whose identity period is that of the lattice projection itself. In general, then, the projection of a lattice complex containing a screw axis, on a plane containing the axis, is not a plane lattice, but rather is a complex whose reciprocal is a plane lattice except for a single lattice line through the origin and parallel to the screw. Along this unique line, the translation is 1, 2, 3, 4, or 6 times the translation of other parallel lattice lines.

The possible plane complex reciprocal patterns, their distribution, and several interpretations among the lattices are illustrated in Fig. 252. In this illustration, the lattice lines other than the unique one are taken from the n-layer. The pattern of the zero layer may have additional doubled translations if glide planes are present parallel to the reciprocal lattice plane. If the symmetry of the crystal is such that the possible screw axis is twofold, and if a glide plane has already been indicated by methods already outlined, then the presence of a twofold screw axis parallel to the glide plane may be neglected for space-group determinative purposes; its presence gives rise to no new diffraction effects, and its presence or absence does not change the space-group symbol. In other cases, however, the zero-layer, unique line should be compared with an n-layer (most conveniently, second-layer) pattern or the equivalent to obtain the normal reciprocal pattern for that set of planes.

The complex pattern illustrated in the upper diagram of Fig. 252 can be seen in an actual example in the central part of Fig. 107.

It should be noted that the introduction of a twofold screw axis into a body-centered lattice does not change the original reciprocal lattice, Fig. 252.

Procedure for space-group determination. The presence of glide planes and screw axes may be determined in several ways. The most direct way is to compare, for each possible glide plane or screw axis, a zero-layer photograph with an n-layer (ordinarily second-layer) photograph in order to detect doubled, etc., translations. Photographs of neighboring layers have sufficiently close ξ scales to bring these out on superficial comparison. This method unfortunately requires many level photographs ordinarily, one zero- and one or more n-layer, in addition to at least one rotation photograph necessary for setting the layer screen first. For the same crystal, only three axial equatorial and one or two axial n-layer photographs are really needed, in addition to one rotation. The rotation plus the n-layer provides the three axial identity periods. The lattice type of the reciprocal cell is also provided by the n-layer patterns. It is an easy matter to predict the normal n-layer patterns of other axial level photographs from this information, and therefore to detect multiple translations on the zero-layer axial photographs.

DIFFRACTION SYMBOLS

The geometrical method very strikingly brings out the limitations of using diffraction data in the determination of space groups. The chief limitation is that only a glide plane or a screw axis has a characteristic effect on the reciprocal lattice; neither a reflection plane nor a rotation axis has an effect apart from that of the lattice itself. Thus, if the lattice is primitive monoclinic, and no glide plane or screw axis is indicated by the diffraction effects, the space group may or may not contain a reflection plane or a twofold axis, and the possible interpretations of the diffraction effects are: Pm, $P2$, or $P2/m$. It is desirable to have a symbol for each distinct diffraction effect. Diffraction effects are capable of giving the centrosymmetrical space point-group, the lattice type, characteristic glide planes, and certain characteristic screw axes. The *diffraction symbol* includes this information in conformity with the orientation point-group and space-group symbols. The symbol includes the point group, followed by the lattice type, followed by the detectable glide planes and screw axes, *undetected possibilities being indicated by dashes when the order is important*. In Table 34, examples of such symbols are given together with symbols of the space groups giving the same diffraction symbol. If there are several choices of space groups for a given diffraction effect, the correct one usually can be determined if the space point-group symmetry of the crystal can be learned. The space group cannot always be fixed in this way, however. Thus, $I222$ cannot be distinguished from $I2_12_12_1$ because the screw, 2_1, does not have projected translations different from body-centered lattice, I. For the same reason, $I23$ cannot be distinguished from $I2_13$. Likewise, left-handed and right-handed screw axes cannot be distinguished from each other.

TABLE 34

DIFFRACTION SYMBOLS

NOTE: Except in the cases of isometric crystals, there are several possible alternative orientations for each space group, and correspondingly, there are several possible orientations for the diffraction symbol. In this table, only one arbitrarily selected orientation has been listed for each diffraction symbol. To use the table, one should first determine experimentally the diffraction symbol of the crystal in question, then develop all its possible (or reasonable) permutations. One of these will be found listed below. For this permutation, the several space groups which are indistinguishable by diffraction methods are then found listed to the right of the diffraction symbol, under their appropriate crystal classes. If only one space group is listed for the diffraction symbol, the space group is uniquely determined, but if several are listed, the correct space group can be determined only by first determining (by some means other than diffraction) to which crystal class the crystal belongs.

The distribution of the diffraction symbols among the crystal systems is as follows:

Triclinic	1
Monoclinic	6
Orthorhombic	33
Tetragonal	37
Hexagonal	20
Isometric	23
	120

Triclinic

Diffraction symmetry	C_i $\bar{1}$	
Crystal class / Diffraction symbol	C_1 1	C_i $\bar{1}$
$\bar{1}P\bar{1}$	$P1$	$P\bar{1}$

Monoclinic

Diffraction symmetry	C_{2h} $2/m$		
Crystal class / Diffraction symbol	C_s m	C_2 2	C_{2h} $2/m$
$2/mP$ -/-	Pm	$P2$	$P2/m$
$2/mP$ -/c	Pc		$P2/c$
$2/mP2_1$/-		$P2_1$	$P2_1/m$
$2/mP2_1$/c			$P2_1/c$
$2/mC$ -/-	Cm	$C2$	$C2/m$
$2/mC$ -/c	Cc		$C2/c$

Orthorhombic   TABLE 34 — *Continued*

Diffraction symmetry / Crystal class / Diffraction symbol	D_{2h} mmm		
	C_{2v} $mm2$	D_2 222	D_{2h} mmm
$mmmP$ - - -	$Pmm2$	$P222$	$Pmmm$
$mmmP$ - - 2_1		$P222_1$	
$mmmP2_12_1$ -		$P2_12_12$	
$mmmP2_12_12_1$		$P2_12_12_1$	
$mmmPc$ - -	$\begin{Bmatrix} Pc2m \\ Pcm2 \end{Bmatrix}$		$Pcmm$
$mmmPn$ - -	$Pnm2$		$Pnmm$
$mmmPcc$ -	$Pcc2$		$Pccm$
$mmmPca$ -	$Pca2$		$Pcam$
$mmmPba$ -	$Pba2$		$Pbam$
$mmmPnc$ -	$Pnc2$		$Pncm$
$mmmPna$ -	$Pna2$		$Pnam$
$mmmPnn$ -	$Pnn2$		$Pnnm$
$mmmPcca$			$Pcca$
$mmmPbca$			$Pbca$
$mmmPccn$			$Pccn$
$mmmPban$			$Pban$
$mmmPbcn$			$Pbcn$
$mmmPnna$			$Pnna$
$mmmPnnn$			$Pnnn$
$mmmC$ - - -	$\begin{Bmatrix} Cmm2 \\ Cm2m \end{Bmatrix}$	$C222$	$Cmmm$
$mmmC$ - - 2_1		$C222_1$	
$mmmC$ - c -	$\begin{Bmatrix} Cmc2 \\ C2cm \end{Bmatrix}$		$Cmcm$
$mmmC$ - - a	$C2ma$		$Cmma$
$mmmC$ - ca	$C2ca$		$Cmca$
$mmmCcc$ -	$Ccc2$		$Cccm$
$mmmCcca$			$Ccca$
$mmmI$ - - -	$Imm2$	$\begin{Bmatrix} I222 \\ I2_12_12_1 \end{Bmatrix}$	$Immm$
$mmmI$ - a -	$Ima2$		$Imam$
$mmmIba$ -	$Iba2$		$Ibam$
$mmmIbca$			$Ibca$
$mmmF$ - - -	$Fmm2$	$F222$	$Fmmm$
$mmmFdd$ -	$Fdd2$		
$mmmFddd$			$Fddd$

Tetragonal   TABLE 34 — *Continued*

Diffraction symmetry / Crystal class / Diffraction symbol	C_{4h} $4/m$			D_{4h} $4/mmm$			
	S_4 $\bar{4}$	C_4 4	C_{4h} $4/m$	D_{2d} $\bar{4}2m$	C_{4v} $4mm$	D_4 42	D_{4h} $4/mmm$
$4/mP$ - / -	$P\bar{4}$	$P4$	$P4/m$				
$4/mP4_2/$ -		$P4_2$	$P4_2/m$				
$4/mP4_1/$ -		$\begin{Bmatrix} P4_1 \\ P4_3 \end{Bmatrix}$					
$4/3P$ - $/n$			$P4/n$				
$4/mP4_2/n$			$P4_2/n$				
$4/mI$ - / -	$I\bar{4}$	$I4$	$I4/m$				
$4/mI4_1/$ -		$I4_1$					
$4/mI4_1/a$			$I4_1/a$				
$4/mmmP$ - / - - -				$\begin{Bmatrix} P\bar{4}2m \\ P\bar{4}m2 \end{Bmatrix}$	$P4mm$	$P42$	$P4/mmm$
$4/mmmP4_2/$ - - -						$P4_22$	
$4/mmmP4_1/$ - - -						$\begin{Bmatrix} P4_12 \\ P4_32 \end{Bmatrix}$	
$4/mmmP$ - / - 2_1 -				$P\bar{4}2_1m$		$P42_1$	
$4/mmmP4_2/$ - 2_1 -						$P4_22_1$	
$4/mmmP4_1/$ - 2_1 -						$\begin{Bmatrix} P4_12_1 \\ P4_32_1 \end{Bmatrix}$	
$4/mmmP$ - / - - c				$P\bar{4}2c$	$P4mc$		$P4/mmc$
$4/mmmP$ - / - 2_1c				$P42_1c$			
$4/mmmP$ - / - b -				$P\bar{4}b2$	$P4bm$		$P4/mbm$
$4/mmmP$ - / - bc					$P4bc$		$P4/mbm$
$4/mmmP$ - / - c -				$P\bar{4}c2$	$P4cm$		$P4/mcm$
$4/mmmP$ - / - cc					$P4cc$		$P4/mcc$
$4/mmmP$ - / - n -				$P\bar{4}n2$	$P4nm$		$P4/mnm$
$4/mmmP$ - / - nc					$P4nc$		$P4/mnc$
$4/mmmP$ - / n - -							$P4/nmm$
$4/mmmP$ - / n - c							$P4/nmc$
$4/mmmP$ - / nb -							$P4/nbm$
$4/mmmP$ - / nbc							$P4/nbc$
$4/mmmP$ - / nc -							$P4/ncm$
$4/mmmP$ - / ncc							$P4/ncc$
$4/mmmP$ - / nn -							$P4/nnm$
$4/mmmP$ - / nnc							$P4/nnc$
$4/mmI$ - / - - -				$\begin{Bmatrix} I\bar{4}2m \\ I\bar{4}m2 \end{Bmatrix}$	$I4mm$	$I42$	$I4/mmm$
$4/mmmI4_1/$ - - -						$I4_12$	
$4/mmmI$ - / - c -				$I\bar{4}c2$	$I4cm$		$I4/mcm$
$4/mmmI$ - / - - d				$I\bar{4}2d$	$I4md$		
$4/mmmI$ - / - cd					$I4cd$		
$4/mmmI$ - / a - d							$I4/amd$
$4/mmmI$ - /acd							$I4/acd$

Hexagonal   TABLE 34 — *Continued*

Diffraction symmetry → Crystal class ↓ / Diffraction symbol	C_{3i} $\bar{3}$		D_{3d} $\bar{3}m$		
	C_3 3	C_{3i} $\bar{3}$	C_{3v} $3m$	D_3 32	D_{3d} $\bar{3}m$
$\bar{3}C$ -	$C3$	$C\bar{3}$			
$\bar{3}C3_1$	$\left\{\begin{array}{l}C3_1\\C3_2\end{array}\right\}$				
$\bar{3}R$ -	$R3$	$R\bar{3}$			
$3mC$ - - -			$\left\{\begin{array}{l}C3m1\\C31m\end{array}\right\}$	$\left\{\begin{array}{l}C321\\C312\end{array}\right\}$	$\left\{\begin{array}{l}C\bar{3}m1\\C\bar{3}1m\end{array}\right\}$
$\bar{3}mC3_1$ - -				$\left\{\begin{array}{l}C3_121\\C3_221\\C3_112\\C3_212\end{array}\right\}$	
$\bar{3}mC$ - c -			$C3c1$		$C\bar{3}c1$
$\bar{3}mC$ - - c			$C31c$		$C\bar{3}1c$
$\bar{3}mR$ - -			$R3m$	$R32$	$R\bar{3}m$
$\bar{3}mR$ - c			$R3c$		$R\bar{3}c$
$6/mC$ - / -					
$6/mC6_1/$-					
$6/mC6_2/$-					
$6/mC6_3/$-					
$6/mmmC$ -/- - -					
$6/mmmC6_1/$- - -					
$6/mmmC6_2/$- - -					
$6/mmmC6_3/$- - -					
$6/mmmC$ -/- c -					
$6/mmmC$ -/- - c					
$6/mmmC$ -/- cc					

TABLE 34 — *Continued*

C_{6h} $6/m$			D_{6h} $6/mmm$			
C_6 6	C_{3h} $\bar{6}$	C_{6h} $6/m$	D_{3h} $\bar{6}m2$	C_{6v} $6mm$	D_6 62	D_{6h} $6/mmm$
$C6$ $\left\{\begin{matrix}C6_1\\C6_5\end{matrix}\right\}$ $\left\{\begin{matrix}C6_2\\C6_4\end{matrix}\right\}$ $C6_3$	$C\bar{6}$	$C6/m$ $C6_3/m$				
			$\left\{\begin{matrix}C\bar{6}m2\\C\bar{6}2m\end{matrix}\right\}$	$C6mm$	$C62$ $\left\{\begin{matrix}C6_12\\C6_52\end{matrix}\right\}$ $\left\{\begin{matrix}C6_22\\C6_42\end{matrix}\right\}$ $C6_32$	$C6/mmm$
			$C\bar{6}c2$ $C\bar{6}2c$	$C6cm$ $C6mc$ $C6cc$		$C6/mcm$ $C6/mmc$ $C6/mcc$

Isometric   TABLE 34 — *Continued*

Diffraction symmetry / Crystal class / Diffraction symbol	T_h $m3$		O_h $m3m$		
	T 23	T_h $m3$	T_d $\bar{4}3m$	O 43	O_h $m3m$
$m3P$ - -	$P23$	$Pm3$			
$m3P2_1$ -	$P2_13$				
$m3Pn$ -		$Pn3$			
$m3Pa$ -		$Pa3$			
$m3I$ - -	$\begin{cases} I23 \\ I2_13 \end{cases}$	$Im3$			
$m3Ia$ -		$Ia3$			
$m3F$ - -	$F23$	$Fm3$			
$m3Fd$ -		$Fd3$			
$m3mP$ - - -			$P\bar{4}3m$	$P43$	$Pm3m$
$m3mP4_1$ - -				$\begin{cases} P4_13 \\ P4_33 \end{cases}$	
$m3mP4_2$ - -				$P4_23$	
$m3mPn$ - -					$Pn3m$
$m3mP$ - - n			$P\bar{4}3n$		$Pm3n$
$m3mPn$ - m					$Pn3n$
$m3mI$ - - -			$I\bar{4}3m$	$I43$	$Im3m$
$m3mI4_1$ - -				$I4_13$	
$m3mI$ - - d			$I\bar{4}3d$		
$m3mIa$ - d					$Ia3d$
$m3mF$ - - -			$F\bar{4}3m$	$F43$	$Fm3m$
$m3mF4_1$ - -				$F4_13$	
$m3mFd$ - -					$Fd3m$
$m3mF$ - - c			$F\bar{4}3c$		$Fm3c$
$m3mFd$ - c					$Fd3c$

LITERATURE

Plane groups

G. Pólya. Über die Analogie der Kristallsymmetrie in der Ebene. *Z. Krist.* (*A*), 60 (1924), 278–282.

Paul Niggli. Die Flächensymmetrien homogener Diskontinuen. *Z. Krist.* (*A*), 60 (1924), 283–298.

Level theory

D. Crowfoot. The interpretation of Weissenberg photographs in relation to crystal symmetry. *Z. Krist.* (*A*), 90 (1935), 215–236.

M. J. Buerger. The application of plane groups to the interpretation of Weissenberg photographs. *Z. Krist.* (*A*), 91 (1935), 255–289.

M. J. Buerger and William Parrish. The unit cell and space group of tourmaline (an example of the inspective equi-inclination treatment of trigonal crystals). *Am. Mineral.*, 22 (1937), 1139–1150.

Reciprocity theory

A. Bravais. Abhandlung über die Systeme von regelmässig auf einer Ebene oder in Raum vertheilten Punkten (1848). (Translated by C. and E. Blasius, and appearing as No. 90 of Ostwald's *Klassiker der exakten Wissenschaften*; Wilhelm Engelmann, Leipzig, 1897.) Pages 112–139.

Ernest Mallard. Traité de cristallographie géométrique et physique. (Dunod, Paris, 1879.) Vol. I, 305–309.

P. P. Ewald. Das "reziproke" Gitter. *Z. Krist.* (*A*), 56 (1921), 148–150.

Examples of use of level theory

M. J. Buerger. Crystallographic data, unit cell, and space group for berthierite (FeSb₂S₄). *Am. Mineral.*, 21 (1936), 442–448.

M. J. Buerger. Crystals of the realgar type: the symmetry, unit cell, and space group of nitrogen sulfide. *Am. Mineral.*, 21 (1936), 575–583.

M. J. Buerger. The symmetry and crystal structure of the minerals of the arsenopyrite group. *Z. Krist.* (*A*), 95 (1936), 83–113.

M. J. Buerger. The symmetry and crystal structure of manganite, Mn(OH)O. *Z. Krist.* (*A*), 95 (1936), 163–174.

M. J. Buerger and Sterling B. Hendricks. The crystal structure of valentinite (orthorhombic Sb₂O₃). *Z. Krist.* (*A*), 98 (1937), 1–30.

M. J. Buerger. The unit cell and space group of cubanite. *Am. Mineral.*, 11 (1937), 1117–1120.

Newton W. Buerger. The unit cell and space group of sternbergite, AgFe₂S₃. *Am. Mineral.*, 22 (1937), 847–854.

M. J. Buerger and William Parrish. The unit cell and space group of tourmaline (an example of the inspective equi-inclination treatment of trigonal crystals). *Am. Mineral.*, 22 (1937), 1139–1150.

M. J. Buerger. The crystal structure of gudmundite (FeSbS) and its bearing on the existence field of the arsenopyrite structural type. *Z. Krist.* (*A*), 101 (1939), 290–316.

INDEX